APEX Calculus II
Version 3.0

Gregory Hartman, Ph.D.
Department of Applied Mathematics
Virginia Military Institute

Contributing Authors

Troy Siemers, Ph.D.
Department of Applied Mathematics
Virginia Military Institute

Brian Heinold, Ph.D.
Department of Mathematics and Computer Science
Mount Saint Mary's University

Dimplekumar Chalishajar, Ph.D.
Department of Applied Mathematics
Virginia Military Institute

Editor

Jennifer Bowen, Ph.D.
Department of Mathematics and Computer Science
The College of Wooster

 Copyright © 2015 Gregory Hartman
Licensed to the public under Creative Commons
Attribution-Noncommercial 4.0 International Public
License

Contents

Table of Contents   iii

Preface   v

5   Integration   **189**
    5.1   Antiderivatives and Indefinite Integration 189
    5.2   The Definite Integral 199
    5.3   Riemann Sums 210
    5.4   The Fundamental Theorem of Calculus 228
    5.5   Numerical Integration 240

6   Techniques of Antidifferentiation   **255**
    6.1   Substitution 255
    6.2   Integration by Parts 275
    6.3   Trigonometric Integrals 286
    6.4   Trigonometric Substitution 296
    6.5   Partial Fraction Decomposition 305
    6.6   Hyperbolic Functions 313
    6.7   L'Hôpital's Rule 324
    6.8   Improper Integration 333

7   Applications of Integration   **345**
    7.1   Area Between Curves 346
    7.2   Volume by Cross-Sectional Area; Disk and Washer Methods . . . 353
    7.3   The Shell Method 361
    7.4   Arc Length and Surface Area 369
    7.5   Work 378
    7.6   Fluid Forces 388

8   Sequences and Series   **397**
    8.1   Sequences 397
    8.2   Infinite Series 411
    8.3   Integral and Comparison Tests 426
    8.4   Ratio and Root Tests 435

	8.5	Alternating Series and Absolute Convergence	441
	8.6	Power Series	452
	8.7	Taylor Polynomials	465
	8.8	Taylor Series	477

A Solutions To Selected Problems — **A.1**

Index — **A.11**

Preface
A Note on Using this Text

Thank you for reading this short preface. Allow us to share a few key points about the text so that you may better understand what you will find beyond this page.

This text is Part II of a three–text series on Calculus. The first part covers material taught in many "Calc 1" courses: limits, derivatives, and the basics of integration, found in Chapters 1 through 6.1. The second text covers material often taught in "Calc 2:" integration and its applications, along with an introduction to sequences, series and Taylor Polynomials, found in Chapters 5 through 8. The third text covers topics common in "Calc 3" or "multivariable calc:" parametric equations, polar coordinates, vector–valued functions, and functions of more than one variable, found in Chapters 9 through 13. All three are available separately for free at www.vmi.edu/APEX. These three texts are intended to work together and make one cohesive text, *APEX Calculus*, which can also be downloaded from the website.

Printing the entire text as one volume makes for a large, heavy, cumbersome book. One can certainly only print the pages they currently need, but some prefer to have a nice, bound copy of the text. Therefore this text has been split into these three manageable parts, each of which can be purchased for under $15 at Amazon.com.

A result of this splitting is that sometimes a concept is said to be explored in an "earlier/later section," though that section does not actually appear in this particular text. Also, the index makes reference to topics, and page numbers, that do not appear in this text. This is done intentionally to show the reader what topics are available for study. Downloading the .pdf of *APEX Calculus* will ensure that you have all the content.

For Students: How to Read this Text

Mathematics textbooks have a reputation for being hard to read. High–level mathematical writing often seeks to say much with few words, and this style often seeps into texts of lower–level topics. This book was written with the goal of being easier to read than many other calculus textbooks, without becoming too verbose.

Each chapter and section starts with an introduction of the coming material, hopefully setting the stage for "why you should care," and ends with a look ahead to see how the just–learned material helps address future problems.

Please read the text; it is written to explain the concepts of Calculus. There are numerous examples to demonstrate the meaning of definitions, the truth of theorems, and the application of mathematical techniques. When you encounter a sentence you don't understand, read it again. If it still doesn't make sense, read on anyway, as sometimes confusing sentences are explained by later sentences.

You don't have to read every equation. The examples generally show "all" the steps needed to solve a problem. Sometimes reading through each step is helpful; sometimes it is confusing. When the steps are illustrating a new technique, one probably should follow each step closely to learn the new technique. When the steps are showing the mathematics needed to find a number to be used later, one can usually skip ahead and see how that number is being used, instead of getting bogged down in reading how the number was found.

Most proofs have been omitted. In mathematics, *proving* something is always true is extremely important, and entails much more than testing to see if it works twice. However, students often are confused by the details of a proof, or become concerned that they should have been able to construct this proof on their own. To alleviate this potential problem, we do not include the proofs to most theorems in the text. The interested reader is highly encouraged to find proofs online or from their instructor. In most cases, one is very capable of understanding what a theorem *means* and *how to apply it* without knowing fully *why* it is true.

Interactive, 3D Graphics

New to Version 3.0 is the addition of interactive, 3D graphics in the .pdf version. Nearly all graphs of objects in space can be rotated, shifted, and zoomed in/out so the reader can better understand the object illustrated.

As of this writing, the only pdf viewers that support these 3D graphics are Adobe Reader & Acrobat (and only the versions for PC/Mac/Unix/Linux computers, not tablets or smartphones). To activate the interactive mode, click on the image. Once activated, one can click/drag to rotate the object and use the scroll wheel on a mouse to zoom in/out. (A great way to investigate an image is to first zoom in on the page of the pdf viewer so the graphic itself takes up much of the screen, then zoom inside the graphic itself.) A CTRL-click/drag pans the object left/right or up/down. By right-clicking on the graph one can access a menu of other options, such as changing the lighting scheme or perspective. One can also revert the graph back to its default view. If you wish to deactive the interactivity, one can right-click and choose the "Disable Content" option.

Thanks

There are many people who deserve recognition for the important role they have played in the development of this text. First, I thank Michelle for her support and encouragement, even as this "project from work" occupied my time and attention at home. Many thanks to Troy Siemers, whose most important contributions extend far beyond the sections he wrote or the 227 figures he coded in Asymptote for 3D interaction. He provided incredible support, advice and encouragement for which I am very grateful. My thanks to Brian Heinold and Dimplekumar Chalishajar for their contributions and to Jennifer Bowen for reading through so much material and providing great feedback early on. Thanks to Troy, Lee Dewald, Dan Joseph, Meagan Herald, Bill Lowe, John David, Vonda Walsh, Geoff Cox, Jessica Libertini and other faculty of VMI who have given me numerous suggestions and corrections based on their experience with teaching from the text. (Special thanks to Troy, Lee & Dan for their patience in teaching Calc III while I was still writing the Calc III material.) Thanks to Randy Cone for encouraging his tutors of VMI's Open Math Lab to read through the text and check the solutions, and thanks to the tutors for spending their time doing so. A very special thanks to Kristi Brown and Paul Janiczek who took this opportunity far above & beyond what I expected, meticulously checking every solution and carefully reading every example. Their comments have been extraordinarily helpful. I am also thankful for the support provided by Wane Schneiter, who as my Dean provided me with extra time to work on this project. I am blessed to have so many people give of their time to make this book better.

APEX – Affordable Print and Electronic teXts

APEX is a consortium of authors who collaborate to produce high–quality, low–cost textbooks. The current textbook–writing paradigm is facing a potential revolution as desktop publishing and electronic formats increase in popularity. However, writing a good textbook is no easy task, as the time requirements alone are substantial. It takes countless hours of work to produce text, write examples and exercises, edit and publish. Through collaboration, however, the cost to any individual can be lessened, allowing us to create texts that we freely distribute electronically and sell in printed form for an incredibly low cost. Having said that, nothing is entirely free; someone always bears some cost. This text "cost" the authors of this book their time, and that was not enough. *APEX Calculus* would not exist had not the Virginia Military Institute, through a generous Jackson–Hope grant, given the lead author significant time away from teaching so he could focus on this text.

Each text is available as a free .pdf, protected by a Creative Commons Attribution - Noncommercial 4.0 copyright. That means you can give the .pdf to anyone you like, print it in any form you like, and even edit the original content and redistribute it. If you do the latter, you must clearly reference this work and you cannot sell your edited work for money.

We encourage others to adapt this work to fit their own needs. One might add sections that are "missing" or remove sections that your students won't need. The source files can be found at github.com/APEXCalculus.

You can learn more at www.vmi.edu/APEX.

5: INTEGRATION

We have spent considerable time considering the derivatives of a function and their applications. In the following chapters, we are going to starting thinking in "the other direction." That is, given a function $f(x)$, we are going to consider functions $F(x)$ such that $F'(x) = f(x)$. There are numerous reasons this will prove to be useful: these functions will help us compute areas, volumes, mass, force, pressure, work, and much more.

5.1 Antiderivatives and Indefinite Integration

Given a function $y = f(x)$, a *differential equation* is one that incorporates y, x, and the derivatives of y. For instance, a simple differential equation is:

$$y' = 2x.$$

Solving a differential equation amounts to finding a function y that satisfies the given equation. Take a moment and consider that equation; can you find a function y such that $y' = 2x$?

Can you find another?

And yet another?

Hopefully one was able to come up with at least one solution: $y = x^2$. "Finding another" may have seemed impossible until one realizes that a function like $y = x^2 + 1$ also has a derivative of $2x$. Once that discovery is made, finding "yet another" is not difficult; the function $y = x^2 + 123,456,789$ also has a derivative of $2x$. The differential equation $y' = 2x$ has many solutions. This leads us to some definitions.

Definition 19   **Antiderivatives and Indefinite Integrals**

Let a function $f(x)$ be given. An **antiderivative** of $f(x)$ is a function $F(x)$ such that $F'(x) = f(x)$.

The set of all antiderivatives of $f(x)$ is the **indefinite integral of f**, denoted by

$$\int f(x)\, dx.$$

Make a note about our definition: we refer to *an* antiderivative of f, as opposed to *the* antiderivative of f, since there is *always* an infinite number of them.

We often use upper-case letters to denote antiderivatives.

Knowing one antiderivative of f allows us to find infinitely more, simply by adding a constant. Not only does this give us *more* antiderivatives, it gives us *all* of them.

Theorem 34   **Antiderivative Forms**

Let $F(x)$ and $G(x)$ be antiderivatives of $f(x)$. Then there exists a constant C such that
$$G(x) = F(x) + C.$$

Given a function f and one of its antiderivatives F, we know *all* antiderivatives of f have the form $F(x) + C$ for some constant C. Using Definition 19, we can say that
$$\int f(x)\, dx = F(x) + C.$$

Let's analyze this indefinite integral notation.

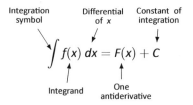

Figure 5.1: Understanding the indefinite integral notation.

Figure 5.1 shows the typical notation of the indefinite integral. The integration symbol, $\int$, is in reality an "elongated S," representing "take the sum." We will later see how *sums* and *antiderivatives* are related.

The function we want to find an antiderivative of is called the *integrand*. It contains the differential of the variable we are integrating with respect to. The $\int$ symbol and the differential dx are not "bookends" with a function sandwiched in between; rather, the symbol $\int$ means "find all antiderivatives of what follows," and the function $f(x)$ and dx are multiplied together; the dx does not "just sit there."

Let's practice using this notation.

Example 109   **Evaluating indefinite integrals**

Evaluate $\int \sin x\, dx$.

Notes:

5.1 Antiderivatives and Indefinite Integration

SOLUTION We are asked to find all functions $F(x)$ such that $F'(x) = \sin x$. Some thought will lead us to one solution: $F(x) = -\cos x$, because $\frac{d}{dx}(-\cos x) = \sin x$.

The indefinite integral of $\sin x$ is thus $-\cos x$, plus a constant of integration. So:

$$\int \sin x \, dx = -\cos x + C.$$

A commonly asked question is "What happened to the *dx*?" The unenlightened response is "Don't worry about it. It just goes away." A full understanding includes the following.

This process of *antidifferentiation* is really solving a *differential* question. The integral

$$\int \sin x \, dx$$

presents us with a differential, $dy = \sin x \, dx$. It is asking: "What is *y*?" We found lots of solutions, all of the form $y = -\cos x + C$.

Letting $dy = \sin x \, dx$, rewrite

$$\int \sin x \, dx \quad \text{as} \quad \int dy.$$

This is asking: "What functions have a differential of the form *dy*?" The answer is "Functions of the form $y + C$, where *C* is a constant." What is *y*? We have lots of choices, all differing by a constant; the simplest choice is $y = -\cos x$.

Understanding all of this is more important later as we try to find antiderivatives of more complicated functions. In this section, we will simply explore the rules of indefinite integration, and one can succeed for now with answering "What happened to the *dx*?" with "It went away."

Let's practice once more before stating integration rules.

Example 110 **Evaluating indefinite integrals**
Evaluate $\int (3x^2 + 4x + 5) \, dx$.

SOLUTION We seek a function $F(x)$ whose derivative is $3x^2 + 4x + 5$. When taking derivatives, we can consider functions term–by–term, so we can likely do that here.

What functions have a derivative of $3x^2$? Some thought will lead us to a cubic, specifically $x^3 + C_1$, where C_1 is a constant.

What functions have a derivative of $4x$? Here the *x* term is raised to the first power, so we likely seek a quadratic. Some thought should lead us to $2x^2 + C_2$, where C_2 is a constant.

Notes:

Finally, what functions have a derivative of 5? Functions of the form $5x + C_3$, where C_3 is a constant.

Our answer appears to be

$$\int (3x^2 + 4x + 5)\, dx = x^3 + C_1 + 2x^2 + C_2 + 5x + C_3.$$

We do not need three separate constants of integration; combine them as one constant, giving the final answer of

$$\int (3x^2 + 4x + 5)\, dx = x^3 + 2x^2 + 5x + C.$$

It is easy to verify our answer; take the derivative of $x^3 + 2x^3 + 5x + C$ and see we indeed get $3x^2 + 4x + 5$.

This final step of "verifying our answer" is important both practically and theoretically. In general, taking derivatives is easier than finding antiderivatives so checking our work is easy and vital as we learn.

We also see that taking the derivative of our answer returns the function in the integrand. Thus we can say that:

$$\frac{d}{dx}\left(\int f(x)\, dx\right) = f(x).$$

Differentiation "undoes" the work done by antidifferentiation.

Theorem 24 gave a list of the derivatives of common functions we had learned at that point. We restate part of that list here to stress the relationship between derivatives and antiderivatives. This list will also be useful as a glossary of common antiderivatives as we learn.

Notes:

Theorem 35 Derivatives and Antiderivatives

Common Differentiation Rules

1. $\frac{d}{dx}\big(cf(x)\big) = c \cdot f'(x)$
2. $\frac{d}{dx}\big(f(x) \pm g(x)\big) = f'(x) \pm g'(x)$
3. $\frac{d}{dx}(C) = 0$
4. $\frac{d}{dx}(x) = 1$
5. $\frac{d}{dx}(x^n) = n \cdot x^{n-1}$
6. $\frac{d}{dx}(\sin x) = \cos x$
7. $\frac{d}{dx}(\cos x) = -\sin x$
8. $\frac{d}{dx}(\tan x) = \sec^2 x$
9. $\frac{d}{dx}(\csc x) = -\csc x \cot x$
10. $\frac{d}{dx}(\sec x) = \sec x \tan x$
11. $\frac{d}{dx}(\cot x) = -\csc^2 x$
12. $\frac{d}{dx}(e^x) = e^x$
13. $\frac{d}{dx}(a^x) = \ln a \cdot a^x$
14. $\frac{d}{dx}(\ln x) = \frac{1}{x}$

Common Indefinite Integral Rules

1. $\int c \cdot f(x)\, dx = c \cdot \int f(x)\, dx$
2. $\int \big(f(x) \pm g(x)\big)\, dx = \int f(x)\, dx \pm \int g(x)\, dx$
3. $\int 0 \, dx = C$
4. $\int 1 \, dx = \int dx = x + C$
5. $\int x^n\, dx = \frac{1}{n+1} x^{n+1} + C \quad (n \ne -1)$
6. $\int \cos x \, dx = \sin x + C$
7. $\int \sin x \, dx = -\cos x + C$
8. $\int \sec^2 x \, dx = \tan x + C$
9. $\int \csc x \cot x \, dx = -\csc x + C$
10. $\int \sec x \tan x \, dx = \sec x + C$
11. $\int \csc^2 x \, dx = -\cot x + C$
12. $\int e^x \, dx = e^x + C$
13. $\int a^x \, dx = \frac{1}{\ln a} \cdot a^x + C$
14. $\int \frac{1}{x} \, dx = \ln |x| + C$

We highlight a few important points from Theorem 35:

- Rule #1 states $\int c \cdot f(x)\, dx = c \cdot \int f(x)\, dx$. This is the Constant Multiple Rule: we can temporarily ignore constants when finding antiderivatives, just as we did when computing derivatives (i.e., $\frac{d}{dx}(3x^2)$ is just as easy to compute as $\frac{d}{dx}(x^2)$). An example:

$$\int 5\cos x \, dx = 5 \cdot \int \cos x \, dx = 5 \cdot (\sin x + C) = 5\sin x + C.$$

In the last step we can consider the constant as also being multiplied by

Notes:

5, but "5 times a constant" is still a constant, so we just write "C".

- Rule #2 is the Sum/Difference Rule: we can split integrals apart when the integrand contains terms that are added/subtracted, as we did in Example 110. So:

$$\int (3x^2 + 4x + 5)\, dx = \int 3x^2\, dx + \int 4x\, dx + \int 5\, dx$$
$$= 3\int x^2\, dx + 4\int x\, dx + \int 5\, dx$$
$$= 3 \cdot \frac{1}{3}x^3 + 4 \cdot \frac{1}{2}x^2 + 5x + C$$
$$= x^3 + 2x^2 + 5x + C$$

In practice we generally do not write out all these steps, but we demonstrate them here for completeness.

- Rule #5 is the Power Rule of indefinite integration. There are two important things to keep in mind:

  1. Notice the restriction that $n \neq -1$. This is important: $\int \frac{1}{x}\, dx \neq$ "$\frac{1}{0}x^0 + C$"; rather, see Rule #14.

  2. We are presenting antidifferentiation as the "inverse operation" of differentiation. Here is a useful quote to remember:

     "Inverse operations do the opposite things in the opposite order."

     When taking a derivative using the Power Rule, we **first** *multiply* by the power, then **second** *subtract* 1 from the power. To find the antiderivative, do the opposite things in the opposite order: **first** *add* one to the power, then **second** *divide* by the power.

- Note that Rule #14 incorporates the absolute value of x. The exercises will work the reader through why this is the case; for now, know the absolute value is important and cannot be ignored.

Initial Value Problems

In Section 2.3 we saw that the derivative of a position function gave a velocity function, and the derivative of a velocity function describes acceleration. We can now go "the other way:" the antiderivative of an acceleration function gives a velocity function, etc. While there is just one derivative of a given function, there are infinite antiderivatives. Therefore we cannot ask "What is *the* velocity of an object whose acceleration is -32ft/s^2?", since there is more than one answer.

Notes:

We can find *the* answer if we provide more information with the question, as done in the following example. Often the additional information comes in the form of an *initial value*, a value of the function that one knows beforehand.

Example 111   Solving initial value problems
The acceleration due to gravity of a falling object is -32 ft/s^2. At time $t = 3$, a falling object had a velocity of -10 ft/s. Find the equation of the object's velocity.

SOLUTION   We want to know a velocity function, $v(t)$. We know two things:

- The acceleration, i.e., $v'(t) = -32$, and
- the velocity at a specific time, i.e., $v(3) = -10$.

Using the first piece of information, we know that $v(t)$ is an antiderivative of $v'(t) = -32$. So we begin by finding the indefinite integral of -32:

$$\int (-32)\, dt = -32t + C = v(t).$$

Now we use the fact that $v(3) = -10$ to find C:

$$v(t) = -32t + C$$
$$v(3) = -10$$
$$-32(3) + C = -10$$
$$C = 86$$

Thus $v(t) = -32t + 86$. We can use this equation to understand the motion of the object: when $t = 0$, the object had a velocity of $v(0) = 86$ ft/s. Since the velocity is positive, the object was moving upward.

When did the object begin moving down? Immediately after $v(t) = 0$:

$$-32t + 86 = 0 \quad \Rightarrow \quad t = \frac{43}{16} \approx 2.69\text{s}.$$

Recognize that we are able to determine quite a bit about the path of the object knowing just its acceleration and its velocity at a single point in time.

Example 112   Solving initial value problems
Find $f(t)$, given that $f''(t) = \cos t$, $f'(0) = 3$ and $f(0) = 5$.

SOLUTION   We start by finding $f'(t)$, which is an antiderivative of $f''(t)$:

$$\int f''(t)\, dt = \int \cos t\, dt = \sin t + C = f'(t).$$

Notes:

So $f'(t) = \sin t + C$ for the correct value of C. We are given that $f'(0) = 3$, so:
$$f'(0) = 3 \quad \Rightarrow \quad \sin 0 + C = 3 \quad \Rightarrow \quad C = 3.$$
Using the initial value, we have found $f'(t) = \sin t + 3$.

We now find $f(t)$ by integrating again.
$$f(t) = \int f'(t)\, dt = \int (\sin t + 3)\, dt = -\cos t + 3t + C.$$
We are given that $f(0) = 5$, so
$$-\cos 0 + 3(0) + C = 5$$
$$-1 + C = 5$$
$$C = 6$$
Thus $f(t) = -\cos t + 3t + 6$.

This section introduced antiderivatives and the indefinite integral. We found they are needed when finding a function given information about its derivative(s). For instance, we found a position function given a velocity function.

In the next section, we will see how position and velocity are unexpectedly related by the areas of certain regions on a graph of the velocity function. Then, in Section 5.4, we will see how areas and antiderivatives are closely tied together.

Notes:

Exercises 5.1

Terms and Concepts

1. Define the term "antiderivative" in your own words.

2. Is it more accurate to refer to "the" antiderivative of $f(x)$ or "an" antiderivative of $f(x)$?

3. Use your own words to define the indefinite integral of $f(x)$.

4. Fill in the blanks: "Inverse operations do the _____ things in the _____ order."

5. What is an "initial value problem"?

6. The derivative of a position function is a _____ function.

7. The antiderivative of an acceleration function is a _____ function.

Problems

In Exercises 8 – 26, evaluate the given indefinite integral.

8. $\int 3x^3\, dx$

9. $\int x^8\, dx$

10. $\int (10x^2 - 2)\, dx$

11. $\int dt$

12. $\int 1\, ds$

13. $\int \frac{1}{3t^2}\, dt$

14. $\int \frac{3}{t^2}\, dt$

15. $\int \frac{1}{\sqrt{x}}\, dx$

16. $\int \sec^2\theta\, d\theta$

17. $\int \sin\theta\, d\theta$

18. $\int (\sec x \tan x + \csc x \cot x)\, dx$

19. $\int 5e^\theta\, d\theta$

20. $\int 3^t\, dt$

21. $\int \frac{5^t}{2}\, dt$

22. $\int (2t+3)^2\, dt$

23. $\int (t^2+3)(t^3-2t)\, dt$

24. $\int x^2 x^3\, dx$

25. $\int e^\pi\, dx$

26. $\int a\, dx$

27. This problem investigates why Theorem 35 states that $\int \frac{1}{x}\, dx = \ln|x| + C$.

   (a) What is the domain of $y = \ln x$?
   (b) Find $\frac{d}{dx}(\ln x)$.
   (c) What is the domain of $y = \ln(-x)$?
   (d) Find $\frac{d}{dx}(\ln(-x))$.
   (e) You should find that $1/x$ has two types of antiderivatives, depending on whether $x > 0$ or $x < 0$. In one expression, give a formula for $\int \frac{1}{x}\, dx$ that takes these different domains into account, and explain your answer.

In Exercises 28 – 38, find $f(x)$ described by the given initial value problem.

28. $f'(x) = \sin x$ and $f(0) = 2$

29. $f'(x) = 5e^x$ and $f(0) = 10$

30. $f'(x) = 4x^3 - 3x^2$ and $f(-1) = 9$

31. $f'(x) = \sec^2 x$ and $f(\pi/4) = 5$

32. $f'(x) = 7^x$ and $f(2) = 1$

33. $f''(x) = 5$ and $f'(0) = 7, f(0) = 3$

34. $f''(x) = 7x$ and $f'(1) = -1, f(1) = 10$

35. $f''(x) = 5e^x$ and $f'(0) = 3, f(0) = 5$

36. $f''(\theta) = \sin\theta$ and $f'(\pi) = 2, f(\pi) = 4$

37. $f''(x) = 24x^2 + 2^x - \cos x$ and $f'(0) = 5, f(0) = 0$

38. $f''(x) = 0$ and $f'(1) = 3, f(1) = 1$

Review

39. Use information gained from the first and second derivatives to sketch $f(x) = \dfrac{1}{e^x + 1}$.

40. Given $y = x^2 e^x \cos x$, find dy.

5.2 The Definite Integral

We start with an easy problem. An object travels in a straight line at a constant velocity of 5 ft/s for 10 seconds. How far away from its starting point is the object?

We approach this problem with the familiar "Distance = Rate × Time" equation. In this case, Distance = 5ft/s × 10s = 50 feet.

It is interesting to note that this solution of 50 feet can be represented graphically. Consider Figure 5.2, where the constant velocity of 5ft/s is graphed on the axes. Shading the area under the line from $t = 0$ to $t = 10$ gives a rectangle with an area of 50 square units; when one considers the units of the axes, we can say this area represents 50 ft.

Now consider a slightly harder situation (and not particularly realistic): an object travels in a straight line with a constant velocity of 5ft/s for 10 seconds, then instantly reverses course at a rate of 2ft/s for 4 seconds. (Since the object is traveling in the opposite direction when reversing course, we say the velocity is a constant -2ft/s.) How far away from the starting point is the object – what is its *displacement*?

Here we use "Distance = Rate$_1$ × Time$_1$ + Rate$_2$ × Time$_2$," which is

$$\text{Distance} = 5 \cdot 10 + (-2) \cdot 4 = 42 \text{ ft.}$$

Hence the object is 42 feet from its starting location.

We can again depict this situation graphically. In Figure 5.3 we have the velocities graphed as straight lines on $[0, 10]$ and $[10, 14]$, respectively. The displacement of the object is

"Area above the t–axis   −   Area below the t–axis,"

which is easy to calculate as $50 - 8 = 42$ feet.

Now consider a more difficult problem.

Example 113   Finding position using velocity
The velocity of an object moving straight up/down under the acceleration of gravity is given as $v(t) = -32t + 48$, where time t is given in seconds and velocity is in ft/s. When $t = 0$, the object had a height of 0 ft.

1. What was the initial velocity of the object?

2. What was the maximum height of the object?

3. What was the height of the object at time $t = 2$?

Solution   It is straightforward to find the initial velocity; at time $t = 0$, $v(0) = -32 \cdot 0 + 48 = 48$ ft/s.

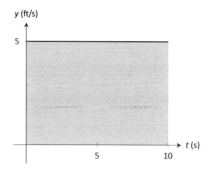

Figure 5.2: The area under a constant velocity function corresponds to distance traveled.

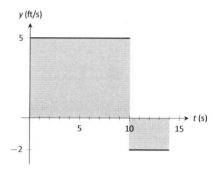

Figure 5.3: The total displacement is the area above the t–axis minus the area below the t–axis.

Notes:

To answer questions about the height of the object, we need to find the object's position function $s(t)$. This is an initial value problem, which we studied in the previous section. We are told the initial height is 0, i.e., $s(0) = 0$. We know $s'(t) = v(t) = -32t + 48$. To find s, we find the indefinite integral of $v(t)$:

$$\int v(t)\, dt = \int (-32t + 48)\, dt = -16t^2 + 48t + C = s(t).$$

Since $s(0) = 0$, we conclude that $C = 0$ and $s(t) = -16t^2 + 48t$.

To find the maximum height of the object, we need to find the maximum of s. Recalling our work finding extreme values, we find the critical points of s by setting its derivative equal to 0 and solving for t:

$$s'(t) = -32t + 48 = 0 \quad \Rightarrow \quad t = 48/32 = 1.5\text{s}.$$

(Notice how we ended up just finding when the velocity was 0ft/s!) The first derivative test shows this is a maximum, so the maximum height of the object is found at

$$s(1.5) = -16(1.5)^2 + 48(1.5) = 36\text{ft}.$$

The height at time $t = 2$ is now straightforward to compute: it is $s(2) = 32$ft.

While we have answered all three questions, let's look at them again graphically, using the concepts of area that we explored earlier.

Figure 5.4 shows a graph of $v(t)$ on axes from $t = 0$ to $t = 3$. It is again straightforward to find $v(0)$. How can we use the graph to find the maximum height of the object?

Recall how in our previous work that the displacement of the object (in this case, its height) was found as the area under the velocity curve, as shaded in the figure. Moreover, the area between the curve and the t–axis that is below the t–axis counted as "negative" area. That is, it represents the object coming back toward its starting position. So to find the maximum distance from the starting point – the maximum height – we find the area under the velocity line that is above the t–axis, i.e., from $t = 0$ to $t = 1.5$. This region is a triangle; its area is

$$\text{Area} = \frac{1}{2}\text{Base} \times \text{Height} = \frac{1}{2} \times 1.5\text{s} \times 48\text{ft/s} = 36\text{ft},$$

which matches our previous calculation of the maximum height.

Finally, we find the total *signed* area under the velocity function from $t = 0$ to $t = 2$ to find the $s(2)$, the height at $t = 2$, which is a displacement, the distance from the current position to the starting position. That is,

Displacement = Area above the t–axis − Area below t–axis.

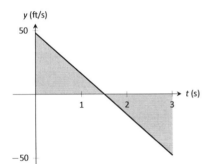

Figure 5.4: A graph of $v(t) = -32t + 48$; the shaded areas help determine displacement.

Notes:

The regions are triangles, and we find
$$\text{Displacement} = \frac{1}{2}(1.5s)(48\text{ft/s}) - \frac{1}{2}(.5s)(16\text{ft/s}) = 32\text{ft}.$$
This also matches our previous calculation of the height at $t = 2$.

Notice how we answered each question in this example in two ways. Our first method was to manipulate equations using our understanding of antiderivatives and derivatives. Our second method was geometric: we answered questions looking at a graph and finding the areas of certain regions of this graph.

The above example does not *prove* a relationship between area under a velocity function and displacement, but it does imply a relationship exists. Section 5.4 will fully establish fact that the area under a velocity function is displacement.

Given a graph of a function $y = f(x)$, we will find that there is great use in computing the area between the curve $y = f(x)$ and the *x*-axis. Because of this, we need to define some terms.

Definition 20   **The Definite Integral, Total Signed Area**

Let $y = f(x)$ be defined on a closed interval $[a, b]$. The **total signed area from $x = a$ to $x = b$ under f** is:
(area under f and above the *x*–axis on $[a, b]$) − (area above f and under the *x*–axis on $[a, b]$).

The **definite integral of f on $[a, b]$** is the total signed area of f on $[a, b]$, denoted
$$\int_a^b f(x)\, dx,$$
where a and b are the **bounds of integration.**

By our definition, the definite integral gives the "signed area under f." We usually drop the word "signed" when talking about the definite integral, and simply say the definite integral gives "the area under f" or, more commonly, "the area under the curve."

The previous section introduced the indefinite integral, which related to antiderivatives. We have now defined the definite integral, which relates to areas under a function. The two are very much related, as we'll see when we learn the Fundamental Theorem of Calculus in Section 5.4. Recall that earlier we said that the "$\int$" symbol was an "elongated S" that represented finding a "sum." In the context of the definite integral, this notation makes a bit more sense, as we are adding up areas under the function f.

Notes:

Chapter 5 Integration

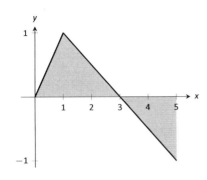

Figure 5.5: A graph of $f(x)$ in Example 114.

We practice using this notation.

Example 114   **Evaluating definite integrals**
Consider the function f given in Figure 5.5.

Find:

1. $\int_0^3 f(x)\, dx$
2. $\int_3^5 f(x)\, dx$
3. $\int_0^5 f(x)\, dx$
4. $\int_0^3 5f(x)\, dx$
5. $\int_1^1 f(x)\, dx$

SOLUTION

1. $\int_0^3 f(x)\, dx$ is the area under f on the interval $[0, 3]$. This region is a triangle, so the area is $\int_0^3 f(x)\, dx = \frac{1}{2}(3)(1) = 1.5$.

2. $\int_3^5 f(x)\, dx$ represents the area of the triangle found under the x–axis on $[3, 5]$. The area is $\frac{1}{2}(2)(1) = 1$; since it is found *under* the x–axis, this is "negative area." Therefore $\int_3^5 f(x)\, dx = -1$.

3. $\int_0^5 f(x)\, dx$ is the total signed area under f on $[0, 5]$. This is $1.5 + (-1) = 0.5$.

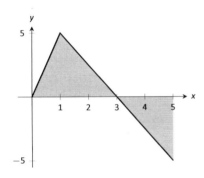

Figure 5.6: A graph of $5f$ in Example 114. (Yes, it looks just like the graph of f in Figure 5.5, just with a different y-scale.)

4. $\int_0^3 5f(x)\, dx$ is the area under $5f$ on $[0, 3]$. This is sketched in Figure 5.6. Again, the region is a triangle, with height 5 times that of the height of the original triangle. Thus the area is $\int_0^3 5f(x)\, dx = 15/2 = 7.5$.

5. $\int_1^1 f(x)\, dx$ is the area under f on the "interval" $[1, 1]$. This describes a line segment, not a region; it has no width. Therefore the area is 0.

This example illustrates some of the properties of the definite integral, given here.

Notes:

5.2 The Definite Integral

Theorem 36   Properties of the Definite Integral

Let f and g be defined on a closed interval I that contains the values a, b and c, and let k be a constant. The following hold:

1. $\displaystyle\int_a^a f(x)\,dx = 0$

2. $\displaystyle\int_a^b f(x)\,dx + \int_b^c f(x)\,dx = \int_a^c f(x)\,dx$

3. $\displaystyle\int_a^b f(x)\,dx = -\int_b^a f(x)\,dx$

4. $\displaystyle\int_a^b \bigl(f(x) \pm g(x)\bigr)\,dx = \int_a^b f(x)\,dx \pm \int_a^b g(x)\,dx$

5. $\displaystyle\int_a^b k \cdot f(x)\,dx = k \cdot \int_a^b f(x)\,dx$

We give a brief justification of Theorem 36 here.

1. As demonstrated in Example 114, there is no "area under the curve" when the region has no width; hence this definite integral is 0.

2. This states that total area is the sum of the areas of subregions. It is easily considered when we let $a < b < c$. We can break the interval $[a,c]$ into two subintervals, $[a,b]$ and $[b,c]$. The total area over $[a,c]$ is the area over $[a,b]$ plus the area over $[b,c]$.

   It is important to note that this still holds true even if $a < b < c$ is not true. We discuss this in the next point.

3. This property can be viewed a merely a convention to make other properties work well. (Later we will see how this property has a justification all its own, not necessarily in support of other properties.) Suppose $b < a < c$. The discussion from the previous point clearly justifies

$$\int_b^a f(x)\,dx + \int_a^c f(x)\,dx = \int_b^c f(x)\,dx. \tag{5.1}$$

However, we still claim that, as originally stated,

$$\int_a^b f(x)\,dx + \int_b^c f(x)\,dx = \int_a^c f(x)\,dx. \tag{5.2}$$

Notes:

How do Equations (5.1) and (5.2) relate? Start with Equation (5.1):

$$\int_b^a f(x)\,dx + \int_a^c f(x)\,dx = \int_b^c f(x)\,dx$$

$$\int_a^c f(x)\,dx = -\int_b^a f(x)\,dx + \int_b^c f(x)\,dx$$

Property (3) justifies changing the sign and switching the bounds of integration on the $-\int_b^a f(x)\,dx$ term; when this is done, Equations (5.1) and (5.2) are equivalent.

The conclusion is this: by adopting the convention of Property (3), Property (2) holds no matter the order of a, b and c. Again, in the next section we will see another justification for this property.

4,5. Each of these may be non–intuitive. Property (5) states that when one scales a function by, for instance, 7, the area of the enclosed region also is scaled by a factor of 7. Both Properties (4) and (5) can be proved using geometry. The details are not complicated but are not discussed here.

Example 115 Evaluating definite integrals using Theorem 36.
Consider the graph of a function $f(x)$ shown in Figure 5.7. Answer the following:

1. Which value is greater: $\int_a^b f(x)\,dx$ or $\int_b^c f(x)\,dx$?

2. Is $\int_a^c f(x)\,dx$ greater or less than 0?

3. Which value is greater: $\int_a^b f(x)\,dx$ or $\int_c^b f(x)\,dx$?

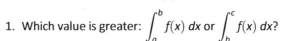

Figure 5.7: A graph of a function in Example 115.

Solution

1. $\int_a^b f(x)\,dx$ has a positive value (since the area is above the x–axis) whereas $\int_b^c f(x)\,dx$ has a negative value. Hence $\int_a^b f(x)\,dx$ is bigger.

2. $\int_a^c f(x)\,dx$ is the total signed area under f between $x = a$ and $x = c$. Since the region below the x–axis looks to be larger than the region above, we conclude that the definite integral has a value less than 0.

3. Note how the second integral has the bounds "reversed." Therefore $\int_c^b f(x)\,dx$ represents a positive number, greater than the area described by the first definite integral. Hence $\int_c^b f(x)\,dx$ is greater.

Notes:

5.2 The Definite Integral

The area definition of the definite integral allows us to use geometry compute the definite integral of some simple functions.

Example 116   Evaluating definite integrals using geometry
Evaluate the following definite integrals:

1. $\int_{-2}^{5} (2x-4)\,dx$ 2. $\int_{-3}^{3} \sqrt{9-x^2}\,dx$.

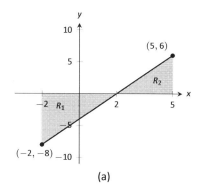

SOLUTION

1. It is useful to sketch the function in the integrand, as shown in Figure 5.8(a). We see we need to compute the areas of two regions, which we have labeled R_1 and R_2. Both are triangles, so the area computation is straightforward:

$$R_1 : \frac{1}{2}(4)(8) = 16 \qquad R_2 : \frac{1}{2}(3)6 = 9.$$

Region R_1 lies under the x–axis, hence it is counted as negative area (we can think of the triangle's height as being "-8"), so

$$\int_{-2}^{5} (2x-4)\,dx = -16 + 9 = -7.$$

2. Recognize that the integrand of this definite integral describes a half circle, as sketched in Figure 5.8(b), with radius 3. Thus the area is:

$$\int_{-3}^{3} \sqrt{9-x^2}\,dx = \frac{1}{2}\pi r^2 = \frac{9}{2}\pi.$$

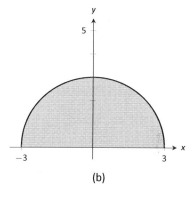

Figure 5.8: A graph of $f(x) = 2x - 4$ in (a) and $f(x) = \sqrt{9-x^2}$ in (b), from Example 116.

Example 117   Understanding motion given velocity
Consider the graph of a velocity function of an object moving in a straight line, given in Figure 5.9, where the numbers in the given regions gives the area of that region. Assume that the definite integral of a velocity function gives displacement. Find the maximum speed of the object and its maximum displacement from its starting position.

SOLUTION   Since the graph gives velocity, finding the maximum speed is simple: it looks to be 15ft/s.
At time $t = 0$, the displacement is 0; the object is at its starting position. At time $t = a$, the object has moved backward 11 feet. Between times $t = a$ and

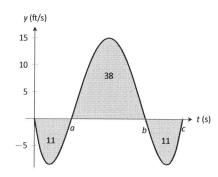

Figure 5.9: A graph of a velocity in Example 117.

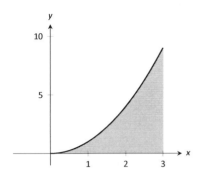

Figure 5.10: What is the area below $y = x^2$ on $[0, 3]$? The region is not a usual geometric shape.

$t = b$, the object moves forward 38 feet, bringing it into a position 27 feet forward of its starting position. From $t = b$ to $t = c$ the object is moving backwards again, hence its maximum displacement is 27 feet from its starting position.

In our examples, we have either found the areas of regions that have nice geometric shapes (such as rectangles, triangles and circles) or the areas were given to us. Consider Figure 5.10, where a region below $y = x^2$ is shaded. What is its area? The function $y = x^2$ is relatively simple, yet the shape it defines has an area that is not simple to find geometrically.

In the next section we will explore how to find the areas of such regions.

Exercises 5.2

Terms and Concepts

1. What is "total signed area"?

2. What is "displacement"?

3. What is $\int_3^3 \sin x\, dx$?

4. Give a single definite integral that has the same value as
$$\int_0^1 (2x+3)\, dx + \int_1^2 (2x+3)\, dx.$$

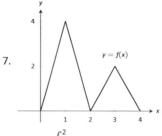

7.

(a) $\int_0^2 f(x)\, dx$

(b) $\int_2^4 f(x)\, dx$

(c) $\int_2^4 2f(x)\, dx$

(d) $\int_0^1 4x\, dx$

(e) $\int_2^3 (2x-4)\, dx$

(f) $\int_2^3 (4x-8)\, dx$

Problems

In Exercises 5 – 9, a graph of a function $f(x)$ is given. Using the geometry of the graph, evaluate the definite integrals.

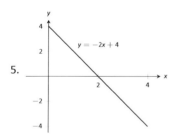

5.

(a) $\int_0^1 (-2x+4)\, dx$

(b) $\int_0^2 (-2x+4)\, dx$

(c) $\int_0^3 (-2x+4)\, dx$

(d) $\int_1^3 (-2x+4)\, dx$

(e) $\int_2^4 (-2x+4)\, dx$

(f) $\int_0^1 (-6x+12)\, dx$

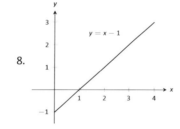

8.

(a) $\int_0^1 (x-1)\, dx$

(b) $\int_0^2 (x-1)\, dx$

(c) $\int_0^3 (x-1)\, dx$

(d) $\int_2^3 (x-1)\, dx$

(e) $\int_1^4 (x-1)\, dx$

(f) $\int_1^4 \bigl((x-1)+1\bigr)\, dx$

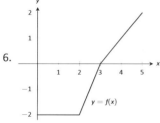

6.

(a) $\int_0^2 f(x)\, dx$

(b) $\int_0^3 f(x)\, dx$

(c) $\int_0^5 f(x)\, dx$

(d) $\int_2^5 f(x)\, dx$

(e) $\int_5^3 f(x)\, dx$

(f) $\int_0^3 -2f(x)\, dx$

9. $f(x) = \sqrt{4-(x-2)^2}$

(a) $\int_0^2 f(x)\, dx$

(b) $\int_2^4 f(x)\, dx$

(c) $\int_0^4 f(x)\, dx$

(d) $\int_0^4 5f(x)\, dx$

In Exercises 10 – 13, a graph of a function $f(x)$ is given; the numbers inside the shaded regions give the area of that region. Evaluate the definite integrals using this area information.

10.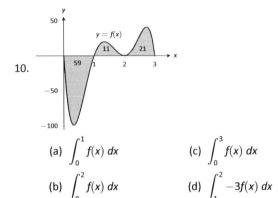

(a) $\int_0^1 f(x)\,dx$

(b) $\int_0^2 f(x)\,dx$

(c) $\int_0^3 f(x)\,dx$

(d) $\int_1^2 -3f(x)\,dx$

11.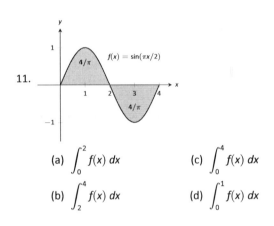

(a) $\int_0^2 f(x)\,dx$

(b) $\int_2^4 f(x)\,dx$

(c) $\int_0^4 f(x)\,dx$

(d) $\int_0^1 f(x)\,dx$

12.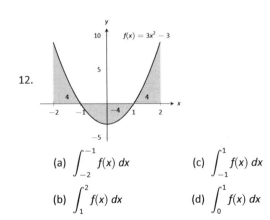

(a) $\int_{-2}^{-1} f(x)\,dx$

(b) $\int_1^2 f(x)\,dx$

(c) $\int_{-1}^1 f(x)\,dx$

(d) $\int_0^1 f(x)\,dx$

13.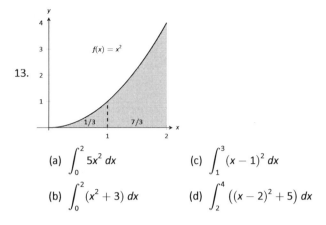

(a) $\int_0^2 5x^2\,dx$

(b) $\int_0^2 (x^2 + 3)\,dx$

(c) $\int_1^3 (x - 1)^2\,dx$

(d) $\int_2^4 \left((x - 2)^2 + 5\right) dx$

In Exercises 14 – 15, a graph of the velocity function of an object moving in a straight line is given. Answer the questions based on that graph.

14.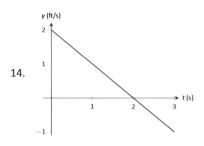

(a) What is the object's maximum velocity?

(b) What is the object's maximum displacement?

(c) What is the object's total displacement on $[0, 3]$?

15.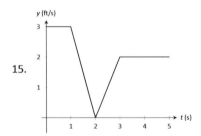

(a) What is the object's maximum velocity?

(b) What is the object's maximum displacement?

(c) What is the object's total displacement on $[0, 5]$?

16. An object is thrown straight up with a velocity, in ft/s, given by $v(t) = -32t + 64$, where t is in seconds, from a height of 48 feet.

(a) What is the object's maximum velocity?

(b) What is the object's maximum displacement?

(c) When does the maximum displacement occur?

(d) When will the object reach a height of 0? (Hint: find when the displacement is -48ft.)

17. An object is thrown straight up with a velocity, in ft/s, given by $v(t) = -32t + 96$, where t is in seconds, from a height of 64 feet.

    (a) What is the object's initial velocity?

    (b) When is the object's displacement 0?

    (c) How long does it take for the object to return to its initial height?

    (d) When will the object reach a height of 210 feet?

In Exercises 18 – 21, let

- $\int_0^2 f(x)\,dx = 5$,
- $\int_0^3 f(x)\,dx = 7$,
- $\int_0^2 g(x)\,dx = -3$, **and**
- $\int_2^3 g(x)\,dx = 5$.

Use these values to evaluate the given definite integrals.

18. $\int_0^2 \big(f(x) + g(x)\big)\,dx$

19. $\int_0^3 \big(f(x) - g(x)\big)\,dx$

20. $\int_2^3 \big(3f(x) + 2g(x)\big)\,dx$

21. Find values for a and b such that
$$\int_0^3 \big(af(x) + bg(x)\big)\,dx = 0$$

In Exercises 22 – 25, let

- $\int_0^3 s(t)\,dt = 10$,
- $\int_3^5 s(t)\,dt = 8$,
- $\int_3^5 r(t)\,dt = -1$, **and**
- $\int_0^5 r(t)\,dt = 11$.

Use these values to evaluate the given definite integrals.

22. $\int_0^3 \big(s(t) + r(t)\big)\,dt$

23. $\int_5^0 \big(s(t) - r(t)\big)\,dt$

24. $\int_3^3 \big(\pi s(t) - 7r(t)\big)\,dt$

25. Find values for a and b such that
$$\int_0^5 \big(ar(t) + bs(t)\big)\,dt = 0$$

Review

In Exercises 26 – 29, evaluate the given indefinite integral.

26. $\int \big(x^3 - 2x^2 + 7x - 9\big)\,dx$

27. $\int \big(\sin x - \cos x + \sec^2 x\big)\,dx$

28. $\int \big(\sqrt[3]{t} + \dfrac{1}{t^2} + 2^t\big)\,dt$

29. $\int \left(\dfrac{1}{x} - \csc x \cot x\right) dx$

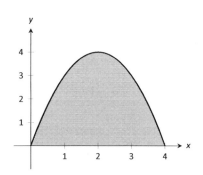

Figure 5.11: A graph of $f(x) = 4x - x^2$. What is the area of the shaded region?

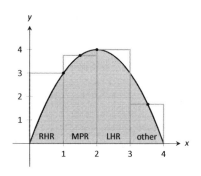

Figure 5.12: Approximating $\int_0^4 (4x - x^2)\, dx$ using rectangles. The heights of the rectangles are determined using different rules.

5.3 Riemann Sums

In the previous section we defined the definite integral of a function on $[a, b]$ to be the signed area between the curve and the x–axis. Some areas were simple to compute; we ended the section with a region whose area was not simple to compute. In this section we develop a technique to find such areas.

A fundamental calculus technique is to first answer a given problem with an approximation, then refine that approximation to make it better, then use limits in the refining process to find the exact answer. That is exactly what we will do here.

Consider the region given in Figure 5.11, which is the area under $y = 4x - x^2$ on $[0, 4]$. What is the signed area of this region – i.e., what is $\int_0^4 (4x - x^2)\, dx$?

We start by approximating. We can surround the region with a rectangle with height and width of 4 and find the area is approximately 16 square units. This is obviously an *over–approximation*; we are including area in the rectangle that is not under the parabola.

We have an approximation of the area, using one rectangle. How can we refine our approximation to make it better? The key to this section is this answer: *use more rectangles*.

Let's use 4 rectangles of equal width of 1. This *partitions* the interval $[0, 4]$ into 4 *subintervals*, $[0, 1]$, $[1, 2]$, $[2, 3]$ and $[3, 4]$. On each subinterval we will draw a rectangle.

There are three common ways to determine the height of these rectangles: the **Left Hand Rule**, the **Right Hand Rule**, and the **Midpoint Rule**. The **Left Hand Rule** says to evaluate the function at the left–hand endpoint of the subinterval and make the rectangle that height. In Figure 5.12, the rectangle drawn on the interval $[2, 3]$ has height determined by the Left Hand Rule; it has a height of $f(2)$. (The rectangle is labeled "LHR.")

The **Right Hand Rule** says the opposite: on each subinterval, evaluate the function at the right endpoint and make the rectangle that height. In the figure, the rectangle drawn on $[0, 1]$ is drawn using $f(1)$ as its height; this rectangle is labeled "RHR.".

The **Midpoint Rule** says that on each subinterval, evaluate the function at the midpoint and make the rectangle that height. The rectangle drawn on $[1, 2]$ was made using the Midpoint Rule, with a height of $f(1.5)$. That rectangle is labeled "MPR."

These are the three most common rules for determining the heights of approximating rectangles, but one is not forced to use one of these three methods. The rectangle on $[3, 4]$ has a height of approximately $f(3.53)$, very close to the Midpoint Rule. It was chosen so that the area of the rectangle is *exactly* the area of the region under f on $[3, 4]$. (Later you'll be able to figure how to do this, too.)

The following example will approximate the value of $\int_0^4 (4x - x^2)\, dx$ using

Notes:

these rules.

Example 118   **Using the Left Hand, Right Hand and Midpoint Rules**
Approximate the value of $\int_0^4 (4x - x^2)\,dx$ using the Left Hand Rule, the Right Hand Rule, and the Midpoint Rule, using 4 equally spaced subintervals.

SOLUTION   We break the interval $[0, 4]$ into four subintervals as before. In Figure 5.13 we see 4 rectangles drawn on $f(x) = 4x - x^2$ using the Left Hand Rule. (The areas of the rectangles are given in each figure.)
Note how in the first subinterval, $[0, 1]$, the rectangle has height $f(0) = 0$. We add up the areas of each rectangle (height $\times$ width) for our Left Hand Rule approximation:

$$f(0) \cdot 1 + f(1) \cdot 1 + f(2) \cdot 1 + f(3) \cdot 1 =$$
$$0 + 3 + 4 + 3 = 10.$$

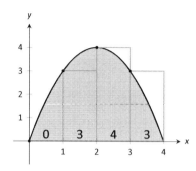

Figure 5.13: Approximating $\int_0^4 (4x - x^2)\,dx$ using the Left Hand Rule in Example 118.

Figure 5.14 shows 4 rectangles drawn under f using the Right Hand Rule; note how the $[3, 4]$ subinterval has a rectangle of height 0.
In this example, these rectangle seem to be the mirror image of those found in Figure 5.13. (This is because of the symmetry of our shaded region.) Our approximation gives the same answer as before, though calculated a different way:

$$f(1) \cdot 1 + f(2) \cdot 1 + f(3) \cdot 1 + f(4) \cdot 1 =$$
$$3 + 4 + 3 + 0 = 10.$$

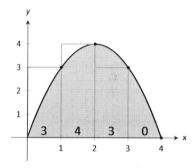

Figure 5.14: Approximating $\int_0^4 (4x - x^2)\,dx$ using the Right Hand Rule in Example 118.

Figure 5.15 shows 4 rectangles drawn under f using the Midpoint Rule. This gives an approximation of $\int_0^4 (4x - x^2)\,dx$ as:

$$f(0.5) \cdot 1 + f(1.5) \cdot 1 + f(2.5) \cdot 1 + f(3.5) \cdot 1 =$$
$$1.75 + 3.75 + 3.75 + 1.75 = 11.$$

Our three methods provide two approximations of $\int_0^4 (4x - x^2)\,dx$: 10 and 11.

Summation Notation

It is hard to tell at this moment which is a better approximation: 10 or 11? We can continue to refine our approximation by using more rectangles. The notation can become unwieldy, though, as we add up longer and longer lists of numbers. We introduce **summation notation** to ameliorate this problem.

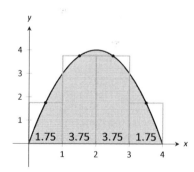

Figure 5.15: Approximating $\int_0^4 (4x - x^2)\,dx$ using the Midpoint Rule in Example 118.

Notes:

Suppose we wish to add up a list of numbers $a_1, a_2, a_3, ..., a_9$. Instead of writing
$$a_1 + a_2 + a_3 + a_4 + a_5 + a_6 + a_7 + a_8 + a_9,$$
we use summation notation and write

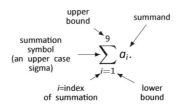

Figure 5.16: Understanding summation notation.

The upper case sigma represents the term "sum." The index of summation in this example is i; any symbol can be used. By convention, the index takes on only the integer values between (and including) the lower and upper bounds.

Let's practice using this notation.

Example 119   **Using summation notation**
Let the numbers $\{a_i\}$ be defined as $a_i = 2i - 1$ for integers i, where $i \geq 1$. So $a_1 = 1, a_2 = 3, a_3 = 5$, etc. (The output is the positive odd integers). Evaluate the following summations:

1. $\displaystyle\sum_{i=1}^{6} a_i$ 2. $\displaystyle\sum_{i=3}^{7} (3a_i - 4)$ 3. $\displaystyle\sum_{i=1}^{4} (a_i)^2$

SOLUTION

1.
$$\sum_{i=1}^{6} a_i = a_1 + a_2 + a_3 + a_4 + a_5 + a_6$$
$$= 1 + 3 + 5 + 7 + 9 + 11$$
$$= 36.$$

2. Note the starting value is different than 1:
$$\sum_{i=3}^{7} a_i = (3a_3 - 4) + (3a_4 - 4) + (3a_5 - 4) + (3a_6 - 4) + (3a_7 - 4)$$
$$= 11 + 17 + 23 + 29 + 35$$
$$= 115.$$

Notes:

3.
$$\sum_{i=1}^{4}(a_i)^2 = (a_1)^2 + (a_2)^2 + (a_3)^2 + (a_4)^2$$
$$= 1^2 + 3^2 + 5^2 + 7^2$$
$$= 84$$

It might seem odd to stress a new, concise way of writing summations only to write each term out as we add them up. It is. The following theorem gives some of the properties of summations that allow us to work with them without writing individual terms. Examples will follow.

Theorem 37   **Properties of Summations**

1. $\sum_{i=1}^{n} c = c \cdot n$, where c is a constant.

2. $\sum_{i=m}^{n} (a_i \pm b_i) = \sum_{i=m}^{n} a_i \pm \sum_{i=m}^{n} b_i$

3. $\sum_{i=m}^{n} c \cdot a_i = c \cdot \sum_{i=m}^{n} a_i$

4. $\sum_{i=m}^{j} a_i + \sum_{i=j+1}^{n} a_i = \sum_{i=m}^{n} a_i$

5. $\sum_{i=1}^{n} i = \dfrac{n(n+1)}{2}$

6. $\sum_{i=1}^{n} i^2 = \dfrac{n(n+1)(2n+1)}{6}$

7. $\sum_{i=1}^{n} i^3 = \left(\dfrac{n(n+1)}{2}\right)^2$

Example 120   **Evaluating summations using Theorem 37**
Revisit Example 119 and, using Theorem 37, evaluate

$$\sum_{i=1}^{6} a_i = \sum_{i=1}^{6} (2i - 1).$$

Notes:

Solution

$$\sum_{i=1}^{6}(2i-1) = \sum_{i=1}^{6} 2i - \sum_{i=1}^{6}(1)$$
$$= \left(2\sum_{i=1}^{6} i\right) - 6$$
$$= 2\frac{6(6+1)}{2} - 6$$
$$= 42 - 6 = 36$$

We obtained the same answer without writing out all six terms. When dealing with small sizes of n, it may be faster to write the terms out by hand. However, Theorem 37 is incredibly important when dealing with large sums as we'll soon see.

Riemann Sums

Consider again $\int_0^4 (4x - x^2)\,dx$. We will approximate this definite integral using 16 equally spaced subintervals and the Right Hand Rule in Example 121. Before doing so, it will pay to do some careful preparation.

Figure 5.17 shows a number line of $[0, 4]$ divided into 16 equally spaced subintervals. We denote 0 as x_1; we have marked the values of x_5, x_9, x_{13} and x_{17}. We could mark them all, but the figure would get crowded. While it is easy to figure that $x_{10} = 2.25$, in general, we want a method of determining the value of x_i without consulting the figure. Consider:

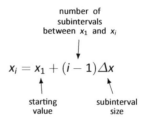

So $x_{10} = x_1 + 9(4/16) = 2.25$.

If we had partitioned $[0, 4]$ into 100 equally spaced subintervals, each subinterval would have length $\Delta x = 4/100 = 0.04$. We could compute x_{32} as

$$x_{32} = x_1 + 31(4/100) = 1.24.$$

(That was far faster than creating a sketch first.)

Figure 5.17: Dividing $[0, 4]$ into 16 equally spaced subintervals.

Notes:

Given any subdivision of $[0, 4]$, the first subinterval is $[x_1, x_2]$; the second is $[x_2, x_3]$; the ith subinterval is $[x_i, x_{i+1}]$.

When using the Left Hand Rule, the height of the ith rectangle will be $f(x_i)$.

When using the Right Hand Rule, the height of the ith rectangle will be $f(x_{i+1})$.

When using the Midpoint Rule, the height of the ith rectangle will be $f\left(\dfrac{x_i + x_{i+1}}{2}\right)$.

Thus approximating $\int_0^4 (4x - x^2)\, dx$ with 16 equally spaced subintervals can be expressed as follows, where $\Delta x = 4/16 = 1/4$:

Left Hand Rule: $\displaystyle\sum_{i=1}^{16} f(x_i)\Delta x$

Right Hand Rule: $\displaystyle\sum_{i=1}^{16} f(x_{i+1})\Delta x$

Midpoint Rule: $\displaystyle\sum_{i=1}^{16} f\left(\dfrac{x_i + x_{i+1}}{2}\right)\Delta x$

We use these formulas in the next two examples. The following example lets us practice using the Right Hand Rule and the summation formulas introduced in Theorem 37.

Example 121   **Approximating definite integrals using sums**
Approximate $\int_0^4 (4x - x^2)\, dx$ using the Right Hand Rule and summation formulas with 16 and 1000 equally spaced intervals.

SOLUTION   Using the formula derived before, using 16 equally spaced intervals and the Right Hand Rule, we can approximate the definite integral as

$$\sum_{i=1}^{16} f(x_{i+1})\Delta x.$$

We have $\Delta x = 4/16 = 0.25$. Since $x_i = 0 + (i-1)\Delta x$, we have

$$\begin{aligned} x_{i+1} &= 0 + \big((i+1) - 1\big)\Delta x \\ &= i\Delta x \end{aligned}$$

Notes:

Using the summation formulas, consider:

$$\int_0^4 (4x - x^2)\,dx \approx \sum_{i=1}^{16} f(x_{i+1})\Delta x$$

$$= \sum_{i=1}^{16} f(i\Delta x)\Delta x$$

$$= \sum_{i=1}^{16} \left(4i\Delta x - (i\Delta x)^2\right)\Delta x$$

$$= \sum_{i=1}^{16} (4i\Delta x^2 - i^2 \Delta x^3)$$

$$= (4\Delta x^2)\sum_{i=1}^{16} i - \Delta x^3 \sum_{i=1}^{16} i^2 \qquad (5.3)$$

$$= (4\Delta x^2)\frac{16 \cdot 17}{2} - \Delta x^3 \frac{16(17)(33)}{6}$$

$$= 4 \cdot 0.25^2 \cdot 136 - 0.25^3 \cdot 1496$$

$$= 10.625$$

We were able to sum up the areas of 16 rectangles with very little computation. In Figure 5.18 the function and the 16 rectangles are graphed. While some rectangles over–approximate the area, other under–approximate the area (by about the same amount). Thus our approximate area of 10.625 is likely a fairly good approximation.

Notice Equation (5.3); by changing the 16's to 1,000's (and appropriately changing the value of Δx), we can use that equation to sum up 1000 rectangles! We do so here, skipping from the original summand to the equivalent of Equation (5.3) to save space. Note that $\Delta x = 4/1000 = 0.004$.

$$\int_0^4 (4x - x^2)\,dx \approx \sum_{i=1}^{1000} f(x_{i+1})\Delta x$$

$$= (4\Delta x^2)\sum_{i=1}^{1000} i - \Delta x^3 \sum_{i=1}^{1000} i^2$$

$$= (4\Delta x^2)\frac{1000 \cdot 1001}{2} - \Delta x^3 \frac{1000(1001)(2001)}{6}$$

$$= 4 \cdot 0.004^2 \cdot 500500 - 0.004^3 \cdot 333,833,500$$

$$= 10.666656$$

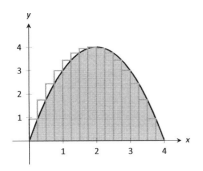

Figure 5.18: Approximating $\int_0^4 (4x-x^2)\,dx$ with the Right Hand Rule and 16 evenly spaced subintervals.

Notes:

Using many, many rectangles, we have a likely good approximation of $\int_0^4 (4x - x^2) \Delta x$. That is,

$$\int_0^4 (4x - x^2)\, dx \approx 10.666656.$$

Before the above example, we stated what the summations for the Left Hand, Right Hand and Midpoint Rules looked like. Each had the same basic structure, which was:

1. each rectangle has the same width, which we referred to as Δx, and

2. each rectangle's height is determined by evaluating f at a particular point in each subinterval. For instance, the Left Hand Rule states that each rectangle's height is determined by evaluating f at the left hand endpoint of the subinterval the rectangle lives on.

One could partition an interval $[a, b]$ with subintervals that did not have the same size. We refer to the length of the first subinterval as Δx_1, the length of the second subinterval as Δx_2, and so on, giving the length of the ith subinterval as Δx_i. Also, one could determine each rectangle's height by evaluating f at *any* point in the ith subinterval. We refer to the point picked in the first subinterval as c_1, the point picked in the second subinterval as c_2, and so on, with c_i representing the point picked in the ith subinterval. Thus the height of the ith subinterval would be $f(c_i)$, and the area of the ith rectangle would be $f(c_i)\Delta x_i$.

Summations of rectangles with area $f(c_i)\Delta x_i$ are named after mathematician Georg Friedrich Bernhard Riemann, as given in the following definition.

Definition 21   **Riemann Sum**

Let f be defined on the closed interval $[a, b]$ and let Δx be a partition of $[a, b]$, with

$$a = x_1 < x_2 < \ldots < x_n < x_{n+1} = b.$$

Let Δx_i denote the length of the ith subinterval $[x_i, x_{i+1}]$ and let c_i denote any value in the ith subinterval.
The sum

$$\sum_{i=1}^n f(c_i)\Delta x_i$$

is a **Riemann sum** of f on $[a, b]$.

Figure 5.19 shows the approximating rectangles of a Riemann sum of $\int_0^4 (4x - x^2)\, dx$. While the rectangles in this example do not approximate well the shaded area, they demonstrate that the subinterval widths may vary and the heights of the rectangles can be determined without following a particular rule.

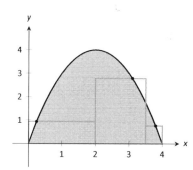

Figure 5.19: An example of a general Riemann sum to approximate $\int_0^4 (4x - x^2)\, dx$.

Notes:

Chapter 5 Integration

"Usually" Riemann sums are calculated using one of the three methods we have introduced. The uniformity of construction makes computations easier. Before working another example, let's summarize some of what we have learned in a convenient way.

Key Idea 8   **Riemann Sum Concepts**

Consider $\int_a^b f(x)\, dx \approx \sum_{i=1}^{n} f(c_i) \Delta x_i$.

1. When the n subintervals have equal length, $\Delta x_i = \Delta x = \dfrac{b-a}{n}$.

2. The ith term of the partition is $x_i = a + (i-1)\Delta x$. (This makes $x_{n+1} = b$.)

3. The Left Hand Rule summation is: $\sum_{i=1}^{n} f(x_i) \Delta x$.

4. The Right Hand Rule summation is: $\sum_{i=1}^{n} f(x_{i+1}) \Delta x$.

5. The Midpoint Rule summation is: $\sum_{i=1}^{n} f\left(\dfrac{x_i + x_{x+1}}{2}\right) \Delta x$.

Let's do another example.

Example 122   **Approximating definite integrals with sums**
Approximate $\int_{-2}^{3} (5x + 2)\, dx$ using the Midpoint Rule and 10 equally spaced intervals.

Solution   Following Key Idea 8, we have

$$\Delta x = \frac{3 - (-2)}{10} = 1/2 \quad \text{and} \quad x_i = (-2) + (1/2)(i - 1) = i/2 - 5/2.$$

As we are using the Midpoint Rule, we will also need x_{i+1} and $\dfrac{x_i + x_{i+1}}{2}$. Since $x_i = i/2 - 5/2$, $x_{i+1} = (i+1)/2 - 5/2 = i/2 - 2$. This gives

$$\frac{x_i + x_{i+1}}{2} = \frac{(i/2 - 5/2) + (i/2 - 2)}{2} = \frac{i - 9/2}{2} = i/2 - 9/4.$$

Notes:

We now construct the Riemann sum and compute its value using summation formulas.

$$\int_{-2}^{3}(5x+2)\,dx \approx \sum_{i=1}^{10} f\left(\frac{x_i + x_{i+1}}{2}\right)\Delta x$$

$$= \sum_{i=1}^{10} f(i/2 - 9/4)\Delta x$$

$$= \sum_{i=1}^{10} \left(5(i/2 - 9/4) + 2\right)\Delta x$$

$$= \Delta x \sum_{i=1}^{10}\left[\left(\frac{5}{2}\right)i - \frac{37}{4}\right]$$

$$= \Delta x \left(\frac{5}{2}\sum_{i=1}^{10}(i) - \sum_{i=1}^{10}\left(\frac{37}{4}\right)\right)$$

$$= \frac{1}{2}\left(\frac{5}{2}\cdot\frac{10(11)}{2} - 10\cdot\frac{37}{4}\right)$$

$$= \frac{45}{2} = 22.5$$

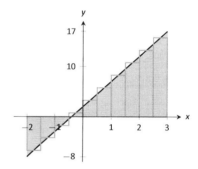

Figure 5.20: Approximating $\int_{-2}^{3}(5x+2)\,dx$ using the Midpoint Rule and 10 evenly spaced subintervals in Example 122.

Note the graph of $f(x) = 5x + 2$ in Figure 5.20. The regions whose area is computed by the definite integral are triangles, meaning we can find the exact answer without summation techniques. We find that the exact answer is indeed 22.5. One of the strengths of the Midpoint Rule is that often each rectangle includes area that should not be counted, but misses other area that should. When the partition size is small, these two amounts are about equal and these errors almost "cancel each other out." In this example, since our function is a line, these errors are exactly equal and they do cancel each other out, giving us the exact answer.

Note too that when the function is negative, the rectangles have a "negative" height. When we compute the area of the rectangle, we use $f(c_i)\Delta x$; when f is negative, the area is counted as negative.

Notice in the previous example that while we used 10 equally spaced intervals, the number "10" didn't play a big role in the calculations until the very end. Mathematicians love to abstract ideas; let's approximate the area of another region using n subintervals, where we do not specify a value of n until the very end.

Example 123 **Approximating definite integrals with a formula, using sums**
Revisit $\int_0^4(4x-x^2)\,dx$ yet again. Approximate this definite integral using the Right

Hand Rule with *n* equally spaced subintervals.

SOLUTION Using Key Idea 8, we know $\Delta x = \frac{4-0}{n} = 4/n$. We also find $x_i = 0 + \Delta x(i-1) = 4(i-1)/n$. The Right Hand Rule uses x_{i+1}, which is $x_{i+1} = 4i/n$.

We construct the Right Hand Rule Riemann sum as follows. Be sure to follow each step carefully. If you get stuck, and do not understand how one line proceeds to the next, you may skip to the result and consider how this result is used. You should come back, though, and work through each step for full understanding.

$$\int_0^4 (4x - x^2)\, dx \approx \sum_{i=1}^n f(x_{i+1})\Delta x$$

$$= \sum_{i=1}^n f\left(\frac{4i}{n}\right)\Delta x$$

$$= \sum_{i=1}^n \left[4\frac{4i}{n} - \left(\frac{4i}{n}\right)^2\right]\Delta x$$

$$= \sum_{i=1}^n \left(\frac{16\Delta x}{n}\right)i - \sum_{i=1}^n \left(\frac{16\Delta x}{n^2}\right)i^2$$

$$= \left(\frac{16\Delta x}{n}\right)\sum_{i=1}^n i - \left(\frac{16\Delta x}{n^2}\right)\sum_{i=1}^n i^2$$

$$= \left(\frac{16\Delta x}{n}\right)\cdot\frac{n(n+1)}{2} - \left(\frac{16\Delta x}{n^2}\right)\frac{n(n+1)(2n+1)}{6} \quad \left(\begin{array}{c}\text{recall}\\ \Delta x = 4/n\end{array}\right)$$

$$= \frac{32(n+1)}{n} - \frac{32(n+1)(2n+1)}{3n^2} \quad \text{(now simplify)}$$

$$= \frac{32}{3}\left(1 - \frac{1}{n^2}\right)$$

The result is an amazing, easy to use formula. To approximate the definite integral with 10 equally spaced subintervals and the Right Hand Rule, set $n = 10$ and compute

$$\int_0^4 (4x - x^2)\, dx \approx \frac{32}{3}\left(1 - \frac{1}{10^2}\right) = 10.56.$$

Recall how earlier we approximated the definite integral with 4 subintervals; with $n = 4$, the formula gives 10, our answer as before.

It is now easy to approximate the integral with 1,000,000 subintervals! Handheld calculators will round off the answer a bit prematurely giving an answer of

Notes:

10.66666667. (The actual answer is 10.666666666656.)

We now take an important leap. Up to this point, our mathematics has been limited to geometry and algebra (finding areas and manipulating expressions). Now we apply *calculus*. For any *finite n*, we know that

$$\int_0^4 (4x - x^2)\, dx \approx \frac{32}{3}\left(1 - \frac{1}{n^2}\right).$$

Both common sense and high–level mathematics tell us that as *n* gets large, the approximation gets better. In fact, if we take the *limit* as $n \to \infty$, we get the *exact area* described by $\int_0^4 (4x - x^2)\, dx$. That is,

$$\begin{aligned}
\int_0^4 (4x - x^2)\, dx &= \lim_{n \to \infty} \frac{32}{3}\left(1 - \frac{1}{n^2}\right) \\
&= \frac{32}{3}(1 - 0) \\
&= \frac{32}{3} = 10.\overline{6}
\end{aligned}$$

This is a fantastic result. By considering *n* equally–spaced subintervals, we obtained a formula for an approximation of the definite integral that involved our variable *n*. As *n* grows large – without bound – the error shrinks to zero and we obtain the exact area.

This section started with a fundamental calculus technique: make an approximation, refine the approximation to make it better, then use limits in the refining process to get an exact answer. That is precisely what we just did.

Let's practice this again.

Example 124   Approximating definite integrals with a formula, using sums
Find a formula that approximates $\int_{-1}^5 x^3\, dx$ using the Right Hand Rule and *n* equally spaced subintervals, then take the limit as $n \to \infty$ to find the exact area.

SOLUTION   Following Key Idea 8, we have $\Delta x = \frac{5-(-1)}{n} = 6/n$. We have $x_i = (-1) + (i-1)\Delta x$; as the Right Hand Rule uses x_{i+1}, we have $x_{i+1} = (-1) + i\Delta x$.

The Riemann sum corresponding to the Right Hand Rule is (followed by sim-

Notes:

plifications):

$$\int_{-1}^{5} x^3\, dx \approx \sum_{i=1}^{n} f(x_{i+1})\Delta x$$

$$= \sum_{i=1}^{n} f(-1 + i\Delta x)\Delta x$$

$$= \sum_{i=1}^{n} (-1 + i\Delta x)^3 \Delta x$$

$$= \sum_{i=1}^{n} \left((i\Delta x)^3 - 3(i\Delta x)^2 + 3i\Delta x - 1\right)\Delta x \quad \text{(now distribute } \Delta x\text{)}$$

$$= \sum_{i=1}^{n} \left(i^3 \Delta x^4 - 3i^2 \Delta x^3 + 3i\Delta x^2 - \Delta x\right) \quad \text{(now split up summation)}$$

$$= \Delta x^4 \sum_{i=1}^{n} i^3 - 3\Delta x^3 \sum_{i=1}^{n} i^2 + 3\Delta x^2 \sum_{i=1}^{n} i - \sum_{i=1}^{n} \Delta x$$

$$= \Delta x^4 \left(\frac{n(n+1)}{2}\right)^2 - 3\Delta x^3 \frac{n(n+1)(2n+1)}{6} + 3\Delta x^2 \frac{n(n+1)}{2} - n\Delta x$$

(use $\Delta x = 6/n$)

$$= \frac{1296}{n^4} \cdot \frac{n^2(n+1)^2}{4} - 3\frac{216}{n^3} \cdot \frac{n(n+1)(2n+1)}{6} + 3\frac{36}{n^2} \cdot \frac{n(n+1)}{2} - 6$$

(now do a sizable amount of algebra to simplify)

$$= 156 + \frac{378}{n} + \frac{216}{n^2}$$

Once again, we have found a compact formula for approximating the definite integral with n equally spaced subintervals and the Right Hand Rule. Using 10 subintervals, we have an approximation of 195.96 (these rectangles are shown in Figure 5.21). Using $n = 100$ gives an approximation of 159.802.

Now find the exact answer using a limit:

$$\int_{-1}^{5} x^3\, dx = \lim_{n\to\infty} \left(156 + \frac{378}{n} + \frac{216}{n^2}\right) = 156.$$

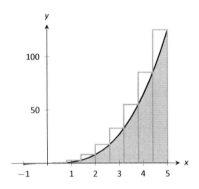

Figure 5.21: Approximating $\int_{-1}^{5} x^3\, dx$ using the Right Hand Rule and 10 evenly spaced subintervals.

Limits of Riemann Sums

We have used limits to evaluate exactly given definite limits. Will this always work? We will show, given not–very–restrictive conditions, that yes, it will always work.

Notes:

The previous two examples demonstrated how an expression such as

$$\sum_{i=1}^{n} f(x_{i+1})\Delta x$$

can be rewritten as an expression explicitly involving n, such as $32/3(1 - 1/n^2)$.

Viewed in this manner, we can think of the summation as a function of n. An n value is given (where n is a positive integer), and the sum of areas of n equally spaced rectangles is returned, using the Left Hand, Right Hand, or Midpoint Rules.

Given a definite integral $\int_a^b f(x)\, dx$, let:

- $S_L(n) = \sum_{i=1}^{n} f(x_i)\Delta x$, the sum of equally spaced rectangles formed using the Left Hand Rule,

- $S_R(n) = \sum_{i=1}^{n} f(x_{i+1})\Delta x$, the sum of equally spaced rectangles formed using the Right Hand Rule, and

- $S_M(n) = \sum_{i=1}^{n} f\left(\frac{x_i + x_{i+1}}{2}\right)\Delta x$, the sum of equally spaced rectangles formed using the Midpoint Rule.

Recall the definition of a limit as $n \to \infty$: $\lim_{n \to \infty} S_L(n) = K$ if, given any $\varepsilon > 0$, there exists $N > 0$ such that

$$|S_L(n) - K| < \varepsilon \quad \text{when} \quad n \geq N.$$

The following theorem states that we can use any of our three rules to find the exact value of a definite integral $\int_a^b f(x)\, dx$. It also goes two steps further. The theorem states that the height of each rectangle doesn't have to be determined following a specific rule, but could be $f(c_i)$, where c_i is any point in the ith subinterval, as discussed before Riemann Sums where defined in Definition 21.

The theorem goes on to state that the rectangles do not need to be of the same width. Using the notation of Definition 21, let Δx_i denote the length of the ith subinterval in a partition of $[a, b]$. Now let $||\Delta x||$ represent the length of the largest subinterval in the partition: that is, $||\Delta x||$ is the largest of all the Δx_i's. If $||\Delta x||$ is small, then $[a, b]$ must be partitioned into many subintervals, since all subintervals must have small lengths. "Taking the limit as $||\Delta x||$ goes to zero" implies that the number n of subintervals in the partition is growing to

Notes:

infinity, as the largest subinterval length is becoming arbitrarily small. We then interpret the expression

$$\lim_{||\Delta x|| \to 0} \sum_{i=1}^{n} f(c_i) \Delta x_i$$

as "the limit of the sum of rectangles, where the width of each rectangle can be different but getting small, and the height of each rectangle is not necessarily determined by a particular rule." The theorem states that this Riemann Sum also gives the value of the definite integral of f over $[a, b]$.

Theorem 38   **Definite Integrals and the Limit of Riemann Sums**

Let f be continuous on the closed interval $[a, b]$ and let $S_L(n)$, $S_R(n)$ and $S_M(n)$ be defined as before. Then:

1. $\lim_{n \to \infty} S_L(n) = \lim_{n \to \infty} S_R(n) = \lim_{n \to \infty} S_M(n) = \lim_{n \to \infty} \sum_{i=1}^{n} f(c_i) \Delta x,$

2. $\lim_{n \to \infty} \sum_{i=1}^{n} f(c_i) \Delta x = \int_a^b f(x)\, dx$, and

3. $\lim_{||\Delta x|| \to 0} \sum_{i=1}^{n} f(c_i) \Delta x_i = \int_a^b f(x)\, dx.$

We summarize what we have learned over the past few sections here.

- Knowing the "area under the curve" can be useful. One common example is: the area under a velocity curve is displacement.

- We have defined the definite integral, $\int_a^b f(x)\, dx$, to be the signed area under f on the interval $[a, b]$.

- While we can approximate a definite integral many ways, we have focused on using rectangles whose heights can be determined using: the Left Hand Rule, the Right Hand Rule and the Midpoint Rule.

- Sums of rectangles of this type are called Riemann sums.

- The exact value of the definite integral can be computed using the limit of a Riemann sum. We generally use one of the above methods as it makes the algebra simpler.

Notes:

We first learned of derivatives through limits then learned rules that made the process simpler. We know of a way to evaluate a definite integral using limits; in the next section we will see how the Fundamental Theorem of Calculus makes the process simpler. The key feature of this theorem is its connection between the indefinite integral and the definite integral.

Notes:

Exercises 5.3

Terms and Concepts

1. A fundamental calculus technique is to use _____ to refine approximations to get an exact answer.

2. What is the upper bound in the summation $\sum_{i=7}^{14}(48i - 201)$?

3. This section approximates definite integrals using what geometric shape?

4. T/F: A sum using the Right Hand Rule is an example of a Riemann Sum.

Problems

In Exercises 5 – 11, write out each term of the summation and compute the sum.

5. $\sum_{i=2}^{4} i^2$

6. $\sum_{i=-1}^{3} (4i - 2)$

7. $\sum_{i=-2}^{2} \sin(\pi i/2)$

8. $\sum_{i=1}^{5} \frac{1}{i}$

9. $\sum_{i=1}^{6} (-1)^i i$

10. $\sum_{i=1}^{4} \left(\frac{1}{i} - \frac{1}{i+1}\right)$

11. $\sum_{i=0}^{5} (-1)^i \cos(\pi i)$

In Exercises 12 – 15, write each sum in summation notation.

12. $3 + 6 + 9 + 12 + 15$

13. $-1 + 0 + 3 + 8 + 15 + 24 + 35 + 48 + 63$

14. $\frac{1}{2} + \frac{2}{3} + \frac{3}{4} + \frac{4}{5}$

15. $1 - e + e^2 - e^3 + e^4$

In Exercises 16 – 22, evaluate the summation using Theorem 37.

16. $\sum_{i=1}^{25} i$

17. $\sum_{i=1}^{10} (3i^2 - 2i)$

18. $\sum_{i=1}^{15} (2i^3 - 10)$

19. $\sum_{i=1}^{10} (-4i^3 + 10i^2 - 7i + 11)$

20. $\sum_{i=1}^{10} (i^3 - 3i^2 + 2i + 7)$

21. $1 + 2 + 3 + \ldots + 99 + 100$

22. $1 + 4 + 9 + \ldots + 361 + 400$

Theorem 37 states

$$\sum_{i=1}^{n} a_i = \sum_{i=1}^{k} a_i + \sum_{i=k+1}^{n} a_i, \text{ so}$$

$$\sum_{i=k+1}^{n} a_i = \sum_{i=1}^{n} a_i - \sum_{i=1}^{k} a_i.$$

Use this fact, along with other parts of Theorem 37, to evaluate the summations given in Exercises 23 – 26.

23. $\sum_{i=11}^{20} i$

24. $\sum_{i=16}^{25} i^3$

25. $\sum_{i=7}^{12} 4$

26. $\sum_{i=5}^{10} 4i^3$

In Exercises 27 – 32, a definite integral
$$\int_a^b f(x)\, dx \text{ is given.}$$

(a) Graph $f(x)$ on $[a, b]$.

(b) Add to the sketch rectangles using the provided rule.

(c) Approximate $\int_a^b f(x)\, dx$ by summing the areas of the rectangles.

27. $\int_{-3}^{3} x^2\, dx$, with 6 rectangles using the Left Hand Rule.

28. $\int_{0}^{2} (5 - x^2)\, dx$, with 4 rectangles using the Midpoint Rule.

29. $\int_{0}^{\pi} \sin x\, dx$, with 6 rectangles using the Right Hand Rule.

30. $\int_{0}^{3} 2^x\, dx$, with 5 rectangles using the Left Hand Rule.

31. $\int_{1}^{2} \ln x\, dx$, with 3 rectangles using the Midpoint Rule.

32. $\int_{1}^{9} \frac{1}{x}\, dx$, with 4 rectangles using the Right Hand Rule.

In Exercises 33 – 38, a definite integral
$$\int_a^b f(x)\, dx \text{ is given. As demonstrated in Examples 123 and 124, do the following.}$$

(a) Find a formula to approximate $\int_a^b f(x)\, dx$ using n subintervals and the provided rule.

(b) Evaluate the formula using $n = 10$, 100 and $1{,}000$.

(c) Find the limit of the formula, as $n \to \infty$, to find the exact value of $\int_a^b f(x)\, dx$.

33. $\int_{0}^{1} x^3\, dx$, using the Right Hand Rule.

34. $\int_{-1}^{1} 3x^2\, dx$, using the Left Hand Rule.

35. $\int_{-1}^{3} (3x - 1)\, dx$, using the Midpoint Rule.

36. $\int_{1}^{4} (2x^2 - 3)\, dx$, using the Left Hand Rule.

37. $\int_{-10}^{10} (5 - x)\, dx$, using the Right Hand Rule.

38. $\int_{0}^{1} (x^3 - x^2)\, dx$, using the Right Hand Rule.

Review

In Exercises 39 – 44, find an antiderivative of the given function.

39. $f(x) = 5 \sec^2 x$

40. $f(x) = \dfrac{7}{x}$

41. $g(t) = 4t^5 - 5t^3 + 8$

42. $g(t) = 5 \cdot 8^t$

43. $g(t) = \cos t + \sin t$

44. $f(x) = \dfrac{1}{\sqrt{x}}$

5.4 The Fundamental Theorem of Calculus

Let $f(t)$ be a continuous function defined on $[a,b]$. The definite integral $\int_a^b f(x)\,dx$ is the "area under f" on $[a,b]$. We can turn this concept into a function by letting the upper (or lower) bound vary.

Let $F(x) = \int_a^x f(t)\,dt$. It computes the area under f on $[a,x]$ as illustrated in Figure 5.22. We can study this function using our knowledge of the definite integral. For instance, $F(a) = 0$ since $\int_a^a f(t)\,dt = 0$.

We can also apply calculus ideas to $F(x)$; in particular, we can compute its derivative. While this may seem like an innocuous thing to do, it has far–reaching implications, as demonstrated by the fact that the result is given as an important theorem.

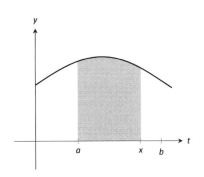

Figure 5.22: The area of the shaded region is $F(x) = \int_a^x f(t)\,dt$.

Theorem 39   **The Fundamental Theorem of Calculus, Part 1**

Let f be continuous on $[a,b]$ and let $F(x) = \int_a^x f(t)\,dt$. Then F is a differentiable function on (a,b), and

$$F'(x) = f(x).$$

Initially this seems simple, as demonstrated in the following example.

Example 125   **Using the Fundamental Theorem of Calculus, Part 1**
Let $F(x) = \displaystyle\int_{-5}^{x} (t^2 + \sin t)\,dt$. What is $F'(x)$?

SOLUTION   Using the Fundamental Theorem of Calculus, we have $F'(x) = x^2 + \sin x$.

This simple example reveals something incredible: $F(x)$ is an antiderivative of $x^2 + \sin x$! Therefore, $F(x) = \frac{1}{3}x^3 - \cos x + C$ for some value of C. (We can find C, but generally we do not care. We know that $F(-5) = 0$, which allows us to compute C. In this case, $C = \cos(-5) + \frac{125}{3}$.)

We have done more than found a complicated way of computing an antiderivative. Consider a function f defined on an open interval containing a, b and c. Suppose we want to compute $\int_a^b f(t)\,dt$. First, let $F(x) = \int_c^x f(t)\,dt$. Using

Notes:

5.4 The Fundamental Theorem of Calculus

the properties of the definite integral found in Theorem 36, we know

$$\int_a^b f(t)\,dt = \int_a^c f(t)\,dt + \int_c^b f(t)\,dt$$
$$= -\int_c^a f(t)\,dt + \int_c^b f(t)\,dt$$
$$= -F(a) + F(b)$$
$$= F(b) - F(a).$$

We now see how indefinite integrals and definite integrals are related: we can evaluate a definite integral using antiderivatives! This is the second part of the Fundamental Theorem of Calculus.

Theorem 40   **The Fundamental Theorem of Calculus, Part 2**

Let f be continuous on $[a, b]$ and let F be *any* antiderivative of f. Then

$$\int_a^b f(x)\,dx = F(b) - F(a).$$

Example 126   **Using the Fundamental Theorem of Calculus, Part 2**
We spent a great deal of time in the previous section studying $\int_0^4 (4x - x^2)\,dx$. Using the Fundamental Theorem of Calculus, evaluate this definite integral.

SOLUTION   We need an antiderivative of $f(x) = 4x - x^2$. All antiderivatives of f have the form $F(x) = 2x^2 - \frac{1}{3}x^3 + C$; for simplicity, choose $C = 0$.

The Fundamental Theorem of Calculus states

$$\int_0^4 (4x - x^2)\,dx = F(4) - F(0) = \left(2(4)^2 - \frac{1}{3}4^3\right) - (0 - 0) = 32 - \frac{64}{3} = 32/3.$$

This is the same answer we obtained using limits in the previous section, just with much less work.

Notation: A special notation is often used in the process of evaluating definite integrals using the Fundamental Theorem of Calculus. Instead of explicitly writing $F(b) - F(a)$, the notation $F(x)\Big|_a^b$ is used. Thus the solution to Example 126 would be written as:

$$\int_0^4 (4x - x^2)\,dx = \left(2x^2 - \frac{1}{3}x^3\right)\Big|_0^4 = \left(2(4)^2 - \frac{1}{3}4^3\right) - (0 - 0) = 32/3.$$

Notes:

The Constant C: *Any antiderivative $F(x)$ can be chosen when using the Fundamental Theorem of Calculus to evaluate a definite integral, meaning any value of C can be picked. The constant always cancels out of the expression when evaluating $F(b) - F(a)$, so it does not matter what value is picked. This being the case, we might as well let $C = 0$.*

Example 127   **Using the Fundamental Theorem of Calculus, Part 2**
Evaluate the following definite integrals.

1. $\displaystyle\int_{-2}^{2} x^3 \, dx$ 2. $\displaystyle\int_{0}^{\pi} \sin x \, dx$ 3. $\displaystyle\int_{0}^{5} e^t \, dt$ 4. $\displaystyle\int_{4}^{9} \sqrt{u} \, du$ 5. $\displaystyle\int_{1}^{5} 2 \, dx$

SOLUTION

1. $\displaystyle\int_{-2}^{2} x^3 \, dx = \frac{1}{4}x^4 \Big|_{-2}^{2} = \left(\frac{1}{4}2^4\right) - \left(\frac{1}{4}(-2)^4\right) = 0.$

2. $\displaystyle\int_{0}^{\pi} \sin x \, dx = -\cos x \Big|_{0}^{\pi} = -\cos \pi - (-\cos 0) = 1 + 1 = 2.$

   (This is interesting; it says that the area under one "hump" of a sine curve is 2.)

3. $\displaystyle\int_{0}^{5} e^t \, dt = e^t \Big|_{0}^{5} = e^5 - e^0 = e^5 - 1 \approx 147.41.$

4. $\displaystyle\int_{4}^{9} \sqrt{u} \, du = \int_{4}^{9} u^{\frac{1}{2}} \, du = \frac{2}{3} u^{\frac{3}{2}} \Big|_{4}^{9} = \frac{2}{3}\left(9^{\frac{3}{2}} - 4^{\frac{3}{2}}\right) = \frac{2}{3}(27 - 8) = \frac{38}{3}.$

5. $\displaystyle\int_{1}^{5} 2 \, dx = 2x \Big|_{1}^{5} = 2(5) - 2 = 2(5 - 1) = 8.$

   This integral is interesting; the integrand is a constant function, hence we are finding the area of a rectangle with width $(5 - 1) = 4$ and height 2. Notice how the evaluation of the definite integral led to $2(4) = 8$.

   In general, if c is a constant, then $\int_a^b c \, dx = c(b - a)$.

Understanding Motion with the Fundamental Theorem of Calculus

We established, starting with Key Idea 1, that the derivative of a position function is a velocity function, and the derivative of a velocity function is an acceleration function. Now consider definite integrals of velocity and acceleration functions. Specifically, if $v(t)$ is a velocity function, what does $\displaystyle\int_a^b v(t) \, dt$ mean?

Notes:

5.4 The Fundamental Theorem of Calculus

The Fundamental Theorem of Calculus states that

$$\int_a^b v(t)\, dt = V(b) - V(a),$$

where $V(t)$ is any antiderivative of $v(t)$. Since $v(t)$ is a velocity function, $V(t)$ must be a position function, and $V(b) - V(a)$ measures a change in position, or **displacement**.

Example 128 **Finding displacement**
A ball is thrown straight up with velocity given by $v(t) = -32t + 20$ ft/s, where t is measured in seconds. Find, and interpret, $\int_0^1 v(t)\, dt$.

Solution Using the Fundamental Theorem of Calculus, we have

$$\int_0^1 v(t)\, dt = \int_0^1 (-32t + 20)\, dt$$
$$= -16t^2 + 20t \Big|_0^1$$
$$= 4.$$

Thus if a ball is thrown straight up into the air with velocity $v(t) = -32t + 20$, the height of the ball, 1 second later, will be 4 feet above the initial height. (Note that the ball has *traveled* much farther. It has gone up to its peak and is falling down, but the difference between its height at $t = 0$ and $t = 1$ is 4ft.)

Integrating a rate of change function gives total change. Velocity is the rate of position change; integrating velocity gives the total change of position, i.e., displacement.

Integrating a speed function gives a similar, though different, result. Speed is also the rate of position change, but does not account for direction. So integrating a speed function gives total change of position, without the possibility of "negative position change." Hence the integral of a speed function gives *distance traveled*.

As acceleration is the rate of velocity change, integrating an acceleration function gives total change in velocity. We do not have a simple term for this analogous to displacement. If $a(t) = 5$ miles/h^2 and t is measured in hours, then

$$\int_0^3 a(t)\, dt = 15$$

means the velocity has increased by 15m/h from $t = 0$ to $t = 3$.

Notes:

The Fundamental Theorem of Calculus and the Chain Rule

Part 1 of the Fundamental Theorem of Calculus (FTC) states that given $F(x) = \int_a^x f(t)\, dt$, $F'(x) = f(x)$. Using other notation, $\dfrac{d}{dx}(F(x)) = f(x)$. While we have just practiced evaluating definite integrals, sometimes finding antiderivatives is impossible and we need to rely on other techniques to approximate the value of a definite integral. Functions written as $F(x) = \int_a^x f(t)\, dt$ are useful in such situations.

It may be of further use to compose such a function with another. As an example, we may compose $F(x)$ with $g(x)$ to get

$$F(g(x)) = \int_a^{g(x)} f(t)\, dt.$$

What is the derivative of such a function? The Chain Rule can be employed to state

$$\frac{d}{dx}\Big(F(g(x))\Big) = F'(g(x))g'(x) = f(g(x))g'(x).$$

An example will help us understand this.

Example 129   **The FTC, Part 1, and the Chain Rule**

Find the derivative of $F(x) = \displaystyle\int_2^{x^2} \ln t\, dt$.

Solution   We can view $F(x)$ as being the function $G(x) = \displaystyle\int_2^x \ln t\, dt$ composed with $g(x) = x^2$; that is, $F(x) = G(g(x))$. The Fundamental Theorem of Calculus states that $G'(x) = \ln x$. The Chain Rule gives us

$$\begin{aligned}
F'(x) &= G'(g(x))g'(x) \\
&= \ln(g(x))g'(x) \\
&= \ln(x^2)2x \\
&= 2x\ln x^2
\end{aligned}$$

Normally, the steps defining $G(x)$ and $g(x)$ are skipped.

Practice this once more.

Example 130   **The FTC, Part 1, and the Chain Rule**

Find the derivative of $F(x) = \displaystyle\int_{\cos x}^{5} t^3\, dt$.

Notes:

5.4 The Fundamental Theorem of Calculus

SOLUTION   Note that $F(x) = -\int_5^{\cos x} t^3\, dt$. Viewed this way, the derivative of F is straightforward:

$$F'(x) = \sin x \cos^3 x.$$

Area Between Curves

Consider continuous functions $f(x)$ and $g(x)$ defined on $[a,b]$, where $f(x) \geq g(x)$ for all x in $[a,b]$, as demonstrated in Figure 5.23. What is the area of the shaded region bounded by the two curves over $[a,b]$?

The area can be found by recognizing that this area is "the area under f — the area under g." Using mathematical notation, the area is

$$\int_a^b f(x)\, dx - \int_a^b g(x)\, dx.$$

Properties of the definite integral allow us to simplify this expression to

$$\int_a^b \bigl(f(x) - g(x)\bigr)\, dx.$$

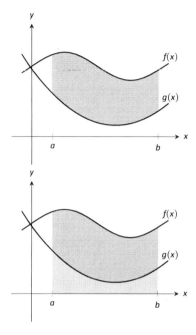

Figure 5.23: Finding the area bounded by two functions on an interval; it is found by subtracting the area under g from the area under f.

Theorem 41   **Area Between Curves**

Let $f(x)$ and $g(x)$ be continuous functions defined on $[a,b]$ where $f(x) \geq g(x)$ for all x in $[a,b]$. The area of the region bounded by the curves $y = f(x)$, $y = g(x)$ and the lines $x = a$ and $x = b$ is

$$\int_a^b \bigl(f(x) - g(x)\bigr)\, dx.$$

Example 131   **Finding area between curves**
Find the area of the region enclosed by $y = x^2 + x - 5$ and $y = 3x - 2$.

SOLUTION   It will help to sketch these two functions, as done in Figure 5.24. The region whose area we seek is completely bounded by these two functions; they seem to intersect at $x = -1$ and $x = 3$. To check, set $x^2 + x - 5 =$

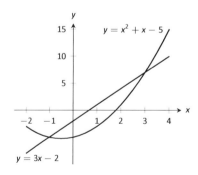

Figure 5.24: Sketching the region enclosed by $y = x^2 + x - 5$ and $y = 3x - 2$ in Example 131.

Notes:

Chapter 5 Integration

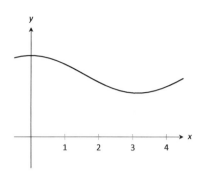

Figure 5.25: A graph of a function f to introduce the Mean Value Theorem.

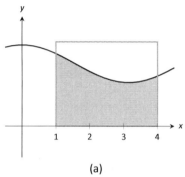

(a)

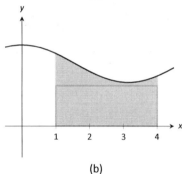

(b)

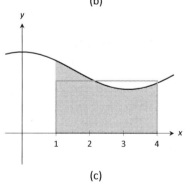

(c)

Figure 5.26: Differently sized rectangles give upper and lower bounds on $\int_1^4 f(x)\,dx$; the last rectangle matches the area exactly.

$3x - 2$ and solve for x:

$$x^2 + x - 5 = 3x - 2$$
$$(x^2 + x - 5) - (3x - 2) = 0$$
$$x^2 - 2x - 3 = 0$$
$$(x - 3)(x + 1) = 0$$
$$x = -1, 3.$$

Following Theorem 41, the area is

$$\int_{-1}^{3} \left(3x - 2 - (x^2 + x - 5)\right) dx = \int_{-1}^{3} (-x^2 + 2x + 3)\, dx$$
$$= \left(-\frac{1}{3}x^3 + x^2 + 3x\right)\Bigg|_{-1}^{3}$$
$$= -\frac{1}{3}(27) + 9 + 9 - \left(\frac{1}{3} + 1 - 3\right)$$
$$= 10\frac{2}{3} = 10.\overline{6}$$

The Mean Value Theorem and Average Value

Consider the graph of a function f in Figure 5.25 and the area defined by $\int_1^4 f(x)\,dx$. Three rectangles are drawn in Figure 5.26; in (a), the height of the rectangle is greater than f on $[1, 4]$, hence the area of this rectangle is is greater than $\int_0^4 f(x)\,dx$.

In (b), the height of the rectangle is smaller than f on $[1, 4]$, hence the area of this rectangle is less than $\int_1^4 f(x)\,dx$.

Finally, in (c) the height of the rectangle is such that the area of the rectangle is *exactly* that of $\int_0^4 f(x)\,dx$. Since rectangles that are "too big", as in (a), and rectangles that are "too little," as in (b), give areas greater/lesser than $\int_1^4 f(x)\,dx$, it makes sense that there is a rectangle, whose top intersects $f(x)$ somewhere on $[1, 4]$, whose area is *exactly* that of the definite integral.

We state this idea formally in a theorem.

Notes:

5.4 The Fundamental Theorem of Calculus

Theorem 42   **The Mean Value Theorem of Integration**

Let f be continuous on $[a, b]$. There exists a value c in $[a, b]$ such that
$$\int_a^b f(x)\,dx = f(c)(b - a).$$

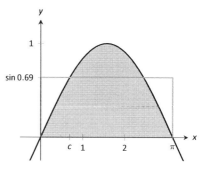

Figure 5.27: A graph of $y = \sin x$ on $[0, \pi]$ and the rectangle guaranteed by the Mean Value Theorem.

This is an *existential* statement; c exists, but we do not provide a method of finding it. Theorem 42 is directly connected to the Mean Value Theorem of Differentiation, given as Theorem 27; we leave it to the reader to see how.

We demonstrate the principles involved in this version of the Mean Value Theorem in the following example.

Example 132   **Using the Mean Value Theorem**
Consider $\int_0^\pi \sin x\,dx$. Find a value c guaranteed by the Mean Value Theorem.

SOLUTION   We first need to evaluate $\int_0^\pi \sin x\,dx$. (This was previously done in Example 127.)
$$\int_0^\pi \sin x\,dx = -\cos x\Big|_0^\pi = 2.$$

Thus we seek a value c in $[0, \pi]$ such that $\pi \sin c = 2$.
$$\pi \sin c = 2 \;\Rightarrow\; \sin c = 2/\pi \;\Rightarrow\; c = \arcsin(2/\pi) \approx 0.69.$$

In Figure 5.27 $\sin x$ is sketched along with a rectangle with height $\sin(0.69)$. The area of the rectangle is the same as the area under $\sin x$ on $[0, \pi]$.

Let f be a function on $[a, b]$ with c such that $f(c)(b-a) = \int_a^b f(x)\,dx$. Consider $\int_a^b \big(f(x) - f(c)\big)\,dx$:

$$\int_a^b \big(f(x) - f(c)\big)\,dx = \int_a^b f(x) - \int_a^b f(c)\,dx$$
$$= f(c)(b - a) - f(c)(b - a)$$
$$= 0.$$

When $f(x)$ is shifted by $-f(c)$, the amount of area under f above the x–axis on $[a, b]$ is the same as the amount of area below the x–axis above f; see Figure 5.28 for an illustration of this. In this sense, we can say that $f(c)$ is the *average value* of f on $[a, b]$.

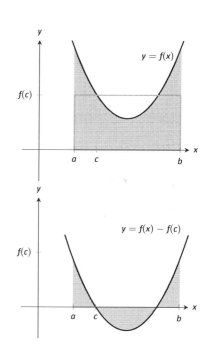

Figure 5.28: On top, a graph of $y = f(x)$ and the rectangle guaranteed by the Mean Value Theorem. Below, $y = f(x)$ is shifted down by $f(c)$; the resulting "area under the curve" is 0.

Notes:

The value $f(c)$ is the average value in another sense. First, recognize that the Mean Value Theorem can be rewritten as

$$f(c) = \frac{1}{b-a} \int_a^b f(x)\, dx,$$

for some value of c in $[a, b]$. Next, partition the interval $[a, b]$ into n equally spaced subintervals, $a = x_1 < x_2 < \ldots < x_{n+1} = b$ and choose any c_i in $[x_i, x_{i+1}]$. The average of the numbers $f(c_1), f(c_2), \ldots, f(c_n)$ is:

$$\frac{1}{n}\Big(f(c_1) + f(c_2) + \ldots + f(c_n)\Big) = \frac{1}{n}\sum_{i=1}^n f(c_i).$$

Multiply this last expression by 1 in the form of $\frac{(b-a)}{(b-a)}$:

$$\frac{1}{n}\sum_{i=1}^n f(c_i) = \sum_{i=1}^n f(c_i)\frac{1}{n}$$

$$= \sum_{i=1}^n f(c_i)\frac{1}{n}\frac{(b-a)}{(b-a)}$$

$$= \frac{1}{b-a}\sum_{i=1}^n f(c_i)\frac{b-a}{n}$$

$$= \frac{1}{b-a}\sum_{i=1}^n f(c_i)\Delta x \quad \text{(where } \Delta x = (b-a)/n\text{)}$$

Now take the limit as $n \to \infty$:

$$\lim_{n\to\infty} \frac{1}{b-a}\sum_{i=1}^n f(c_i)\Delta x = \frac{1}{b-a}\int_a^b f(x)\, dx = f(c).$$

This tells us this: when we evaluate f at n (somewhat) equally spaced points in $[a, b]$, the average value of these samples is $f(c)$ as $n \to \infty$.

This leads us to a definition.

Definition 22 **The Average Value of f on $[a, b]$**

Let f be continuous on $[a, b]$. The **average value of** f **on** $[a, b]$ is $f(c)$, where c is a value in $[a, b]$ guaranteed by the Mean Value Theorem. I.e.,

$$\text{Average Value of } f \text{ on } [a, b] = \frac{1}{b-a}\int_a^b f(x)\, dx.$$

Notes:

An application of this definition is given in the following example.

Example 133   **Finding the average value of a function**
An object moves back and forth along a straight line with a velocity given by $v(t) = (t-1)^2$ on $[0, 3]$, where t is measured in seconds and $v(t)$ is measured in ft/s.
What is the average velocity of the object?

SOLUTION   By our definition, the average velocity is:

$$\frac{1}{3-0}\int_0^3 (t-1)^2\, dt = \frac{1}{3}\int_0^3 (t^2 - 2t + 1)\, dt = \frac{1}{3}\left(\frac{1}{3}t^3 - t^2 + t\right)\Big|_0^3 = 1 \text{ ft/s}.$$

We can understand the above example through a simpler situation. Suppose you drove 100 miles in 2 hours. What was your average speed? The answer is simple: displacement/time = 100 miles/2 hours = 50 mph.

What was the displacement of the object in Example 133? We calculate this by integrating its velocity function: $\int_0^3 (t-1)^2\, dt = 3$ ft. Its final position was 3 feet from its initial position after 3 seconds: its average velocity was 1 ft/s.

This section has laid the groundwork for a lot of great mathematics to follow. The most important lesson is this: definite integrals can be evaluated using antiderivatives. Since the previous section established that definite integrals are the limit of Riemann sums, we can later create Riemann sums to approximate values other than "area under the curve," convert the sums to definite integrals, then evaluate these using the Fundamental Theorem of Calculus. This will allow us to compute the work done by a variable force, the volume of certain solids, the arc length of curves, and more.

The downside is this: generally speaking, computing antiderivatives is much more difficult than computing derivatives. The next chapter is devoted to techniques of finding antiderivatives so that a wide variety of definite integrals can be evaluated. Before that, the next section explores techniques of approximating the value of definite integrals beyond using the Left Hand, Right Hand and Midpoint Rules.

Notes:

Exercises 5.4

Terms and Concepts

1. How are definite and indefinite integrals related?

2. What constant of integration is most commonly used when evaluating definite integrals?

3. T/F: If f is a continuous function, then $F(x) = \int_a^x f(t)\, dt$ is also a continuous function.

4. The definite integral can be used to find "the area under a curve." Give two other uses for definite integrals.

Problems

In Exercises 5 – 28, evaluate the definite integral.

5. $\int_1^3 (3x^2 - 2x + 1)\, dx$

6. $\int_0^4 (x-1)^2\, dx$

7. $\int_{-1}^1 (x^3 - x^5)\, dx$

8. $\int_{\pi/2}^{\pi} \cos x\, dx$

9. $\int_0^{\pi/4} \sec^2 x\, dx$

10. $\int_1^e \frac{1}{x}\, dx$

11. $\int_{-1}^1 5^x\, dx$

12. $\int_{-2}^{-1} (4 - 2x^3)\, dx$

13. $\int_0^{\pi} (2\cos x - 2\sin x)\, dx$

14. $\int_1^3 e^x\, dx$

15. $\int_0^4 \sqrt{t}\, dt$

16. $\int_9^{25} \frac{1}{\sqrt{t}}\, dt$

17. $\int_1^8 \sqrt[3]{x}\, dx$

18. $\int_1^2 \frac{1}{x}\, dx$

19. $\int_1^2 \frac{1}{x^2}\, dx$

20. $\int_1^2 \frac{1}{x^3}\, dx$

21. $\int_0^1 x\, dx$

22. $\int_0^1 x^2\, dx$

23. $\int_0^1 x^3\, dx$

24. $\int_0^1 x^{100}\, dx$

25. $\int_{-4}^4 dx$

26. $\int_{-10}^{-5} 3\, dx$

27. $\int_{-2}^2 0\, dx$

28. $\int_{\pi/6}^{\pi/3} \csc x \cot x\, dx$

29. Explain why:

    (a) $\int_{-1}^1 x^n\, dx = 0$, when n is a positive, odd integer, and

    (b) $\int_{-1}^1 x^n\, dx = 2\int_0^1 x^n\, dx$ when n is a positive, even integer.

In Exercises 30 – 33, find a value c guaranteed by the Mean Value Theorem.

30. $\int_0^2 x^2\, dx$

31. $\int_{-2}^2 x^2\, dx$

32. $\int_0^1 e^x\, dx$

33. $\int_0^{16} \sqrt{x}\, dx$

In Exercises 34 – 39, find the average value of the function on the given interval.

34. $f(x) = \sin x$ on $[0, \pi/2]$

35. $y = \sin x$ on $[0, \pi]$

36. $y = x$ on $[0, 4]$

37. $y = x^2$ on $[0, 4]$

38. $y = x^3$ on $[0, 4]$

39. $g(t) = 1/t$ on $[1, e]$

In Exercises 40 – 44, a velocity function of an object moving along a straight line is given. Find the displacement of the object over the given time interval.

40. $v(t) = -32t + 20$ ft/s on $[0, 5]$

41. $v(t) = -32t + 200$ ft/s on $[0, 10]$

42. $v(t) = 2^t$ mph on $[-1, 1]$

43. $v(t) = \cos t$ ft/s on $[0, 3\pi/2]$

44. $v(t) = \sqrt[4]{t}$ ft/s on $[0, 16]$

In Exercises 45 – 48, an acceleration function of an object moving along a straight line is given. Find the change of the object's velocity over the given time interval.

45. $a(t) = -32$ft/s^2 on $[0, 2]$

46. $a(t) = 10$ft/s^2 on $[0, 5]$

47. $a(t) = t$ ft/s^2 on $[0, 2]$

48. $a(t) = \cos t$ ft/s^2 on $[0, \pi]$

In Exercises 49 – 52, sketch the given functions and find the area of the enclosed region.

49. $y = 2x$, $y = 5x$, and $x = 3$.

50. $y = -x + 1$, $y = 3x + 6$, $x = 2$ and $x = -1$.

51. $y = x^2 - 2x + 5$, $y = 5x - 5$.

52. $y = 2x^2 + 2x - 5$, $y = x^2 + 3x + 7$.

In Exercises 53 – 56, find $F'(x)$.

53. $F(x) = \displaystyle\int_2^{x^3+x} \frac{1}{t}\,dt$

54. $F(x) = \displaystyle\int_{x^3}^{0} t^3\,dt$

55. $F(x) = \displaystyle\int_x^{x^2} (t+2)\,dt$

56. $F(x) = \displaystyle\int_{\ln x}^{e^x} \sin t\,dt$

5.5 Numerical Integration

The Fundamental Theorem of Calculus gives a concrete technique for finding the exact value of a definite integral. That technique is based on computing antiderivatives. Despite the power of this theorem, there are still situations where we must *approximate* the value of the definite integral instead of finding its exact value. The first situation we explore is where we *cannot* compute the antiderivative of the integrand. The second case is when we actually do not know the integrand, but only its value when evaluated at certain points.

An **elementary function** is any function that is a combination of polynomials, n^{th} roots, rational, exponential, logarithmic and trigonometric functions. We can compute the derivative of any elementary function, but there are many elementary functions of which we cannot compute an antiderivative. For example, the following functions do not have antiderivatives that we can express with elementary functions:

$$e^{-x^2}, \quad \sin(x^3) \quad \text{and} \quad \frac{\sin x}{x}.$$

The simplest way to refer to the antiderivatives of e^{-x^2} is to simply write $\int e^{-x^2}\,dx$.

This section outlines three common methods of approximating the value of definite integrals. We describe each as a systematic method of approximating area under a curve. By approximating this area accurately, we find an accurate approximation of the corresponding definite integral.

We will apply the methods we learn in this section to the following definite integrals:

$$\int_0^1 e^{-x^2}\,dx, \quad \int_{-\frac{\pi}{4}}^{\frac{\pi}{2}} \sin(x^3)\,dx, \quad \text{and} \quad \int_{0.5}^{4\pi} \frac{\sin(x)}{x}\,dx,$$

as pictured in Figure 5.29.

The Left and Right Hand Rule Methods

In Section 5.3 we addressed the problem of evaluating definite integrals by approximating the area under the curve using rectangles. We revisit those ideas here before introducing other methods of approximating definite integrals.

We start with a review of notation. Let f be a continuous function on the interval $[a, b]$. We wish to approximate $\int_a^b f(x)\,dx$. We partition $[a, b]$ into n equally spaced subintervals, each of length $\Delta x = \dfrac{b-a}{n}$. The endpoints of these

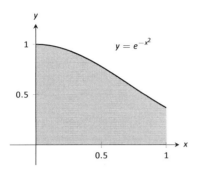

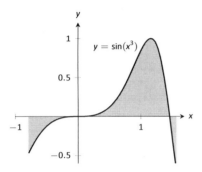

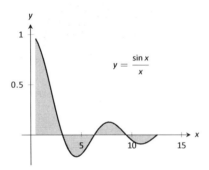

Figure 5.29: Graphically representing three definite integrals that cannot be evaluated using antiderivatives.

Notes:

subintervals are labeled as

$x_1 = a$, $x_2 = a + \Delta x$, $x_3 = a + 2\Delta x$, ..., $x_i = a + (i-1)\Delta x$, ..., $x_{n+1} = b$.

Key Idea 8 states that to use the Left Hand Rule we use the summation $\sum_{i=1}^{n} f(x_i)\Delta x$ and to use the Right Hand Rule we use $\sum_{i=1}^{n} f(x_{i+1})\Delta x$. We review the use of these rules in the context of examples.

Example 134   **Approximating definite integrals with rectangles**
Approximate $\int_0^1 e^{-x^2}\,dx$ using the Left and Right Hand Rules with 5 equally spaced subintervals.

SOLUTION   We begin by partitioning the interval $[0,1]$ into 5 equally spaced intervals. We have $\Delta x = \frac{1-0}{5} = 1/5 = 0.2$, so

$x_1 = 0$, $x_2 = 0.2$, $x_3 = 0.4$, $x_4 = 0.6$, $x_5 = 0.8$, and $x_6 = 1$.

Using the Left Hand Rule, we have:

$$\sum_{i=1}^{n} f(x_i)\Delta x = \big(f(x_1) + f(x_2) + f(x_3) + f(x_4) + f(x_5)\big)\Delta x$$
$$= \big(f(0) + f(0.2) + f(0.4) + f(0.6) + f(0.8)\big)\Delta x$$
$$\approx \big(1 + 0.961 + 0.852 + 0.698 + 0.527\big)(0.2)$$
$$\approx 0.808.$$

Using the Right Hand Rule, we have:

$$\sum_{i=1}^{n} f(x_{i+1})\Delta x = \big(f(x_2) + f(x_3) + f(x_4) + f(x_5) + f(x_6)\big)\Delta x$$
$$= \big(f(0.2) + f(0.4) + f(0.6) + f(0.8) + f(1)\big)\Delta x$$
$$\approx \big(0.961 + 0.852 + 0.698 + 0.527 + 0.368\big)(0.2)$$
$$\approx 0.681.$$

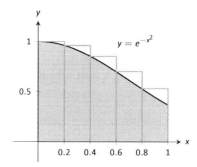
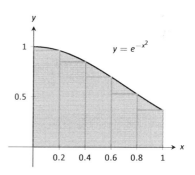

Figure 5.30: Approximating $\int_0^1 e^{-x^2}\,dx$ in Example 134.

Figure 5.30 shows the rectangles used in each method to approximate the definite integral. These graphs show that in this particular case, the Left Hand Rule is an over approximation and the Right Hand Rule is an under approximation. To get a better approximation, we could use more rectangles, as we did in

Notes:

Chapter 5 Integration

x_i	Exact	Approx.	$\sin(x_i^3)$
x_1	$-\pi/4$	-0.785	-0.466
x_2	$-7\pi/40$	-0.550	-0.165
x_3	$-\pi/10$	-0.314	-0.031
x_4	$-\pi/40$	-0.0785	0
x_5	$\pi/20$	0.157	0.004
x_6	$\pi/8$	0.393	0.061
x_7	$\pi/5$	0.628	0.246
x_8	$11\pi/40$	0.864	0.601
x_9	$7\pi/20$	1.10	0.971
x_{10}	$17\pi/40$	1.34	0.690
x_{11}	$\pi/2$	1.57	-0.670

Figure 5.31: Table of values used to approximate $\int_{-\pi/4}^{\pi/2} \sin(x^3)\, dx$ in Example 135.

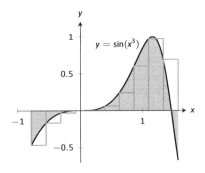

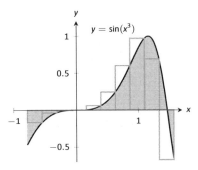

Figure 5.32: Approximating $\int_{-\pi/4}^{\pi/2} \sin(x^3)\, dx$ in Example 135.

Section 5.3. We could also average the Left and Right Hand Rule results together, giving
$$\frac{0.808 + 0.681}{2} = 0.7445.$$
The actual answer, accurate to 4 places after the decimal, is 0.7468, showing our average is a good approximation.

Example 135   **Approximating definite integrals with rectangles**

Approximate $\int_{-\frac{\pi}{4}}^{\frac{\pi}{2}} \sin(x^3)\, dx$ using the Left and Right Hand Rules with 10 equally spaced subintervals.

SOLUTION   We begin by finding Δx:
$$\frac{b-a}{n} = \frac{\pi/2 - (-\pi/4)}{10} = \frac{3\pi}{40} \approx 0.236.$$

It is useful to write out the endpoints of the subintervals in a table; in Figure 5.31, we give the exact values of the endpoints, their decimal approximations, and decimal approximations of $\sin(x^3)$ evaluated at these points.

Once this table is created, it is straightforward to approximate the definite integral using the Left and Right Hand Rules. (Note: the table itself is easy to create, especially with a standard spreadsheet program on a computer. The last two columns are all that are needed.) The Left Hand Rule sums the first 10 values of $\sin(x_i^3)$ and multiplies the sum by Δx; the Right Hand Rule sums the last 10 values of $\sin(x_i^3)$ and multiplies by Δx. Therefore we have:

Left Hand Rule: $\int_{-\frac{\pi}{4}}^{\frac{\pi}{2}} \sin(x^3)\, dx \approx (1.91)(0.236) = 0.451.$

Right Hand Rule: $\int_{-\frac{\pi}{4}}^{\frac{\pi}{2}} \sin(x^3)\, dx \approx (1.71)(0.236) = 0.404.$

Average of the Left and Right Hand Rules: 0.4275.

The actual answer, accurate to 3 places after the decimal, is 0.460. Our approximations were once again fairly good. The rectangles used in each approximation are shown in Figure 5.32. It is clear from the graphs that using more rectangles (and hence, narrower rectangles) should result in a more accurate approximation.

The Trapezoidal Rule

In Example 134 we approximated the value of $\int_0^1 e^{-x^2}\, dx$ with 5 rectangles of equal width. Figure 5.30 shows the rectangles used in the Left and Right Hand

Notes:

Rules. These graphs clearly show that rectangles do not match the shape of the graph all that well, and that accurate approximations will only come by using lots of rectangles.

Instead of using rectangles to approximate the area, we can instead use *trapezoids*. In Figure 5.33, we show the region under $f(x) = e^{-x^2}$ on $[0, 1]$ approximated with 5 trapezoids of equal width; the top "corners" of each trapezoid lies on the graph of $f(x)$. It is clear from this figure that these trapezoids more accurately approximate the area under f and hence should give a better approximation of $\int_0^1 e^{-x^2}\, dx$. (In fact, these trapezoids seem to give a *great* approximation of the area!)

The formula for the area of a trapezoid is given in Figure 5.34. We approximate $\int_0^1 e^{-x^2}\, dx$ with these trapezoids in the following example.

Example 136   **Approximating definite integrals using trapezoids**

Use 5 trapezoids of equal width to approximate $\displaystyle\int_0^1 e^{-x^2}\, dx$.

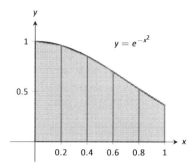

Figure 5.33: Approximating $\int_0^1 e^{-x^2}\, dx$ using 5 trapezoids of equal widths.

Solution   To compute the areas of the 5 trapezoids in Figure 5.33, it will again be useful to create a table of values as shown in Figure 5.35.

The leftmost trapezoid has legs of length 1 and 0.961 and a height of 0.2. Thus, by our formula, the area of the leftmost trapezoid is:

$$\frac{1+0.961}{2}(0.2) = 0.1961.$$

Moving right, the next trapezoid has legs of length 0.961 and 0.852 and a height of 0.2. Thus its area is:

$$\frac{0.961+0.852}{2}(0.2) = 0.1813.$$

The sum of the areas of all 5 trapezoids is:

$$\frac{1+0.961}{2}(0.2) + \frac{0.961+0.852}{2}(0.2) + \frac{0.852+0.698}{2}(0.2) +$$
$$\frac{0.698+0.527}{2}(0.2) + \frac{0.527+0.368}{2}(0.2) = 0.7445.$$

We approximate $\displaystyle\int_0^1 e^{-x^2}\, dx \approx 0.7445$.

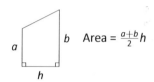

Figure 5.34: The area of a trapezoid.

x_i	$e^{-x_i^2}$
0	1
0.2	0.961
0.4	0.852
0.6	0.698
0.8	0.527
1	0.368

Figure 5.35: A table of values of e^{-x^2}.

There are many things to observe in this example. Note how each term in the final summation was multiplied by both 1/2 and by $\Delta x = 0.2$. We can factor these coefficients out, leaving a more concise summation as:

$$\frac{1}{2}(0.2)\Big[(1+0.961)+(0.961+0.852)+(0.852+0.698)+(0.698+0.527)+(0.527+0.368)\Big].$$

Notes:

Now notice that all numbers except for the first and the last are added twice. Therefore we can write the summation even more concisely as

$$\frac{0.2}{2}\Big[1 + 2(0.961 + 0.852 + 0.698 + 0.527) + 0.368\Big].$$

This is the heart of the **Trapezoidal Rule**, wherein a definite integral $\int_a^b f(x)\,dx$ is approximated by using trapezoids of equal widths to approximate the corresponding area under f. Using n equally spaced subintervals with endpoints x_1, $x_2, \ldots, x_{n+1}$, we again have $\Delta x = \dfrac{b-a}{n}$. Thus:

$$\int_a^b f(x)\,dx \approx \sum_{i=1}^n \frac{f(x_i) + f(x_{i+1})}{2}\Delta x$$

$$= \frac{\Delta x}{2}\sum_{i=1}^n \big(f(x_i) + f(x_{i+1})\big)$$

$$= \frac{\Delta x}{2}\Big[f(x_1) + 2\sum_{i=2}^n f(x_i) + f(x_{n+1})\Big].$$

Example 137   **Using the Trapezoidal Rule**

Revisit Example 135 and approximate $\int_{-\frac{\pi}{4}}^{\frac{\pi}{2}} \sin(x^3)\,dx$ using the Trapezoidal Rule and 10 equally spaced subintervals.

SOLUTION   We refer back to Figure 5.31 for the table of values of $\sin(x^3)$. Recall that $\Delta x = 3\pi/40 \approx 0.236$. Thus we have:

$$\int_{-\frac{\pi}{4}}^{\frac{\pi}{2}} \sin(x^3)\,dx \approx \frac{0.236}{2}\Big[-0.466 + 2\big(-0.165 + (-0.031) + \ldots + 0.69\big) + (-0.67)\Big]$$

$$= 0.4275.$$

Notice how "quickly" the Trapezoidal Rule can be implemented once the table of values is created. This is true for all the methods explored in this section; the real work is creating a table of x_i and $f(x_i)$ values. Once this is completed, approximating the definite integral is not difficult. Again, using technology is wise. Spreadsheets can make quick work of these computations and make using lots of subintervals easy.

Also notice the approximations the Trapezoidal Rule gives. It is the average of the approximations given by the Left and Right Hand Rules! This effectively

Notes:

renders the Left and Right Hand Rules obsolete. They are useful when first learning about definite integrals, but if a real approximation is needed, one is generally better off using the Trapezoidal Rule instead of either the Left or Right Hand Rule.

How can we improve on the Trapezoidal Rule, apart from using more and more trapezoids? The answer is clear once we look back and consider what we have *really* done so far. The Left Hand Rule is not *really* about using rectangles to approximate area. Instead, it approximates a function f with constant functions on small subintervals and then computes the definite integral of these constant functions. The Trapezoidal Rule is really approximating a function f with a linear function on a small subinterval, then computes the definite integral of this linear function. In both of these cases the definite integrals are easy to compute in geometric terms.

So we have a progression: we start by approximating f with a constant function and then with a linear function. What is next? A quadratic function. By approximating the curve of a function with lots of parabolas, we generally get an even better approximation of the definite integral. We call this process **Simpson's Rule**, named after Thomas Simpson (1710-1761), even though others had used this rule as much as 100 years prior.

Simpson's Rule

Given one point, we can create a constant function that goes through that point. Given two points, we can create a linear function that goes through those points. Given three points, we can create a quadratic function that goes through those three points (given that no two have the same x–value).

Consider three points (x_1, y_1), (x_2, y_2) and (x_3, y_3) whose x–values are equally spaced and $x_1 < x_2 < x_3$. Let f be the quadratic function that goes through these three points. It is not hard to show that

$$\int_{x_1}^{x_3} f(x)\, dx = \frac{x_3 - x_1}{6}\left(y_1 + 4y_2 + y_3\right). \tag{5.4}$$

Consider Figure 5.36. A function f goes through the 3 points shown and the parabola g that also goes through those points is graphed with a dashed line. Using our equation from above, we know exactly that

$$\int_1^3 g(x)\, dx = \frac{3-1}{6}\bigl(3 + 4(1) + 2\bigr) = 3.$$

Since g is a good approximation for f on $[1, 3]$, we can state that

$$\int_1^3 f(x)\, dx \approx 3.$$

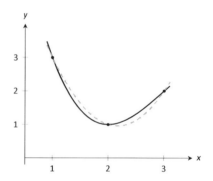

Figure 5.36: A graph of a function f and a parabola that approximates it well on $[1, 3]$.

Notes:

Chapter 5 Integration

Notice how the interval $[1, 3]$ was split into two subintervals as we needed 3 points. Because of this, whenever we use Simpson's Rule, we need to break the interval into an even number of subintervals.

In general, to approximate $\int_a^b f(x)\,dx$ using Simpson's Rule, subdivide $[a, b]$ into n subintervals, where n is even and each subinterval has width $\Delta x = (b - a)/n$. We approximate f with $n/2$ parabolic curves, using Equation (5.4) to compute the area under these parabolas. Adding up these areas gives the formula:

$$\int_a^b f(x)\,dx \approx \frac{\Delta x}{3}\Big[f(x_1)+4f(x_2)+2f(x_3)+4f(x_4)+\ldots+2f(x_{n-1})+4f(x_n)+f(x_{n+1})\Big].$$

Note how the coefficients of the terms in the summation have the pattern 1, 4, 2, 4, 2, 4, ..., 2, 4, 1.

Let's demonstrate Simpson's Rule with a concrete example.

Example 138   **Using Simpson's Rule**

Approximate $\int_0^1 e^{-x^2}\,dx$ using Simpson's Rule and 4 equally spaced subintervals.

SOLUTION   We begin by making a table of values as we have in the past, as shown in Figure 5.37(a). Simpson's Rule states that

$$\int_0^1 e^{-x^2}\,dx \approx \frac{0.25}{3}\Big[1+4(0.939)+2(0.779)+4(0.570)+0.368\Big] = 0.746\overline{83}.$$

Recall in Example 134 we stated that the correct answer, accurate to 4 places after the decimal, was 0.7468. Our approximation with Simpson's Rule, with 4 subintervals, is better than our approximation with the Trapezoidal Rule using 5!

Figure 5.37(b) shows $f(x) = e^{-x^2}$ along with its approximating parabolas, demonstrating how good our approximation is. The approximating curves are nearly indistinguishable from the actual function.

Example 139   **Using Simpson's Rule**

Approximate $\int_{-\pi/4}^{\pi/2} \sin(x^3)\,dx$ using Simpson's Rule and 10 equally spaced intervals.

SOLUTION   Figure 5.38 shows the table of values that we used in the past for this problem, shown here again for convenience. Again, $\Delta x = (\pi/2 + \pi/4)/10 \approx 0.236$.

x_i	$e^{-x_i^2}$
0	1
0.25	0.939
0.5	0.779
0.75	0.570
1	0.368

(a)

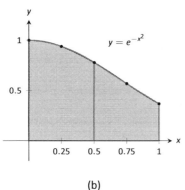

(b)

Figure 5.37: A table of values to approximate $\int_0^1 e^{-x^2}\,dx$, along with a graph of the function.

x_i	$\sin(x_i^3)$
−0.785	−0.466
−0.550	−0.165
−0.314	−0.031
−0.0785	0
0.157	0.004
0.393	0.061
0.628	0.246
0.864	0.601
1.10	0.971
1.34	0.690
1.57	−0.670

Figure 5.38: Table of values used to approximate $\int_{-\pi/4}^{\pi/2} \sin(x^3)\,dx$ in Example 139.

Notes:

5.5 Numerical Integration

Simpson's Rule states that

$$\int_{-\frac{\pi}{4}}^{\frac{\pi}{2}} \sin(x^3)\, dx \approx \frac{0.236}{3}\Big[(-0.466) + 4(-0.165) + 2(-0.031) + \ldots$$
$$\ldots + 2(0.971) + 4(0.69) + (-0.67)\Big]$$
$$= 0.4701$$

Recall that the actual value, accurate to 3 decimal places, is 0.460. Our approximation is within one $1/100^{\text{th}}$ of the correct value. The graph in Figure 5.39 shows how closely the parabolas match the shape of the graph.

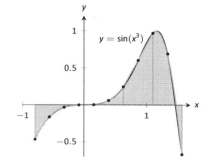

Figure 5.39: Approximating $\int_{-\frac{\pi}{4}}^{\frac{\pi}{2}} \sin(x^3)\, dx$ in Example 139 with Simpson's Rule and 10 equally spaced intervals.

Summary and Error Analysis

We summarize the key concepts of this section thus far in the following Key Idea.

Key Idea 9   Numerical Integration

Let f be a continuous function on $[a, b]$, let n be a positive integer, and let $\Delta x = \dfrac{b-a}{n}$.
Set $x_1 = a$, $x_2 = a + \Delta x$, …, $x_i = a + (i-1)\Delta x$, $x_{n+1} = b$.
Consider $\displaystyle\int_a^b f(x)\, dx$.

Left Hand Rule: $\displaystyle\int_a^b f(x)\, dx \approx \Delta x\Big[f(x_1) + f(x_2) + \ldots + f(x_n)\Big]$.

Right Hand Rule: $\displaystyle\int_a^b f(x)\, dx \approx \Delta x\Big[f(x_2) + f(x_3) + \ldots + f(x_{n+1})\Big]$.

Trapezoidal Rule: $\displaystyle\int_a^b f(x)\, dx \approx \frac{\Delta x}{2}\Big[f(x_1) + 2f(x_2) + 2f(x_3) + \ldots + 2f(x_n) + f(x_{n+1})\Big]$.

Simpson's Rule: $\displaystyle\int_a^b f(x)\, dx \approx \frac{\Delta x}{3}\Big[f(x_1) + 4f(x_2) + 2f(x_3) + \ldots + 4f(x_n) + f(x_{n+1})\Big]$ (n even).

In our examples, we approximated the value of a definite integral using a given method then compared it to the "right" answer. This should have raised several questions in the reader's mind, such as:

1. How was the "right" answer computed?

2. If the right answer can be found, what is the point of approximating?

3. If there is value to approximating, how are we supposed to know if the approximation is any good?

Notes:

These are good questions, and their answers are educational. In the examples, *the* right answer was never computed. Rather, an approximation accurate to a certain number of places after the decimal was given. In Example 134, we do not know the *exact* answer, but we know it starts with 0.7468. These more accurate approximations were computed using numerical integration but with more precision (i.e., more subintervals and the help of a computer).

Since the exact answer cannot be found, approximation still has its place. How are we to tell if the approximation is any good?

"Trial and error" provides one way. Using technology, make an approximation with, say, 10, 100, and 200 subintervals. This likely will not take much time at all, and a trend should emerge. If a trend does not emerge, try using yet more subintervals. Keep in mind that trial and error is never foolproof; you might stumble upon a problem in which a trend will not emerge.

A second method is to use Error Analysis. While the details are beyond the scope of this text, there are some formulas that give *bounds* for how good your approximation will be. For instance, the formula might state that the approximation is within 0.1 of the correct answer. If the approximation is 1.58, then one knows that the correct answer is between 1.48 and 1.68. By using lots of subintervals, one can get an approximation as accurate as one likes. Theorem 43 states what these bounds are.

Theorem 43   **Error Bounds in the Trapezoidal and Simpson's Rules**

1. Let E_T be the error in approximating $\int_a^b f(x)\, dx$ using the Trapezoidal Rule.

    If f has a continuous 2nd derivative on $[a, b]$ and M is any upper bound of $|f''(x)|$ on $[a, b]$, then

    $$E_T \leq \frac{(b-a)^3}{12n^2} M.$$

2. Let E_S be the error in approximating $\int_a^b f(x)\, dx$ using Simpson's Rule.

    If f has a continuous 4th derivative on $[a, b]$ and M is any upper bound of $|f^{(4)}|$ on $[a, b]$, then

    $$E_S \leq \frac{(b-a)^5}{180n^4} M.$$

Notes:

There are some key things to note about this theorem.

1. The larger the interval, the larger the error. This should make sense intuitively.

2. The error shrinks as more subintervals are used (i.e., as *n* gets larger).

3. The error in Simpson's Rule has a term relating to the 4th derivative of *f*. Consider a cubic polynomial: it's 4th derivative is 0. Therefore, the error in approximating the definite integral of a cubic polynomial with Simpson's Rule is 0 – Simpson's Rule computes the exact answer!

We revisit Examples 136 and 138 and compute the error bounds using Theorem 43 in the following example.

Example 140   **Computing error bounds**
Find the error bounds when approximating $\int_0^1 e^{-x^2}\, dx$ using the Trapezoidal Rule and 5 subintervals, and using Simpson's Rule with 4 subintervals.

SOLUTION
Trapezoidal Rule with $n = 5$:
We start by computing the 2nd derivative of $f(x) = e^{-x^2}$:

$$f''(x) = e^{-x^2}(4x^2 - 2).$$

Figure 5.40 shows a graph of $f''(x)$ on $[0, 1]$. It is clear that the largest value of f'', in absolute value, is 2. Thus we let $M = 2$ and apply the error formula from Theorem 43.

$$E_T = \frac{(1-0)^3}{12 \cdot 5^2} \cdot 2 = 0.00\overline{6}.$$

Our error estimation formula states that our approximation of 0.7445 found in Example 136 is within 0.0067 of the correct answer, hence we know that

$$0.7445 - 0.0067 = .7378 \leq \int_0^1 e^{-x^2}\, dx \leq 0.7512 = 0.7445 + 0.0067.$$

We had earlier computed the exact answer, correct to 4 decimal places, to be 0.7468, affirming the validity of Theorem 43.

Simpson's Rule with $n = 4$:
We start by computing the 4th derivative of $f(x) = e^{-x^2}$:

$$f^{(4)}(x) = e^{-x^2}(16x^4 - 48x^2 + 12).$$

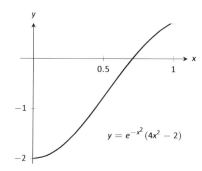

Figure 5.40: Graphing $f''(x)$ in Example 140 to help establish error bounds.

Notes:

Chapter 5 Integration

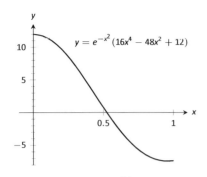

Figure 5.41: Graphing $f^{(4)}(x)$ in Example 140 to help establish error bounds.

Time	Speed (mph)
0	0
1	25
2	22
3	19
4	39
5	0
6	43
7	59
8	54
9	51
10	43
11	35
12	40
13	43
14	30
15	0
16	0
17	28
18	40
19	42
20	40
21	39
22	40
23	23
24	0

Figure 5.42: Speed data collected at 30 second intervals for Example 141.

Figure 5.41 shows a graph of $f^{(4)}(x)$ on $[0, 1]$. It is clear that the largest value of $f^{(4)}$, in absolute value, is 12. Thus we let $M = 12$ and apply the error formula from Theorem 43.

$$E_s = \frac{(1-0)^5}{180 \cdot 4^4} \cdot 12 = 0.00026.$$

Our error estimation formula states that our approximation of $0.746\overline{83}$ found in Example 138 is within 0.00026 of the correct answer, hence we know that

$$0.74683 - 0.00026 = .74657 \leq \int_0^1 e^{-x^2}\, dx \leq 0.74709 = 0.74683 + 0.00026.$$

Once again we affirm the validity of Theorem 43.

At the beginning of this section we mentioned two main situations where numerical integration was desirable. We have considered the case where an antiderivative of the integrand cannot be computed. We now investigate the situation where the integrand is not known. This is, in fact, the most widely used application of Numerical Integration methods. "Most of the time" we observe behavior but do not know "the" function that describes it. We instead collect data about the behavior and make approximations based off of this data. We demonstrate this in an example.

Example 141   **Approximating distance traveled**
One of the authors drove his daughter home from school while she recorded their speed every 30 seconds. The data is given in Figure 5.42. Approximate the distance they traveled.

SOLUTION   Recall that by integrating a speed function we get distance traveled. We have information about $v(t)$; we will use Simpson's Rule to approximate $\int_a^b v(t)\, dt$.

The most difficult aspect of this problem is converting the given data into the form we need it to be in. The speed is measured in miles per hour, whereas the time is measured in 30 second increments.

We need to compute $\Delta x = (b-a)/n$. Clearly, $n = 24$. What are a and b? Since we start at time $t = 0$, we have that $a = 0$. The final recorded time came after 24 periods of 30 seconds, which is 12 minutes or 1/5 of an hour. Thus we have

$$\Delta x = \frac{b-a}{n} = \frac{1/5 - 0}{24} = \frac{1}{120}; \quad \frac{\Delta x}{3} = \frac{1}{360}.$$

Notes:

Thus the distance traveled is approximately:

$$\int_0^{0.2} v(t)\,dt \approx \frac{1}{360}\Big[f(x_1) + 4f(x_2) + 2f(x_3) + \cdots + 4f(x_n) + f(x_{n+1})\Big]$$
$$= \frac{1}{360}\Big[0 + 4\cdot 25 + 2\cdot 22 + \cdots + 2\cdot 40 + 4\cdot 23 + 0\Big]$$
$$\approx 6.2167 \text{ miles.}$$

We approximate the author drove 6.2 miles. (Because we are sure the reader wants to know, the author's odometer recorded the distance as about 6.05 miles.)

We started this chapter learning about antiderivatives and indefinite integrals. We then seemed to change focus by looking at areas between the graph of a function and the *x*-axis. We defined these areas as the definite integral of the function, using a notation very similar to the notation of the indefinite integral. The Fundamental Theorem of Calculus tied these two seemingly separate concepts together: we can find areas under a curve, i.e., we can evaluate a definite integral, using antiderivatives.

We ended the chapter by noting that antiderivatives are sometimes more than difficult to find: they are impossible. Therefore we developed numerical techniques that gave us good approximations of definite integrals.

We used the definite integral to compute areas, and also to compute displacements and distances traveled. There is far more we can do than that. In Chapter 7 we'll see more applications of the definite integral. Before that, in Chapter 6 we'll learn advanced techniques of integration, analogous to learning rules like the Product, Quotient and Chain Rules of differentiation.

Notes:

Exercises 5.5

Terms and Concepts

1. T/F: Simpson's Rule is a method of approximating antiderivatives.

2. What are the two basic situations where approximating the value of a definite integral is necessary?

3. Why are the Left and Right Hand Rules rarely used?

Problems

In Exercises 4 – 11, a definite integral is given.

(a) Approximate the definite integral with the Trapezoidal Rule and $n = 4$.

(b) Approximate the definite integral with Simpson's Rule and $n = 4$.

(c) Find the exact value of the integral.

4. $\int_{-1}^{1} x^2 \, dx$

5. $\int_{0}^{10} 5x \, dx$

6. $\int_{0}^{\pi} \sin x \, dx$

7. $\int_{0}^{4} \sqrt{x} \, dx$

8. $\int_{0}^{3} (x^3 + 2x^2 - 5x + 7) \, dx$

9. $\int_{0}^{1} x^4 \, dx$

10. $\int_{0}^{2\pi} \cos x \, dx$

11. $\int_{-3}^{3} \sqrt{9 - x^2} \, dx$

In Exercises 12 – 19, approximate the definite integral with the Trapezoidal Rule and Simpson's Rule, with $n = 6$.

12. $\int_{0}^{1} \cos(x^2) \, dx$

13. $\int_{-1}^{1} e^{x^2} \, dx$

14. $\int_{0}^{5} \sqrt{x^2 + 1} \, dx$

15. $\int_{0}^{\pi} x \sin x \, dx$

16. $\int_{0}^{\pi/2} \sqrt{\cos x} \, dx$

17. $\int_{1}^{4} \ln x \, dx$

18. $\int_{-1}^{1} \frac{1}{\sin x + 2} \, dx$

19. $\int_{0}^{6} \frac{1}{\sin x + 2} \, dx$

In Exercises 20 – 23, find n such that the error in approximating the given definite integral is less than 0.0001 when using:

(a) the Trapezoidal Rule

(b) Simpson's Rule

20. $\int_{0}^{\pi} \sin x \, dx$

21. $\int_{1}^{4} \frac{1}{\sqrt{x}} \, dx$

22. $\int_{0}^{\pi} \cos(x^2) \, dx$

23. $\int_{0}^{5} x^4 \, dx$

In Exercises 24 – 25, a region is given. Find the area of the region using Simpson's Rule:

(a) where the measurements are in centimeters, taken in 1 cm increments, and

(b) where the measurements are in hundreds of yards, taken in 100 yd increments.

24.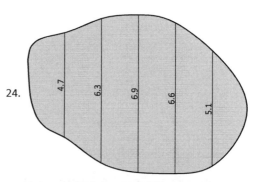

25.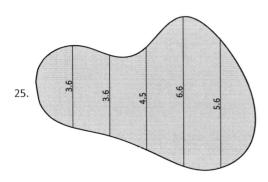

6: Techniques of Antidifferentiation

The previous chapter introduced the antiderivative and connected it to signed areas under a curve through the Fundamental Theorem of Calculus. The next chapter explores more applications of definite integrals than just area. As evaluating definite integrals will become important, we will want to find antiderivatives of a variety of functions.

This chapter is devoted to exploring techniques of antidifferentiation. While not every function has an antiderivative in terms of elementary functions (a concept introduced in the section on Numerical Integration), we can still find antiderivatives of a wide variety of functions.

6.1 Substitution

We motivate this section with an example. Let $f(x) = (x^2 + 3x - 5)^{10}$. We can compute $f'(x)$ using the Chain Rule. It is:

$$f'(x) = 10(x^2 + 3x - 5)^9 \cdot (2x + 3) = (20x + 30)(x^2 + 3x - 5)^9.$$

Now consider this: What is $\int (20x + 30)(x^2 + 3x - 5)^9 \, dx$? We have the answer in front of us;

$$\int (20x + 30)(x^2 + 3x - 5)^9 \, dx = (x^2 + 3x - 5)^{10} + C.$$

How would we have evaluated this indefinite integral without starting with $f(x)$ as we did?

This section explores *integration by substitution*. It allows us to "undo the Chain Rule." Substitution allows us to evaluate the above integral without knowing the original function first.

The underlying principle is to rewrite a "complicated" integral of the form $\int f(x) \, dx$ as a not–so–complicated integral $\int h(u) \, du$. We'll formally establish later how this is done. First, consider again our introductory indefinite integral, $\int (20x + 30)(x^2 + 3x - 5)^9 \, dx$. Arguably the most "complicated" part of the integrand is $(x^2 + 3x - 5)^9$. We wish to make this simpler; we do so through a substitution. Let $u = x^2 + 3x - 5$. Thus

$$(x^2 + 3x - 5)^9 = u^9.$$

We have established u as a function of x, so now consider the differential of u:

$$du = (2x + 3)dx.$$

Keep in mind that $(2x+3)$ and dx are multiplied; the dx is not "just sitting there."
Return to the original integral and do some substitutions through algebra:

$$\int (20x+30)(x^2+3x-5)^9\,dx = \int 10(2x+3)(x^2+3x-5)^9\,dx$$
$$= \int 10(\underbrace{x^2+3x-5}_{u})^9\underbrace{(2x+3)\,dx}_{du}$$
$$= \int 10u^9\,du$$
$$= u^{10} + C \quad \text{(replace } u \text{ with } x^2+3x-5)$$
$$= (x^2+3x-5)^{10} + C$$

One might well look at this and think "I (sort of) followed how that worked, but I could never come up with that on my own," but the process is learnable. This section contains numerous examples through which the reader will gain understanding and mathematical maturity enabling them to regard substitution as a natural tool when evaluating integrals.

We stated before that integration by substitution "undoes" the Chain Rule. Specifically, let $F(x)$ and $g(x)$ be differentiable functions and consider the derivative of their composition:

$$\frac{d}{dx}\Big(F\big(g(x)\big)\Big) = F'\big(g(x)\big)g'(x).$$

Thus

$$\int F'\big(g(x)\big)g'(x)\,dx = F\big(g(x)\big) + C.$$

Integration by substitution works by recognizing the "inside" function $g(x)$ and replacing it with a variable. By setting $u = g(x)$, we can rewrite the derivative as

$$\frac{d}{dx}\Big(F(u)\Big) = F'(u)u'.$$

Since $du = g'(x)dx$, we can rewrite the above integral as

$$\int F'\big(g(x)\big)g'(x)\,dx = \int F'(u)\,du = F(u) + C = F\big(g(x)\big) + C.$$

This concept is important so we restate it in the context of a theorem.

Notes:

6.1 Substitution

> **Theorem 44   Integration by Substitution**
>
> Let F and g be differentiable functions, where the range of g is an interval I contained in the domain of F. Then
>
> $$\int F'(g(x))g'(x)\,dx = F(g(x)) + C.$$
>
> If $u = g(x)$, then $du = g'(x)dx$ and
>
> $$\int F'(g(x))g'(x)\,dx = \int F'(u)\,du = F(u) + C = F(g(x)) + C.$$

The point of substitution is to make the integration step easy. Indeed, the step $\int F'(u)\,du = F(u) + C$ looks easy, as the antiderivative of the derivative of F is just F, plus a constant. The "work" involved is making the proper substitution. There is not a step–by–step process that one can memorize; rather, experience will be one's guide. To gain experience, we now embark on many examples.

Example 142   Integrating by substitution

Evaluate $\int x\sin(x^2 + 5)\,dx$.

SOLUTION   Knowing that substitution is related to the Chain Rule, we choose to let u be the "inside" function of $\sin(x^2 + 5)$. (This is not *always* a good choice, but it is often the best place to start.)

Let $u = x^2 + 5$, hence $du = 2x\,dx$. The integrand has an $x\,dx$ term, but not a $2x\,dx$ term. (Recall that multiplication is commutative, so the x does not physically have to be next to dx for there to be an $x\,dx$ term.) We can divide both sides of the du expression by 2:

$$du = 2x\,dx \quad \Rightarrow \quad \frac{1}{2}du = x\,dx.$$

We can now substitute.

$$\int x\sin(x^2 + 5)\,dx = \int \underbrace{\sin(x^2 + 5)}_{u}\,\underbrace{x\,dx}_{\frac{1}{2}du}$$

$$= \int \frac{1}{2}\sin u\,du$$

Notes:

Chapter 6 Techniques of Antidifferentiation

$$= -\frac{1}{2}\cos u + C \quad \text{(now replace } u \text{ with } x^2 + 5\text{)}$$

$$= -\frac{1}{2}\cos(x^2 + 5) + C.$$

Thus $\int x\sin(x^2 + 5)\,dx = -\frac{1}{2}\cos(x^2 + 5) + C$. We can check our work by evaluating the derivative of the right hand side.

Example 143   **Integrating by substitution**

Evaluate $\int \cos(5x)\,dx$.

Solution   Again let u replace the "inside" function. Letting $u = 5x$, we have $du = 5dx$. Since our integrand does not have a $5dx$ term, we can divide the previous equation by 5 to obtain $\frac{1}{5}du = dx$. We can now substitute.

$$\int \cos(5x)\,dx = \int \cos(\underbrace{5x}_{u})\underbrace{dx}_{\frac{1}{5}du}$$

$$= \int \frac{1}{5}\cos u\,du$$

$$= \frac{1}{5}\sin u + C$$

$$= \frac{1}{5}\sin(5x) + C.$$

We can again check our work through differentiation.

The previous example exhibited a common, and simple, type of substitution. The "inside" function was a linear function (in this case, $y = 5x$). When the inside function is linear, the resulting integration is very predictable, outlined here.

Key Idea 10   **Substitution With A Linear Function**

Consider $\int F'(ax + b)\,dx$, where $a \neq 0$ and b are constants. Letting $u = ax + b$ gives $du = a \cdot dx$, leading to the result

$$\int F'(ax + b)\,dx = \frac{1}{a}F(ax + b) + C.$$

Thus $\int \sin(7x - 4)\,dx = -\frac{1}{7}\cos(7x - 4) + C$. Our next example can use Key Idea 10, but we will only employ it after going through all of the steps.

Notes:

6.1 Substitution

Example 144 Integrating by substituting a linear function

Evaluate $\displaystyle\int \frac{7}{-3x+1}\,dx$.

Solution View this a composition of functions $f(g(x))$, where $f(x) = 7/x$ and $g(x) = -3x+1$. Employing our understanding of substitution, we let $u = -3x+1$, the inside function. Thus $du = -3dx$. The integrand lacks a -3; hence divide the previous equation by -3 to obtain $-du/3 = dx$. We can now evaluate the integral through substitution.

$$\int \frac{7}{-3x+1}\,dx = \int \frac{7}{u}\frac{du}{-3}$$
$$= \frac{-7}{3}\int \frac{du}{u}$$
$$= \frac{-7}{3}\ln|u| + C$$
$$= -\frac{7}{3}\ln|-3x+1| + C.$$

Using Key Idea 10 is faster, recognizing that u is linear and $a = -3$. One may want to continue writing out all the steps until they are comfortable with this particular shortcut.

Not all integrals that benefit from substitution have a clear "inside" function. Several of the following examples will demonstrate ways in which this occurs.

Example 145 Integrating by substitution

Evaluate $\displaystyle\int \sin x \cos x\,dx$.

Solution There is not a composition of function here to exploit; rather, just a product of functions. Do not be afraid to experiment; when given an integral to evaluate, it is often beneficial to think "If I let u be *this*, then du must be *that* ..." and see if this helps simplify the integral at all.

In this example, let's set $u = \sin x$. Then $du = \cos x\,dx$, which we have as part of the integrand! The substitution becomes very straightforward:

$$\int \sin x \cos x\,dx = \int u\,du$$
$$= \frac{1}{2}u^2 + C$$
$$= \frac{1}{2}\sin^2 x + C.$$

Notes:

One would do well to ask "What would happen if we let $u = \cos x$?" The result is just as easy to find, yet looks very different. The challenge to the reader is to evaluate the integral letting $u = \cos x$ and discover why the answer is the same, yet looks different.

Our examples so far have required "basic substitution." The next example demonstrates how substitutions can be made that often strike the new learner as being "nonstandard."

Example 146 **Integrating by substitution**

Evaluate $\int x\sqrt{x+3}\,dx$.

SOLUTION Recognizing the composition of functions, set $u = x + 3$. Then $du = dx$, giving what seems initially to be a simple substitution. But at this stage, we have:

$$\int x\sqrt{x+3}\,dx = \int x\sqrt{u}\,du.$$

We cannot evaluate an integral that has both an x and an u in it. We need to convert the x to an expression involving just u.

Since we set $u = x+3$, we can also state that $u - 3 = x$. Thus we can replace x in the integrand with $u - 3$. It will also be helpful to rewrite $\sqrt{u}$ as $u^{\frac{1}{2}}$.

$$\int x\sqrt{x+3}\,dx = \int (u-3)u^{\frac{1}{2}}\,du$$
$$= \int \left(u^{\frac{3}{2}} - 3u^{\frac{1}{2}}\right) du$$
$$= \frac{2}{5}u^{\frac{5}{2}} - 2u^{\frac{3}{2}} + C$$
$$= \frac{2}{5}(x+3)^{\frac{5}{2}} - 2(x+3)^{\frac{3}{2}} + C.$$

Checking your work is always a good idea. In this particular case, some algebra will be needed to make one's answer match the integrand in the original problem.

Example 147 **Integrating by substitution**

Evaluate $\int \dfrac{1}{x \ln x}\,dx$.

SOLUTION This is another example where there does not seem to be an obvious composition of functions. The line of thinking used in Example 146 is useful here: choose something for u and consider what this implies du must

Notes:

be. If u can be chosen such that du also appears in the integrand, then we have chosen well.

Choosing $u = 1/x$ makes $du = -1/x^2\, dx$; that does not seem helpful. However, setting $u = \ln x$ makes $du = 1/x\, dx$, which is part of the integrand. Thus:

$$\int \frac{1}{x\ln x}\, dx = \int \underbrace{\frac{1}{\ln x}}_{1/u} \underbrace{\frac{1}{x}\, dx}_{du}$$
$$= \int \frac{1}{u}\, du$$
$$= \ln|u| + C$$
$$= \ln|\ln x| + C.$$

The final answer is interesting; the natural log of the natural log. Take the derivative to confirm this answer is indeed correct.

Integrals Involving Trigonometric Functions

Section 6.3 delves deeper into integrals of a variety of trigonometric functions; here we use substitution to establish a foundation that we will build upon.

The next three examples will help fill in some missing pieces of our antiderivative knowledge. We know the antiderivatives of the sine and cosine functions; what about the other standard functions tangent, cotangent, secant and cosecant? We discover these next.

Example 148   **Integration by substitution: antiderivatives of $\tan x$**
Evaluate $\int \tan x\, dx$.

Solution   The previous paragraph established that we did not know the antiderivatives of tangent, hence we must assume that we have learned something in this section that can help us evaluate this indefinite integral.

Rewrite $\tan x$ as $\sin x / \cos x$. While the presence of a composition of functions may not be immediately obvious, recognize that $\cos x$ is "inside" the $1/x$ function. Therefore, we see if setting $u = \cos x$ returns usable results. We have

Notes:

that $du = -\sin x \, dx$, hence $-du = \sin x \, dx$. We can integrate:

$$\int \tan x \, dx = \int \frac{\sin x}{\cos x} \, dx$$
$$= \int \underbrace{\frac{1}{\cos x}}_{u} \underbrace{\sin x \, dx}_{-du}$$
$$= \int \frac{-1}{u} \, du$$
$$= -\ln|u| + C$$
$$= -\ln|\cos x| + C.$$

Some texts prefer to bring the -1 inside the logarithm as a power of $\cos x$, as in:

$$-\ln|\cos x| + C = \ln|(\cos x)^{-1}| + C$$
$$= \ln\left|\frac{1}{\cos x}\right| + C$$
$$= \ln|\sec x| + C.$$

Thus the result they give is $\int \tan x \, dx = \ln|\sec x| + C$. These two answers are equivalent.

Example 149   **Integrating by substitution: antiderivatives of** $\sec x$
Evaluate $\int \sec x \, dx$.

SOLUTION   This example employs a wonderful trick: multiply the integrand by "1" so that we see how to integrate more clearly. In this case, we write "1" as

$$1 = \frac{\sec x + \tan x}{\sec x + \tan x}.$$

This may seem like it came out of left field, but it works beautifully. Consider:

$$\int \sec x \, dx = \int \sec x \cdot \frac{\sec x + \tan x}{\sec x + \tan x} \, dx$$
$$= \int \frac{\sec^2 x + \sec x \tan x}{\sec x + \tan x} \, dx.$$

Notes:

Now let $u = \sec x + \tan x$; this means $du = (\sec x \tan x + \sec^2 x)\, dx$, which is our numerator. Thus:

$$= \int \frac{du}{u}$$
$$= \ln|u| + C$$
$$= \ln|\sec x + \tan x| + C.$$

We can use similar techniques to those used in Examples 148 and 149 to find antiderivatives of $\cot x$ and $\csc x$ (which the reader can explore in the exercises.) We summarize our results here.

Theorem 45   **Antiderivatives of Trigonometric Functions**

1. $\int \sin x\, dx = -\cos x + C$
2. $\int \cos x\, dx = \sin x + C$
3. $\int \tan x\, dx = -\ln|\cos x| + C$
4. $\int \csc x\, dx = -\ln|\csc x + \cot x| + C$
5. $\int \sec x\, dx = \ln|\sec x + \tan x| + C$
6. $\int \cot x\, dx = \ln|\sin x| + C$

We explore one more common trigonometric integral.

Example 150   **Integration by substitution: powers of $\cos x$ and $\sin x$**

Evaluate $\int \cos^2 x\, dx$.

SOLUTION   We have a composition of functions as $\cos^2 x = (\cos x)^2$. However, setting $u = \cos x$ means $du = -\sin x\, dx$, which we do not have in the integral. Another technique is needed.

The process we'll employ is to use a Power Reducing formula for $\cos^2 x$ (perhaps consult the back of this text for this formula), which states

$$\cos^2 x = \frac{1 + \cos(2x)}{2}.$$

The right hand side of this equation is not difficult to integrate. We have:

$$\int \cos^2 x\, dx = \int \frac{1 + \cos(2x)}{2}\, dx$$
$$= \int \left(\frac{1}{2} + \frac{1}{2}\cos(2x) \right) dx.$$

Notes:

Now use Key Idea 10:

$$= \frac{1}{2}x + \frac{1}{2}\frac{\sin(2x)}{2} + C$$
$$= \frac{1}{2}x + \frac{\sin(2x)}{4} + C.$$

We'll make significant use of this power–reducing technique in future sections.

Simplifying the Integrand

It is common to be reluctant to manipulate the integrand of an integral; at first, our grasp of integration is tenuous and one may think that working with the integrand will improperly change the results. Integration by substitution works using a different logic: as long as *equality* is maintained, the integrand can be manipulated so that its *form* is easier to deal with. The next two examples demonstrate common ways in which using algebra first makes the integration easier to perform.

Example 151   **Integration by substitution: simplifying first**
Evaluate $\int \frac{x^3 + 4x^2 + 8x + 5}{x^2 + 2x + 1}\,dx.$

SOLUTION   One may try to start by setting u equal to either the numerator or denominator; in each instance, the result is not workable.

When dealing with rational functions (i.e., quotients made up of polynomial functions), it is an almost universal rule that everything works better when the degree of the numerator is less than the degree of the denominator. Hence we use polynomial division.

We skip the specifics of the steps, but note that when $x^2 + 2x + 1$ is divided into $x^3 + 4x^2 + 8x + 5$, it goes in $x + 2$ times with a remainder of $3x + 3$. Thus

$$\frac{x^3 + 4x^2 + 8x + 5}{x^2 + 2x + 1} = x + 2 + \frac{3x + 3}{x^2 + 2x + 1}.$$

Integrating $x + 2$ is simple. The fraction can be integrated by setting $u = x^2 + 2x + 1$, giving $du = (2x + 2)\,dx$. This is very similar to the numerator. Note that

Notes:

$du/2 = (x+1)\,dx$ and then consider the following:

$$\int \frac{x^3 + 4x^2 + 8x + 5}{x^2 + 2x + 1}\,dx = \int \left(x + 2 + \frac{3x + 3}{x^2 + 2x + 1}\right)\,dx$$

$$= \int (x+2)\,dx + \int \frac{3(x+1)}{x^2 + 2x + 1}\,dx$$

$$= \frac{1}{2}x^2 + 2x + C_1 + \int \frac{3}{u}\frac{du}{2}$$

$$= \frac{1}{2}x^2 + 2x + C_1 + \frac{3}{2}\ln|u| + C_2$$

$$= \frac{1}{2}x^2 + 2x + \frac{3}{2}\ln|x^2 + 2x + 1| + C.$$

In some ways, we "lucked out" in that after dividing, substitution was able to be done. In later sections we'll develop techniques for handling rational functions where substitution is not directly feasible.

Example 152   **Integration by alternate methods**
Evaluate $\int \dfrac{x^2 + 2x + 3}{\sqrt{x}}\,dx$ with, and without, substitution.

SOLUTION   We already know how to integrate this particular example. Rewrite $\sqrt{x}$ as $x^{\frac{1}{2}}$ and simplify the fraction:

$$\frac{x^2 + 2x + 3}{x^{1/2}} = x^{\frac{3}{2}} + 2x^{\frac{1}{2}} + 3x^{-\frac{1}{2}}.$$

We can now integrate using the Power Rule:

$$\int \frac{x^2 + 2x + 3}{x^{1/2}}\,dx = \int \left(x^{\frac{3}{2}} + 2x^{\frac{1}{2}} + 3x^{-\frac{1}{2}}\right)\,dx$$

$$= \frac{2}{5}x^{\frac{5}{2}} + \frac{4}{3}x^{\frac{3}{2}} + 6x^{\frac{1}{2}} + C$$

This is a perfectly fine approach. We demonstrate how this can also be solved using substitution as its implementation is rather clever.
Let $u = \sqrt{x} = x^{\frac{1}{2}}$; therefore

$$du = \frac{1}{2}x^{-\frac{1}{2}}\,dx = \frac{1}{2\sqrt{x}}\,dx \quad \Rightarrow \quad 2\,du = \frac{1}{\sqrt{x}}\,dx.$$

This gives us $\int \dfrac{x^2 + 2x + 3}{\sqrt{x}}\,dx = \int (x^2 + 2x + 3)\cdot 2\,du$. What are we to do with the other x terms? Since $u = x^{\frac{1}{2}}$, $u^2 = x$, etc. We can then replace x^2 and

Notes:

Chapter 6 Techniques of Antidifferentiation

x with appropriate powers of u. We thus have

$$\int \frac{x^2 + 2x + 3}{\sqrt{x}} dx = \int (x^2 + 2x + 3) \cdot 2\, du$$
$$= \int 2(u^4 + 2u^2 + 3)\, du$$
$$= \frac{2}{5}u^5 + \frac{4}{3}u^3 + 6u + C$$
$$= \frac{2}{5}x^{\frac{5}{2}} + \frac{4}{3}x^{\frac{3}{2}} + 6x^{\frac{1}{2}} + C,$$

which is obviously the same answer we obtained before. In this situation, substitution is arguably more work than our other method. The fantastic thing is that it works. It demonstrates how flexible integration is.

Substitution and Inverse Trigonometric Functions

When studying derivatives of inverse functions, we learned that

$$\frac{d}{dx}\left(\tan^{-1} x\right) = \frac{1}{1 + x^2}.$$

Applying the Chain Rule to this is not difficult; for instance,

$$\frac{d}{dx}\left(\tan^{-1} 5x\right) = \frac{5}{1 + 25x^2}.$$

We now explore how Substitution can be used to "undo" certain derivatives that are the result of the Chain Rule applied to Inverse Trigonometric functions. We begin with an example.

Example 153 **Integrating by substitution: inverse trigonometric functions**
Evaluate $\int \frac{1}{25 + x^2}\, dx.$

SOLUTION The integrand looks similar to the derivative of the arctangent function. Note:

$$\frac{1}{25 + x^2} = \frac{1}{25(1 + \frac{x^2}{25})}$$
$$= \frac{1}{25\left(1 + \left(\frac{x}{5}\right)^2\right)}$$
$$= \frac{1}{25} \frac{1}{1 + \left(\frac{x}{5}\right)^2}.$$

Notes:

6.1 Substitution

Thus
$$\int \frac{1}{25+x^2}\,dx = \frac{1}{25}\int \frac{1}{1+\left(\frac{x}{5}\right)^2}\,dx.$$

This can be integrated using Substitution. Set $u = x/5$, hence $du = dx/5$ or $dx = 5du$. Thus

$$\begin{aligned}
\int \frac{1}{25+x^2}\,dx &= \frac{1}{25}\int \frac{1}{1+\left(\frac{x}{5}\right)^2}\,dx \\
&= \frac{1}{5}\int \frac{1}{1+u^2}\,du \\
&= \frac{1}{5}\tan^{-1} u + C \\
&= \frac{1}{5}\tan^{-1}\left(\frac{x}{5}\right) + C
\end{aligned}$$

Example 153 demonstrates a general technique that can be applied to other integrands that result in inverse trigonometric functions. The results are summarized here.

Theorem 46   **Integrals Involving Inverse Trigonomentric Functions**

Let $a > 0$.

1. $\int \dfrac{1}{a^2+x^2}\,dx = \dfrac{1}{a}\tan^{-1}\left(\dfrac{x}{a}\right) + C$

2. $\int \dfrac{1}{\sqrt{a^2-x^2}}\,dx = \sin^{-1}\left(\dfrac{x}{a}\right) + C$

3. $\int \dfrac{1}{x\sqrt{x^2-a^2}}\,dx = \dfrac{1}{a}\sec^{-1}\left(\dfrac{|x|}{a}\right) + C$

Let's practice using Theorem 46.

Example 154   **Integrating by substitution: inverse trigonometric functions**
Evaluate the given indefinite integrals.

$$\int \frac{1}{9+x^2}\,dx, \quad \int \frac{1}{x\sqrt{x^2 - \frac{1}{100}}}\,dx \quad \text{and} \quad \int \frac{1}{\sqrt{5-x^2}}\,dx.$$

Notes:

Chapter 6 Techniques of Antidifferentiation

SOLUTION Each can be answered using a straightforward application of Theorem 46.

$$\int \frac{1}{9+x^2}\, dx = \frac{1}{3}\tan^{-1}\frac{x}{3} + C, \text{ as } a = 3.$$

$$\int \frac{1}{x\sqrt{x^2 - \frac{1}{100}}}\, dx = 10\sec^{-1} 10x + C, \text{ as } a = \frac{1}{10}.$$

$$\int \frac{1}{\sqrt{5-x^2}} = \sin^{-1}\frac{x}{\sqrt{5}} + C, \text{ as } a = \sqrt{5}.$$

Most applications of Theorem 46 are not as straightforward. The next examples show some common integrals that can still be approached with this theorem.

Example 155   Integrating by substitution: completing the square

Evaluate $\int \frac{1}{x^2 - 4x + 13}\, dx$.

SOLUTION Initially, this integral seems to have nothing in common with the integrals in Theorem 46. As it lacks a square root, it almost certainly is not related to arcsine or arcsecant. It is, however, related to the arctangent function.

We see this by *completing the square* in the denominator. We give a brief reminder of the process here.

Start with a quadratic with a leading coefficient of 1. It will have the form of $x^2 + bx + c$. Take 1/2 of b, square it, and add/subtract it back into the expression. I.e.,

$$x^2 + bx + c = \underbrace{x^2 + bx + \frac{b^2}{4}}_{(x+b/2)^2} - \frac{b^2}{4} + c$$

$$= \left(x + \frac{b}{2}\right)^2 + c - \frac{b^2}{4}$$

In our example, we take half of -4 and square it, getting 4. We add/subtract it into the denominator as follows:

$$\frac{1}{x^2 - 4x + 13} = \frac{1}{\underbrace{x^2 - 4x + 4}_{(x-2)^2} - 4 + 13}$$

$$= \frac{1}{(x-2)^2 + 9}$$

Notes:

We can now integrate this using the arctangent rule. Technically, we need to substitute first with $u = x - 2$, but we can employ Key Idea 10 instead. Thus we have

$$\int \frac{1}{x^2 - 4x + 13}\, dx = \int \frac{1}{(x-2)^2 + 9}\, dx = \frac{1}{3} \tan^{-1} \frac{x-2}{3} + C.$$

Example 156 **Integrals requiring multiple methods**

Evaluate $\int \frac{4 - x}{\sqrt{16 - x^2}}\, dx.$

SOLUTION This integral requires two different methods to evaluate it. We get to those methods by splitting up the integral:

$$\int \frac{4 - x}{\sqrt{16 - x^2}}\, dx = \int \frac{4}{\sqrt{16 - x^2}}\, dx - \int \frac{x}{\sqrt{16 - x^2}}\, dx.$$

The first integral is handled using a straightforward application of Theorem 46; the second integral is handled by substitution, with $u = 16 - x^2$. We handle each separately.

$$\int \frac{4}{\sqrt{16 - x^2}}\, dx = 4 \sin^{-1} \frac{x}{4} + C.$$

$\int \frac{x}{\sqrt{16 - x^2}}\, dx$: Set $u = 16 - x^2$, so $du = -2x\,dx$ and $x\,dx = -du/2$. We have

$$\int \frac{x}{\sqrt{16 - x^2}}\, dx = \int \frac{-du/2}{\sqrt{u}}$$
$$= -\frac{1}{2} \int \frac{1}{\sqrt{u}}\, du$$
$$= -\sqrt{u} + C$$
$$= -\sqrt{16 - x^2} + C.$$

Combining these together, we have

$$\int \frac{4 - x}{\sqrt{16 - x^2}}\, dx = 4 \sin^{-1} \frac{x}{4} + \sqrt{16 - x^2} + C.$$

Substitution and Definite Integration

This section has focused on evaluating indefinite integrals as we are learning a new technique for finding antiderivatives. However, much of the time integration is used in the context of a definite integral. Definite integrals that require substitution can be calculated using the following workflow:

Notes:

1. Start with a definite integral $\int_a^b f(x)\,dx$ that requires substitution.

2. Ignore the bounds; use substitution to evaluate $\int f(x)\,dx$ and find an antiderivative $F(x)$.

3. Evaluate $F(x)$ at the bounds; that is, evaluate $F(x)\Big|_a^b = F(b) - F(a)$.

This workflow works fine, but substitution offers an alternative that is powerful and amazing (and a little time saving).

At its heart, (using the notation of Theorem 44) substitution converts integrals of the form $\int F'(g(x))g'(x)\,dx$ into an integral of the form $\int F'(u)\,du$ with the substitution of $u = g(x)$. The following theorem states how the bounds of a definite integral can be changed as the substitution is performed.

> **Theorem 47**   **Substitution with Definite Integrals**
>
> Let F and g be differentiable functions, where the range of g is an interval I that is contained in the domain of F. Then
>
> $$\int_a^b F'\big(g(x)\big)g'(x)\,dx = \int_{g(a)}^{g(b)} F'(u)\,du.$$

In effect, Theorem 47 states that once you convert to integrating with respect to u, you do not need to switch back to evaluating with respect to x. A few examples will help one understand.

Example 157   **Definite integrals and substitution: changing the bounds**

Evaluate $\int_0^2 \cos(3x - 1)\,dx$ using Theorem 47.

SOLUTION   Observing the composition of functions, let $u = 3x - 1$, hence $du = 3dx$. As $3dx$ does not appear in the integrand, divide the latter equation by 3 to get $du/3 = dx$.

By setting $u = 3x - 1$, we are implicitly stating that $g(x) = 3x - 1$. Theorem 47 states that the new lower bound is $g(0) = -1$; the new upper bound is

Notes:

$g(2) = 5$. We now evaluate the definite integral:

$$\int_1^2 \cos(3x-1)\,dx = \int_{-1}^5 \cos u\,\frac{du}{3}$$
$$= \frac{1}{3}\sin u \Big|_{-1}^5$$
$$= \frac{1}{3}\big(\sin 5 - \sin(-1)\big) \approx -0.039.$$

Notice how once we converted the integral to be in terms of u, we never went back to using x.

The graphs in Figure 6.1 tell more of the story. In (a) the area defined by the original integrand is shaded, whereas in (b) the area defined by the new integrand is shaded. In this particular situation, the areas look very similar; the new region is "shorter" but "wider," giving the same area.

Example 158   **Definite integrals and substitution: changing the bounds**
Evaluate $\displaystyle\int_0^{\pi/2} \sin x \cos x\,dx$ using Theorem 47.

SOLUTION   We saw the corresponding indefinite integral in Example 145. In that example we set $u = \sin x$ but stated that we could have let $u = \cos x$. For variety, we do the latter here.

Let $u = g(x) = \cos x$, giving $du = -\sin x\,dx$ and hence $\sin x\,dx = -du$. The new upper bound is $g(\pi/2) = 0$; the new lower bound is $g(0) = 1$. Note how the lower bound is actually larger than the upper bound now. We have

$$\int_0^{\pi/2} \sin x \cos x\,dx = \int_1^0 -u\,du \quad \text{(switch bounds \& change sign)}$$
$$= \int_0^1 u\,du$$
$$= \frac{1}{2}u^2 \Big|_0^1 = 1/2.$$

In Figure 6.2 we have again graphed the two regions defined by our definite integrals. Unlike the previous example, they bear no resemblance to each other. However, Theorem 47 guarantees that they have the same area.

Integration by substitution is a powerful and useful integration technique. The next section introduces another technique, called Integration by Parts. As substitution "undoes" the Chain Rule, integration by parts "undoes" the Product Rule. Together, these two techniques provide a strong foundation on which most other integration techniques are based.

Notes:

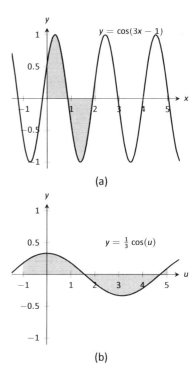

Figure 6.1: Graphing the areas defined by the definite integrals of Example 157.

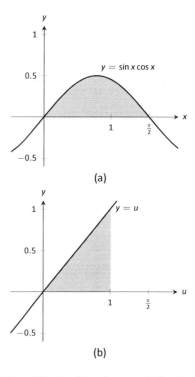

Figure 6.2: Graphing the areas defined by the definite integrals of Example 158.

Exercises 6.1

Terms and Concepts

1. Substitution "undoes" what derivative rule?

2. T/F: One can use algebra to rewrite the integrand of an integral to make it easier to evaluate.

Problems

In Exercises 3 – 14, evaluate the indefinite integral to develop an understanding of Substitution.

3. $\int 3x^2 \left(x^3 - 5\right)^7 dx$

4. $\int (2x - 5) \left(x^2 - 5x + 7\right)^3 dx$

5. $\int x \left(x^2 + 1\right)^8 dx$

6. $\int (12x + 14) \left(3x^2 + 7x - 1\right)^5 dx$

7. $\int \dfrac{1}{2x + 7} dx$

8. $\int \dfrac{1}{\sqrt{2x + 3}} dx$

9. $\int \dfrac{x}{\sqrt{x + 3}} dx$

10. $\int \dfrac{x^3 - x}{\sqrt{x}} dx$

11. $\int \dfrac{e^{\sqrt{x}}}{\sqrt{x}} dx$

12. $\int \dfrac{x^4}{\sqrt{x^5 + 1}} dx$

13. $\int \dfrac{\frac{1}{x} + 1}{x^2} dx$

14. $\int \dfrac{\ln(x)}{x} dx$

In Exercises 15 – 23, use Substitution to evaluate the indefinite integral involving trigonometric functions.

15. $\int \sin^2(x) \cos(x) dx$

16. $\int \cos(3 - 6x) dx$

17. $\int \sec^2(4 - x) dx$

18. $\int \sec(2x) dx$

19. $\int \tan^2(x) \sec^2(x) dx$

20. $\int x \cos\left(x^2\right) dx$

21. $\int \tan^2(x) dx$

22. $\int \cot x \, dx$. Do not just refer to Theorem 45 for the answer; justify it through Substitution.

23. $\int \csc x \, dx$. Do not just refer to Theorem 45 for the answer; justify it through Substitution.

In Exercises 24 – 30, use Substitution to evaluate the indefinite integral involving exponential functions.

24. $\int e^{3x-1} dx$

25. $\int e^{x^3} x^2 dx$

26. $\int e^{x^2 - 2x + 1}(x - 1) dx$

27. $\int \dfrac{e^x + 1}{e^x} dx$

28. $\int \dfrac{e^x - e^{-x}}{e^{2x}} dx$

29. $\int 3^{3x} dx$

30. $\int 4^{2x} dx$

In Exercises 31 – 34, use Substitution to evaluate the indefinite integral involving logarithmic functions.

31. $\int \dfrac{\ln x}{x} dx$

32. $\int \dfrac{(\ln x)^2}{x} dx$

33. $\int \dfrac{\ln\left(x^3\right)}{x} dx$

34. $\int \dfrac{1}{x\ln(x^2)}\,dx$

In Exercises 35 – 40, use Substitution to evaluate the indefinite integral involving rational functions.

35. $\int \dfrac{x^2+3x+1}{x}\,dx$

36. $\int \dfrac{x^3+x^2+x+1}{x}\,dx$

37. $\int \dfrac{x^3-1}{x+1}\,dx$

38. $\int \dfrac{x^2+2x-5}{x-3}\,dx$

39. $\int \dfrac{3x^2-5x+7}{x+1}\,dx$

40. $\int \dfrac{x^2+2x+1}{x^3+3x^2+3x}\,dx$

In Exercises 41 – 50, use Substitution to evaluate the indefinite integral involving inverse trigonometric functions.

41. $\int \dfrac{7}{x^2+7}\,dx$

42. $\int \dfrac{3}{\sqrt{9-x^2}}\,dx$

43. $\int \dfrac{14}{\sqrt{5-x^2}}\,dx$

44. $\int \dfrac{2}{x\sqrt{x^2-9}}\,dx$

45. $\int \dfrac{5}{\sqrt{x^4-16x^2}}\,dx$

46. $\int \dfrac{x}{\sqrt{1-x^4}}\,dx$

47. $\int \dfrac{1}{x^2-2x+8}\,dx$

48. $\int \dfrac{2}{\sqrt{-x^2+6x+7}}\,dx$

49. $\int \dfrac{3}{\sqrt{-x^2+8x+9}}\,dx$

50. $\int \dfrac{5}{x^2+6x+34}\,dx$

In Exercises 51 – 75, evaluate the indefinite integral.

51. $\int \dfrac{x^2}{(x^3+3)^2}\,dx$

52. $\int (3x^2+2x)(5x^3+5x^2+2)^8\,dx$

53. $\int \dfrac{x}{\sqrt{1-x^2}}\,dx$

54. $\int x^2\csc^2(x^3+1)\,dx$

55. $\int \sin(x)\sqrt{\cos(x)}\,dx$

56. $\int \dfrac{1}{x-5}\,dx$

57. $\int \dfrac{7}{3x+2}\,dx$

58. $\int \dfrac{3x^3+4x^2+2x-22}{x^2+3x+5}\,dx$

59. $\int \dfrac{2x+7}{x^2+7x+3}\,dx$

60. $\int \dfrac{9(2x+3)}{3x^2+9x+7}\,dx$

61. $\int \dfrac{-x^3+14x^2-46x-7}{x^2-7x+1}\,dx$

62. $\int \dfrac{x}{x^4+81}\,dx$

63. $\int \dfrac{2}{4x^2+1}\,dx$

64. $\int \dfrac{1}{x\sqrt{4x^2-1}}\,dx$

65. $\int \dfrac{1}{\sqrt{16-9x^2}}\,dx$

66. $\int \dfrac{3x-2}{x^2-2x+10}\,dx$

67. $\int \dfrac{7-2x}{x^2+12x+61}\,dx$

68. $\int \dfrac{x^2+5x-2}{x^2-10x+32}\,dx$

69. $\int \dfrac{x^3}{x^2+9}\,dx$

70. $\int \dfrac{x^3-x}{x^2+4x+9}\,dx$

71. $\int \dfrac{\sin(x)}{\cos^2(x)+1}\,dx$

72. $\int \dfrac{\cos(x)}{\sin^2(x) + 1}\,dx$

73. $\int \dfrac{\cos(x)}{1 - \sin^2(x)}\,dx$

74. $\int \dfrac{3x - 3}{\sqrt{x^2 - 2x - 6}}\,dx$

75. $\int \dfrac{x - 3}{\sqrt{x^2 - 6x + 8}}\,dx$

In Exercises 76 – 83, evaluate the definite integral.

76. $\int_1^3 \dfrac{1}{x - 5}\,dx$

77. $\int_2^6 x\sqrt{x - 2}\,dx$

78. $\int_{-\pi/2}^{\pi/2} \sin^2 x \cos x \, dx$

79. $\int_0^1 2x(1 - x^2)^4 \, dx$

80. $\int_{-2}^{-1} (x + 1)e^{x^2 + 2x + 1}\, dx$

81. $\int_{-1}^1 \dfrac{1}{1 + x^2}\,dx$

82. $\int_2^4 \dfrac{1}{x^2 - 6x + 10}\,dx$

83. $\int_1^{\sqrt{3}} \dfrac{1}{\sqrt{4 - x^2}}\,dx$

6.2 Integration by Parts

Here's a simple integral that we can't yet evaluate:

$$\int x \cos x \, dx.$$

It's a simple matter to take the derivative of the integrand using the Product Rule, but there is no Product Rule for integrals. However, this section introduces *Integration by Parts*, a method of integration that is based on the Product Rule for derivatives. It will enable us to evaluate this integral.

The Product Rule says that if u and v are functions of x, then $(uv)' = u'v + uv'$. For simplicity, we've written u for $u(x)$ and v for $v(x)$. Suppose we integrate both sides with respect to x. This gives

$$\int (uv)' \, dx = \int (u'v + uv') \, dx.$$

By the Fundamental Theorem of Calculus, the left side integrates to uv. The right side can be broken up into two integrals, and we have

$$uv = \int u'v \, dx + \int uv' \, dx.$$

Solving for the second integral we have

$$\int uv' \, dx = uv - \int u'v \, dx.$$

Using differential notation, we can write $du = u'(x)dx$ and $dv = v'(x)dx$ and the expression above can be written as follows:

$$\int u \, dv = uv - \int v \, du.$$

This is the Integration by Parts formula. For reference purposes, we state this in a theorem.

Theorem 48   **Integration by Parts**

Let u and v be differentiable functions of x on an interval I containing a and b. Then

$$\int u \, dv = uv - \int v \, du,$$

and

$$\int_{x=a}^{x=b} u \, dv = uv \Big|_a^b - \int_{x=a}^{x=b} v \, du.$$

Notes:

Let's try an example to understand our new technique.

Example 159   **Integrating using Integration by Parts**
Evaluate $\int x\cos x\,dx$.

SOLUTION   The key to Integration by Parts is to identify part of the integrand as "u" and part as "dv." Regular practice will help one make good identifications, and later we will introduce some principles that help. For now, let $u = x$ and $dv = \cos x\,dx$.

It is generally useful to make a small table of these values as done below. Right now we only know u and dv as shown on the left of Figure 6.3; on the right we fill in the rest of what we need. If $u = x$, then $du = dx$. Since $dv = \cos x\,dx$, v is an antiderivative of $\cos x$. We choose $v = \sin x$.

$$\begin{array}{ll} u = x & v = ? \\ du = ? & dv = \cos x\,dx \end{array} \quad\Rightarrow\quad \begin{array}{ll} u = x & v = \sin x \\ du = dx & dv = \cos x\,dx \end{array}$$

Figure 6.3: Setting up Integration by Parts.

Now substitute all of this into the Integration by Parts formula, giving

$$\int x\cos x\,dx = x\sin x - \int \sin x\,dx.$$

We can then integrate $\sin x$ to get $-\cos x + C$ and overall our answer is

$$\int x\cos x\,dx = x\sin x + \cos x + C.$$

Note how the antiderivative contains a product, $x\sin x$. This product is what makes Integration by Parts necessary.

The example above demonstrates how Integration by Parts works in general. We try to identify u and dv in the integral we are given, and the key is that we usually want to choose u and dv so that du is simpler than u and v is hopefully not too much more complicated than dv. This will mean that the integral on the right side of the Integration by Parts formula, $\int v\,du$ will be simpler to integrate than the original integral $\int u\,dv$.

In the example above, we chose $u = x$ and $dv = \cos x\,dx$. Then $du = dx$ was simpler than u and $v = \sin x$ is no more complicated than dv. Therefore, instead of integrating $x\cos x\,dx$, we could integrate $\sin x\,dx$, which we knew how to do.

A useful mnemonic for helping to determine u is "LIATE," where

L = **L**ogarithmic, I = **I**nverse Trig., A = **A**lgebraic (polynomials),
T = **T**rigonometric, and E = **E**xponential.

Notes:

6.2 Integration by Parts

If the integrand contains both a logarithmic and an algebraic term, in general letting u be the logarithmic term works best, as indicated by L coming before A in LIATE.

We now consider another example.

Example 160 **Integrating using Integration by Parts**

Evaluate $\int xe^x\, dx$.

SOLUTION The integrand contains an **A**lgebraic term (x) and an **E**xponential term (e^x). Our mnemonic suggests letting u be the algebraic term, so we choose $u = x$ and $dv = e^x\, dx$. Then $du = dx$ and $v = e^x$ as indicated by the tables below.

$$
\begin{array}{ll}
u = x & v = ? \\
du = ? & dv = e^x\, dx
\end{array}
\quad\Rightarrow\quad
\begin{array}{ll}
u = x & v = e^x \\
du = dx & dv = e^x\, dx
\end{array}
$$

Figure 6.4: Setting up Integration by Parts.

We see du is simpler than u, while there is no change in going from dv to v. This is good. The Integration by Parts formula gives

$$\int xe^x\, dx = xe^x - \int e^x\, dx.$$

The integral on the right is simple; our final answer is

$$\int xe^x\, dx = xe^x - e^x + C.$$

Note again how the antiderivatives contain a product term.

Example 161 **Integrating using Integration by Parts**

Evaluate $\int x^2 \cos x\, dx$.

SOLUTION The mnemonic suggests letting $u = x^2$ instead of the trigonometric function, hence $dv = \cos x\, dx$. Then $du = 2x\, dx$ and $v = \sin x$ as shown below.

$$
\begin{array}{ll}
u = x^2 & v = ? \\
du = ? & dv = \cos x\, dx
\end{array}
\quad\Rightarrow\quad
\begin{array}{ll}
u = x^2 & v = \sin x \\
du = 2x\, dx & dv = \cos x\, dx
\end{array}
$$

Figure 6.5: Setting up Integration by Parts.

Notes:

The Integration by Parts formula gives

$$\int x^2 \cos x \, dx = x^2 \sin x - \int 2x \sin x \, dx.$$

At this point, the integral on the right is indeed simpler than the one we started with, but to evaluate it, we need to do Integration by Parts again. Here we choose $u = 2x$ and $dv = \sin x$ and fill in the rest below.

$$\begin{array}{ll} u = 2x & v = ? \\ du = ? & dv = \sin x \, dx \end{array} \Rightarrow \begin{array}{ll} u = 2x & v = -\cos x \\ du = 2 \, dx & dv = \sin x \, dx \end{array}$$

Figure 6.6: Setting up Integration by Parts (again).

$$\int x^2 \cos x \, dx = x^2 \sin x - \left(-2x \cos x - \int -2 \cos x \, dx \right).$$

The integral all the way on the right is now something we can evaluate. It evaluates to $-2 \sin x$. Then going through and simplifying, being careful to keep all the signs straight, our answer is

$$\int x^2 \cos x \, dx = x^2 \sin x + 2x \cos x - 2 \sin x + C.$$

Example 162   Integrating using Integration by Parts

Evaluate $\int e^x \cos x \, dx$.

SOLUTION   This is a classic problem. Our mnemonic suggests letting u be the trigonometric function instead of the exponential. In this particular example, one can let u be either $\cos x$ or e^x; to demonstrate that we do not have to follow LIATE, we choose $u = e^x$ and hence $dv = \cos x \, dx$. Then $du = e^x \, dx$ and $v = \sin x$ as shown below.

$$\begin{array}{ll} u = e^x & v = ? \\ du = ? & dv = \cos x \, dx \end{array} \Rightarrow \begin{array}{ll} u = e^x & v = \sin x \\ du = e^x \, dx & dv = \cos x \, dx \end{array}$$

Figure 6.7: Setting up Integration by Parts.

Notice that du is no simpler than u, going against our general rule (but bear with us). The Integration by Parts formula yields

$$\int e^x \cos x \, dx = e^x \sin x - \int e^x \sin x \, dx.$$

Notes:

The integral on the right is not much different than the one we started with, so it seems like we have gotten nowhere. Let's keep working and apply Integration by Parts to the new integral, using $u = e^x$ and $dv = \sin x \, dx$. This leads us to the following:

$$\begin{array}{ll} u = e^x & v = ? \\ du = ? & dv = \sin x \, dx \end{array} \quad \Rightarrow \quad \begin{array}{ll} u = e^x & v = -\cos x \\ du = e^x \, dx & dv = \sin x \, dx \end{array}$$

Figure 6.8: Setting up Integration by Parts (again).

The Integration by Parts formula then gives:

$$\int e^x \cos x \, dx = e^x \sin x - \left(-e^x \cos x - \int -e^x \cos x \, dx \right)$$
$$= e^x \sin x + e^x \cos x - \int e^x \cos x \, dx.$$

It seems we are back right where we started, as the right hand side contains $\int e^x \cos x \, dx$. But this is actually a good thing.

Add $\int e^x \cos x \, dx$ to both sides. This gives

$$2 \int e^x \cos x \, dx = e^x \sin x + e^x \cos x$$

Now divide both sides by 2:

$$\int e^x \cos x \, dx = \frac{1}{2} \left(e^x \sin x + e^x \cos x \right).$$

Simplifying a little and adding the constant of integration, our answer is thus

$$\int e^x \cos x \, dx = \frac{1}{2} e^x \left(\sin x + \cos x \right) + C.$$

Example 163   **Integrating using Integration by Parts: antiderivative of $\ln x$**
Evaluate $\int \ln x \, dx$.

SOLUTION   One may have noticed that we have rules for integrating the familiar trigonometric functions and e^x, but we have not yet given a rule for integrating $\ln x$. That is because $\ln x$ can't easily be integrated with any of the rules we have learned up to this point. But we can find its antiderivative by a

Notes:

clever application of Integration by Parts. Set $u = \ln x$ and $dv = dx$. This is a good, sneaky trick to learn as it can help in other situations. This determines $du = (1/x)\,dx$ and $v = x$ as shown below.

$$\begin{array}{cc} u = \ln x & v = ? \\ du = ? & dv = dx \end{array} \quad \Rightarrow \quad \begin{array}{cc} u = \ln x & v = x \\ du = 1/x\,dx & dv = dx \end{array}$$

Figure 6.9: Setting up Integration by Parts.

Putting this all together in the Integration by Parts formula, things work out very nicely:

$$\int \ln x\,dx = x\ln x - \int x\frac{1}{x}\,dx.$$

The new integral simplifies to $\int 1\,dx$, which is about as simple as things get. Its integral is $x + C$ and our answer is

$$\int \ln x\,dx = x\ln x - x + C.$$

Example 164   Integrating using Int. by Parts: antiderivative of arctan x
Evaluate $\int \arctan x\,dx$.

SOLUTION   The same sneaky trick we used above works here. Let $u = \arctan x$ and $dv = dx$. Then $du = 1/(1 + x^2)\,dx$ and $v = x$. The Integration by Parts formula gives

$$\int \arctan x\,dx = x\arctan x - \int \frac{x}{1+x^2}\,dx.$$

The integral on the right can be solved by substitution. Taking $u = 1 + x^2$, we get $du = 2x\,dx$. The integral then becomes

$$\int \arctan x\,dx = x\arctan x - \frac{1}{2}\int \frac{1}{u}\,du.$$

The integral on the right evaluates to $\ln|u| + C$, which becomes $\ln(1 + x^2) + C$. Therefore, the answer is

$$\int \arctan x\,dx = x\arctan x - \ln(1 + x^2) + C.$$

Notes:

Substitution Before Integration

When taking derivatives, it was common to employ multiple rules (such as using both the Quotient and the Chain Rules). It should then come as no surprise that some integrals are best evaluated by combining integration techniques. In particular, here we illustrate making an "unusual" substitution first before using Integration by Parts.

Example 165 **Integration by Parts after substitution**

Evaluate $\int \cos(\ln x)\, dx$.

SOLUTION The integrand contains a composition of functions, leading us to think Substitution would be beneficial. Letting $u = \ln x$, we have $du = 1/x\, dx$. This seems problematic, as we do not have a $1/x$ in the integrand. But consider:
$$du = \frac{1}{x}\, dx \Rightarrow x \cdot du = dx.$$
Since $u = \ln x$, we can use inverse functions and conclude that $x = e^u$. Therefore we have that
$$\begin{aligned} dx &= x \cdot du \\ &= e^u\, du. \end{aligned}$$
We can thus replace $\ln x$ with u and dx with $e^u\, du$. Thus we rewrite our integral as
$$\int \cos(\ln x)\, dx = \int e^u \cos u\, du.$$
We evaluated this integral in Example 162. Using the result there, we have:
$$\begin{aligned} \int \cos(\ln x)\, dx &= \int e^u \cos u\, du \\ &= \frac{1}{2}e^u\big(\sin u + \cos u\big) + C \\ &= \frac{1}{2}e^{\ln x}\big(\sin(\ln x) + \cos(\ln x)\big) + C \\ &= \frac{1}{2}x\big(\sin(\ln x) + \cos(\ln x)\big) + C. \end{aligned}$$

Definite Integrals and Integration By Parts

So far we have focused only on evaluating indefinite integrals. Of course, we can use Integration by Parts to evaluate definite integrals as well, as Theorem

Notes:

48 states. We do so in the next example.

Example 166   **Definite integration using Integration by Parts**

Evaluate $\int_1^2 x^2 \ln x \, dx$.

SOLUTION   Our mnemonic suggests letting $u = \ln x$, hence $dv = x^2 \, dx$. We then get $du = (1/x) \, dx$ and $v = x^3/3$ as shown below.

$$\begin{array}{ll} u = \ln x & v = ? \\ du = ? & dv = x^2 \, dx \end{array} \Rightarrow \begin{array}{ll} u = \ln x & v = x^3/3 \\ du = 1/x \, dx & dv = x^2 \, dx \end{array}$$

Figure 6.10: Setting up Integration by Parts.

The Integration by Parts formula then gives

$$\int_1^2 x^2 \ln x \, dx = \frac{x^3}{3} \ln x \Big|_1^2 - \int_1^2 \frac{x^3}{3} \frac{1}{x} \, dx$$

$$= \frac{x^3}{3} \ln x \Big|_1^2 - \int_1^2 \frac{x^2}{3} \, dx$$

$$= \frac{x^3}{3} \ln x \Big|_1^2 - \frac{x^3}{9} \Big|_1^2$$

$$= \left(\frac{x^3}{3} \ln x - \frac{x^3}{9} \right) \Big|_1^2$$

$$= \left(\frac{8}{3} \ln 2 - \frac{8}{9} \right) - \left(\frac{1}{3} \ln 1 - \frac{1}{9} \right)$$

$$= \frac{8}{3} \ln 2 - \frac{7}{9}$$

$$\approx 1.07.$$

In general, Integration by Parts is useful for integrating certain products of functions, like $\int xe^x \, dx$ or $\int x^3 \sin x \, dx$. It is also useful for integrals involving logarithms and inverse trigonometric functions.

As stated before, integration is generally more difficult than derivation. We are developing tools for handling a large array of integrals, and experience will tell us when one tool is preferable/necessary over another. For instance, consider the three similar–looking integrals

$$\int xe^x \, dx, \quad \int xe^{x^2} \, dx \quad \text{and} \quad \int xe^{x^3} \, dx.$$

Notes:

While the first is calculated easily with Integration by Parts, the second is best approached with Substitution. Taking things one step further, the third integral has no answer in terms of elementary functions, so none of the methods we learn in calculus will get us the exact answer.

Integration by Parts is a very useful method, second only to substitution. In the following sections of this chapter, we continue to learn other integration techniques. The next section focuses on handling integrals containing trigonometric functions.

Notes:

Exercises 6.2

Terms and Concepts

1. T/F: Integration by Parts is useful in evaluating integrands that contain products of functions.

2. T/F: Integration by Parts can be thought of as the "opposite of the Chain Rule."

3. For what is "LIATE" useful?

Problems

In Exercises 4 – 33, evaluate the given indefinite integral.

4. $\int x \sin x \, dx$

5. $\int xe^{-x} \, dx$

6. $\int x^2 \sin x \, dx$

7. $\int x^3 \sin x \, dx$

8. $\int xe^{x^2} \, dx$

9. $\int x^3 e^x \, dx$

10. $\int xe^{-2x} \, dx$

11. $\int e^x \sin x \, dx$

12. $\int e^{2x} \cos x \, dx$

13. $\int e^{2x} \sin(3x) \, dx$

14. $\int e^{5x} \cos(5x) \, dx$

15. $\int \sin x \cos x \, dx$

16. $\int \sin^{-1} x \, dx$

17. $\int \tan^{-1}(2x) \, dx$

18. $\int x \tan^{-1} x \, dx$

19. $\int \sin^{-1} x \, dx$

20. $\int x \ln x \, dx$

21. $\int (x-2) \ln x \, dx$

22. $\int x \ln(x-1) \, dx$

23. $\int x \ln(x^2) \, dx$

24. $\int x^2 \ln x \, dx$

25. $\int (\ln x)^2 \, dx$

26. $\int (\ln(x+1))^2 \, dx$

27. $\int x \sec^2 x \, dx$

28. $\int x \csc^2 x \, dx$

29. $\int x\sqrt{x-2} \, dx$

30. $\int x\sqrt{x^2-2} \, dx$

31. $\int \sec x \tan x \, dx$

32. $\int x \sec x \tan x \, dx$

33. $\int x \csc x \cot x \, dx$

In Exercises 34 – 38, evaluate the indefinite integral after first making a substitution.

34. $\int \sin(\ln x) \, dx$

35. $\int \sin(\sqrt{x}) \, dx$

36. $\int \ln(\sqrt{x}) \, dx$

37. $\int e^{\sqrt{x}} \, dx$

38. $\int e^{\ln x}\, dx$

In Exercises 39 – 47, evaluate the definite integral. Note: the corresponding indefinite integrals appear in Exercises 4 – 12.

39. $\int_0^{\pi} x \sin x\, dx$

40. $\int_{-1}^{1} xe^{-x}\, dx$

41. $\int_{-\pi/4}^{\pi/4} x^2 \sin x\, dx$

42. $\int_{-\pi/2}^{\pi/2} x^3 \sin x\, dx$

43. $\int_0^{\sqrt{\ln 2}} xe^{x^2}\, dx$

44. $\int_0^1 x^3 e^x\, dx$

45. $\int_1^2 xe^{-2x}\, dx$

46. $\int_0^{\pi} e^x \sin x\, dx$

47. $\int_{-\pi/2}^{\pi/2} e^{2x} \cos x\, dx$

Chapter 6　Techniques of Antidifferentiation

6.3　Trigonometric Integrals

Functions involving trigonometric functions are useful as they are good at describing periodic behavior. This section describes several techniques for finding antiderivatives of certain combinations of trigonometric functions.

Integrals of the form $\int \sin^m x \cos^n x \, dx$

In learning the technique of Substitution, we saw the integral $\int \sin x \cos x \, dx$ in Example 145. The integration was not difficult, and one could easily evaluate the indefinite integral by letting $u = \sin x$ or by letting $u = \cos x$. This integral is easy since the power of both sine and cosine is 1.

We generalize this integral and consider integrals of the form $\int \sin^m x \cos^n x \, dx$, where m, n are nonnegative integers. Our strategy for evaluating these integrals is to use the identity $\cos^2 x + \sin^2 x = 1$ to convert high powers of one trigonometric function into the other, leaving a single sine or cosine term in the integrand. We summarize the general technique in the following Key Idea.

Key Idea 11　**Integrals Involving Powers of Sine and Cosine**

Consider $\int \sin^m x \cos^n x \, dx$, where m, n are nonnegative integers.

1. If m is odd, then $m = 2k + 1$ for some integer k. Rewrite
$$\sin^m x = \sin^{2k+1} x = \sin^{2k} x \sin x = (\sin^2 x)^k \sin x = (1 - \cos^2 x)^k \sin x.$$
Then
$$\int \sin^m x \cos^n x \, dx = \int (1 - \cos^2 x)^k \sin x \cos^n x \, dx = -\int (1 - u^2)^k u^n \, du,$$
where $u = \cos x$ and $du = -\sin x \, dx$.

2. If n is odd, then using substitutions similar to that outlined above we have
$$\int \sin^m x \cos^n x \, dx = \int u^m (1 - u^2)^k \, du,$$
where $u = \sin x$ and $du = \cos x \, dx$.

3. If both m and n are even, use the power–reducing identities
$$\cos^2 x = \frac{1 + \cos(2x)}{2} \quad \text{and} \quad \sin^2 x = \frac{1 - \cos(2x)}{2}$$
to reduce the degree of the integrand. Expand the result and apply the principles of this Key Idea again.

Notes:

286

6.3 Trigonometric Integrals

We practice applying Key Idea 11 in the next examples.

Example 167   **Integrating powers of sine and cosine**

Evaluate $\int \sin^5 x \cos^8 x \, dx$.

SOLUTION   The power of the sine term is odd, so we rewrite $\sin^5 x$ as

$$\sin^5 x = \sin^4 x \sin x = (\sin^2 x)^2 \sin x = (1 - \cos^2 x)^2 \sin x.$$

Our integral is now $\int (1 - \cos^2 x)^2 \cos^8 x \sin x \, dx$. Let $u = \cos x$, hence $du = -\sin x \, dx$. Making the substitution and expanding the integrand gives

$$\int (1-\cos^2)^2 \cos^8 x \sin x \, dx = -\int (1-u^2)^2 u^8 \, du = -\int \left(1 - 2u^2 + u^4\right) u^8 \, du = -\int \left(u^8 - 2u^{10} + u^{12}\right) du.$$

This final integral is not difficult to evaluate, giving

$$-\int \left(u^8 - 2u^{10} + u^{12}\right) du = -\frac{1}{9}u^9 + \frac{2}{11}u^{11} - \frac{1}{13}u^{13} + C$$

$$= -\frac{1}{9}\cos^9 x + \frac{2}{11}\cos^{11} x - \frac{1}{13}\cos^{13} x + C.$$

Example 168   **Integrating powers of sine and cosine**

Evaluate $\int \sin^5 x \cos^9 x \, dx$.

SOLUTION   The powers of both the sine and cosine terms are odd, therefore we can apply the techniques of Key Idea 11 to either power. We choose to work with the power of the cosine term since the previous example used the sine term's power.

We rewrite $\cos^9 x$ as

$$\cos^9 x = \cos^8 x \cos x$$
$$= (\cos^2 x)^4 \cos x$$
$$= (1 - \sin^2 x)^4 \cos x$$
$$= (1 - 4\sin^2 x + 6\sin^4 x - 4\sin^6 x + \sin^8 x) \cos x.$$

We rewrite the integral as

$$\int \sin^5 x \cos^9 x \, dx = \int \sin^5 x \left(1 - 4\sin^2 x + 6\sin^4 x - 4\sin^6 x + \sin^8 x\right) \cos x \, dx.$$

Notes:

Chapter 6 Techniques of Antidifferentiation

Now substitute and integrate, using $u = \sin x$ and $du = \cos x\, dx$.

$$\int \sin^5 x \left(1 - 4\sin^2 x + 6\sin^4 x - 4\sin^6 x + \sin^8 x\right) \cos x\, dx =$$

$$\int u^5 \left(1 - 4u^2 + 6u^4 - 4u^6 + u^8\right) du = \int \left(u^5 - 4u^7 + 6u^9 - 4u^{11} + u^{13}\right) du$$

$$= \frac{1}{6}u^6 - \frac{1}{2}u^8 + \frac{3}{5}u^{10} - \frac{1}{3}u^{12} + \frac{1}{14}u^{14} + C$$

$$= \frac{1}{6}\sin^6 x - \frac{1}{2}\sin^8 x + \frac{3}{5}\sin^{10} x + \ldots$$

$$- \frac{1}{3}\sin^{12} x + \frac{1}{14}\sin^{14} x + C.$$

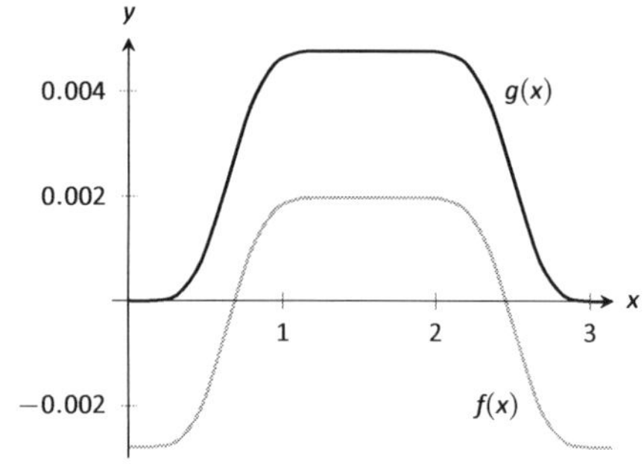

Technology Note: The work we are doing here can be a bit tedious, but the skills developed (problem solving, algebraic manipulation, etc.) are important. Nowadays problems of this sort are often solved using a computer algebra system. The powerful program *Mathematica*® integrates $\int \sin^5 x \cos^9 x\, dx$ as

$$f(x) = -\frac{45\cos(2x)}{16384} - \frac{5\cos(4x)}{8192} + \frac{19\cos(6x)}{49152} + \frac{\cos(8x)}{4096} - \frac{\cos(10x)}{81920} - \frac{\cos(12x)}{24576} - \frac{\cos(14x)}{114688},$$

which clearly has a different form than our answer in Example 168, which is

$$g(x) = \frac{1}{6}\sin^6 x - \frac{1}{2}\sin^8 x + \frac{3}{5}\sin^{10} x - \frac{1}{3}\sin^{12} x + \frac{1}{14}\sin^{14} x.$$

Figure 6.11 shows a graph of f and g; they are clearly not equal, but they differ *only by a constant*. That is $g(x) = f(x) + C$ for some constant C. So we have two different antiderivatives of the same function, meaning both answers are correct.

Figure 6.11: A plot of $f(x)$ and $g(x)$ from Example 168 and the Technology Note.

Example 169 **Integrating powers of sine and cosine**

Evaluate $\int \cos^4 x \sin^2 x\, dx$.

Solution The powers of sine and cosine are both even, so we employ the power–reducing formulas and algebra as follows.

$$\int \cos^4 x \sin^2 x\, dx = \int \left(\frac{1+\cos(2x)}{2}\right)^2 \left(\frac{1-\cos(2x)}{2}\right) dx$$

$$= \int \frac{1 + 2\cos(2x) + \cos^2(2x)}{4} \cdot \frac{1-\cos(2x)}{2}\, dx$$

$$= \int \frac{1}{8}\left(1 + \cos(2x) - \cos^2(2x) - \cos^3(2x)\right) dx$$

The $\cos(2x)$ term is easy to integrate, especially with Key Idea 10. The $\cos^2(2x)$ term is another trigonometric integral with an even power, requiring the power–reducing formula again. The $\cos^3(2x)$ term is a cosine function with an odd power, requiring a substitution as done before. We integrate each in turn below.

Notes:

$$\int \cos(2x)\,dx = \frac{1}{2}\sin(2x) + C.$$

$$\int \cos^2(2x)\,dx = \int \frac{1+\cos(4x)}{2}\,dx = \frac{1}{2}\left(x + \frac{1}{4}\sin(4x)\right) + C.$$

Finally, we rewrite $\cos^3(2x)$ as

$$\cos^3(2x) = \cos^2(2x)\cos(2x) = \left(1 - \sin^2(2x)\right)\cos(2x).$$

Letting $u = \sin(2x)$, we have $du = 2\cos(2x)\,dx$, hence

$$\int \cos^3(2x)\,dx = \int \left(1 - \sin^2(2x)\right)\cos(2x)\,dx$$
$$= \int \frac{1}{2}(1 - u^2)\,du$$
$$= \frac{1}{2}\left(u - \frac{1}{3}u^3\right) + C$$
$$= \frac{1}{2}\left(\sin(2x) - \frac{1}{3}\sin^3(2x)\right) + C$$

Putting all the pieces together, we have

$$\int \cos^4 x \sin^2 x\,dx = \int \frac{1}{8}\left(1 + \cos(2x) - \cos^2(2x) - \cos^3(2x)\right)dx$$
$$= \frac{1}{8}\left[x + \frac{1}{2}\sin(2x) - \frac{1}{2}\left(x + \frac{1}{4}\sin(4x)\right) - \frac{1}{2}\left(\sin(2x) - \frac{1}{3}\sin^3(2x)\right)\right] + C$$
$$= \frac{1}{8}\left[\frac{1}{2}x - \frac{1}{8}\sin(4x) + \frac{1}{6}\sin^3(2x)\right] + C$$

The process above was a bit long and tedious, but being able to work a problem such as this from start to finish is important.

Integrals of the form $\int \sin(mx)\sin(nx)\,dx$, $\int \cos(mx)\cos(nx)\,dx$, **and** $\int \sin(mx)\cos(nx)\,dx$.

Functions that contain products of sines and cosines of differing periods are important in many applications including the analysis of sound waves. Integrals of the form

$$\int \sin(mx)\sin(nx)\,dx, \quad \int \cos(mx)\cos(nx)\,dx \quad \text{and} \quad \int \sin(mx)\cos(nx)\,dx$$

Notes:

are best approached by first applying the Product to Sum Formulas found in the back cover of this text, namely

$$\sin(mx)\sin(nx) = \frac{1}{2}\Big[\cos\big((m-n)x\big) - \cos\big((m+n)x\big)\Big]$$
$$\cos(mx)\cos(nx) = \frac{1}{2}\Big[\cos\big((m-n)x\big) + \cos\big((m+n)x\big)\Big]$$
$$\sin(mx)\cos(nx) = \frac{1}{2}\Big[\sin\big((m-n)x\big) + \sin\big((m+n)x\big)\Big]$$

Example 170   **Integrating products of $\sin(mx)$ and $\cos(nx)$**

Evaluate $\int \sin(5x)\cos(2x)\, dx$.

SOLUTION   The application of the formula and subsequent integration are straightforward:

$$\int \sin(5x)\cos(2x)\, dx = \int \frac{1}{2}\Big[\sin(3x) + \sin(7x)\Big]\, dx$$
$$= -\frac{1}{6}\cos(3x) - \frac{1}{14}\cos(7x) + C$$

Integrals of the form $\int \tan^m x \sec^n x\, dx$.

When evaluating integrals of the form $\int \sin^m x \cos^n x\, dx$, the Pythagorean Theorem allowed us to convert even powers of sine into even powers of cosine, and vise–versa. If, for instance, the power of sine was odd, we pulled out one $\sin x$ and converted the remaining even power of $\sin x$ into a function using powers of $\cos x$, leading to an easy substitution.

The same basic strategy applies to integrals of the form $\int \tan^m x \sec^n x\, dx$, albeit a bit more nuanced. The following three facts will prove useful:

- $\frac{d}{dx}(\tan x) = \sec^2 x$,

- $\frac{d}{dx}(\sec x) = \sec x \tan x$, and

- $1 + \tan^2 x = \sec^2 x$ (the Pythagorean Theorem).

If the integrand can be manipulated to separate a $\sec^2 x$ term with the remaining secant power even, or if a $\sec x \tan x$ term can be separated with the remaining $\tan x$ power even, the Pythagorean Theorem can be employed, leading to a simple substitution. This strategy is outlined in the following Key Idea.

Notes:

6.3 Trigonometric Integrals

Key Idea 12 **Integrals Involving Powers of Tangent and Secant**

Consider $\int \tan^m x \sec^n x \, dx$, where m, n are nonnegative integers.

1. If n is even, then $n = 2k$ for some integer k. Rewrite $\sec^n x$ as
$$\sec^n x = \sec^{2k} x = \sec^{2k-2} x \sec^2 x = (1 + \tan^2 x)^{k-1} \sec^2 x.$$

   Then
$$\int \tan^m x \sec^n x \, dx = \int \tan^m x (1 + \tan^2 x)^{k-1} \sec^2 x \, dx = \int u^m (1 + u^2)^{k-1} \, du,$$
   where $u = \tan x$ and $du = \sec^2 x \, dx$.

2. If m is odd, then $m = 2k + 1$ for some integer k. Rewrite $\tan^m x \sec^n x$ as
$$\tan^m x \sec^n x = \tan^{2k+1} x \sec^n x = \tan^{2k} x \sec^{n-1} x \sec x \tan x = (\sec^2 x - 1)^k \sec^{n-1} x \sec x \tan x.$$

   Then
$$\int \tan^m x \sec^n x \, dx = \int (\sec^2 x - 1)^k \sec^{n-1} x \sec x \tan x \, dx = \int (u^2 - 1)^k u^{n-1} \, du,$$
   where $u = \sec x$ and $du = \sec x \tan x \, dx$.

3. If n is odd and m is even, then $m = 2k$ for some integer k. Convert $\tan^m x$ to $(\sec^2 x - 1)^k$. Expand the new integrand and use Integration By Parts, with $dv = \sec^2 x \, dx$.

4. If m is even and $n = 0$, rewrite $\tan^m x$ as
$$\tan^m x = \tan^{m-2} x \tan^2 x = \tan^{m-2} x (\sec^2 x - 1) = \tan^{m-2} \sec^2 x - \tan^{m-2} x.$$

   So
$$\int \tan^m x \, dx = \underbrace{\int \tan^{m-2} \sec^2 x \, dx}_{\text{apply rule \#1}} - \underbrace{\int \tan^{m-2} x \, dx}_{\text{apply rule \#4 again}}.$$

The techniques described in items 1 and 2 of Key Idea 12 are relatively straightforward, but the techniques in items 3 and 4 can be rather tedious. A few examples will help with these methods.

Notes:

Example 171   **Integrating powers of tangent and secant**

Evaluate $\int \tan^2 x \sec^6 x \, dx$.

Solution   Since the power of secant is even, we use rule #1 from Key Idea 12 and pull out a $\sec^2 x$ in the integrand. We convert the remaining powers of secant into powers of tangent.

$$\int \tan^2 x \sec^6 x \, dx = \int \tan^2 x \sec^4 x \sec^2 x \, dx$$

$$= \int \tan^2 x (1 + \tan^2 x)^2 \sec^2 x \, dx$$

Now substitute, with $u = \tan x$, with $du = \sec^2 x \, dx$.

$$= \int u^2 (1 + u^2)^2 \, du$$

We leave the integration and subsequent substitution to the reader. The final answer is

$$= \frac{1}{3} \tan^3 x + \frac{2}{5} \tan^5 x + \frac{1}{7} \tan^7 x + C.$$

Example 172   **Integrating powers of tangent and secant**

Evaluate $\int \sec^3 x \, dx$.

Solution   We apply rule #3 from Key Idea 12 as the power of secant is odd and the power of tangent is even (0 is an even number). We use Integration by Parts; the rule suggests letting $dv = \sec^2 x \, dx$, meaning that $u = \sec x$.

$$\begin{array}{ll} u = \sec x & v = ? \\ du = ? & dv = \sec^2 x \, dx \end{array} \Rightarrow \begin{array}{ll} u = \sec x & v = \tan x \\ du = \sec x \tan x \, dx & dv = \sec^2 x \, dx \end{array}$$

Figure 6.12: Setting up Integration by Parts.

Employing Integration by Parts, we have

$$\int \sec^3 x \, dx = \int \underbrace{\sec x}_{u} \cdot \underbrace{\sec^2 x \, dx}_{dv}$$

$$= \sec x \tan x - \int \sec x \tan^2 x \, dx.$$

Notes:

6.3 Trigonometric Integrals

This new integral also requires applying rule #3 of Key Idea 12:

$$= \sec x \tan x - \int \sec x \left(\sec^2 x - 1 \right) dx$$

$$= \sec x \tan x - \int \sec^3 x \, dx + \int \sec x \, dx$$

$$= \sec x \tan x - \int \sec^3 x \, dx + \ln|\sec x + \tan x|$$

In previous applications of Integration by Parts, we have seen where the original integral has reappeared in our work. We resolve this by adding $\int \sec^3 x \, dx$ to both sides, giving:

$$2 \int \sec^3 x \, dx = \sec x \tan x + \ln|\sec x + \tan x|$$

$$\int \sec^3 x \, dx = \frac{1}{2} \Big(\sec x \tan x + \ln|\sec x + \tan x| \Big) + C$$

We give one more example.

Example 173 **Integrating powers of tangent and secant**
Evaluate $\int \tan^6 x \, dx$.

SOLUTION We employ rule #4 of Key Idea 12.

$$\int \tan^6 x \, dx = \int \tan^4 x \tan^2 x \, dx$$

$$= \int \tan^4 x \left(\sec^2 x - 1 \right) dx$$

$$= \int \tan^4 x \sec^2 x \, dx - \int \tan^4 x \, dx$$

Integrate the first integral with substitution, $u = \tan x$; integrate the second by employing rule #4 again.

$$= \frac{1}{5} \tan^5 x - \int \tan^2 x \tan^2 x \, dx$$

$$= \frac{1}{5} \tan^5 x - \int \tan^2 x \left(\sec^2 x - 1 \right) dx$$

$$= \frac{1}{5} \tan^5 x - \int \tan^2 x \sec^2 x \, dx + \int \tan^2 x \, dx$$

Notes:

Again, use substitution for the first integral and rule #4 for the second.

$$= \frac{1}{5}\tan^5 x - \frac{1}{3}\tan^3 x + \int \left(\sec^2 x - 1\right) dx$$

$$= \frac{1}{5}\tan^5 x - \frac{1}{3}\tan^3 x + \tan x - x + C.$$

These latter examples were admittedly long, with repeated applications of the same rule. Try to not be overwhelmed by the length of the problem, but rather admire how robust this solution method is. A trigonometric function of a high power can be systematically reduced to trigonometric functions of lower powers until all antiderivatives can be computed.

The next section introduces an integration technique known as Trigonometric Substitution, a clever combination of Substitution and the Pythagorean Theorem.

Notes:

Exercises 6.3

Terms and Concepts

1. T/F: $\int \sin^2 x \cos^2 x \, dx$ cannot be evaluated using the techniques described in this section since both powers of sin x and cos x are even.

2. T/F: $\int \sin^3 x \cos^3 x \, dx$ cannot be evaluated using the techniques described in this section since both powers of sin x and cos x are odd.

3. T/F: This section addresses how to evaluate indefinite integrals such as $\int \sin^5 x \tan^3 x \, dx$.

Problems

In Exercises 4 – 26, evaluate the indefinite integral.

4. $\int \sin x \cos^4 x \, dx$

5. $\int \sin^3 x \cos x \, dx$

6. $\int \sin^3 x \cos^2 x \, dx$

7. $\int \sin^3 x \cos^3 x \, dx$

8. $\int \sin^6 x \cos^5 x \, dx$

9. $\int \sin^2 x \cos^7 x \, dx$

10. $\int \sin^2 x \cos^2 x \, dx$

11. $\int \sin(5x) \cos(3x) \, dx$

12. $\int \sin(x) \cos(2x) \, dx$

13. $\int \sin(3x) \sin(7x) \, dx$

14. $\int \sin(\pi x) \sin(2\pi x) \, dx$

15. $\int \cos(x) \cos(2x) \, dx$

16. $\int \cos\left(\frac{\pi}{2}x\right) \cos(\pi x) \, dx$

17. $\int \tan^4 x \sec^2 x \, dx$

18. $\int \tan^2 x \sec^4 x \, dx$

19. $\int \tan^3 x \sec^4 x \, dx$

20. $\int \tan^3 x \sec^2 x \, dx$

21. $\int \tan^3 x \sec^3 x \, dx$

22. $\int \tan^5 x \sec^5 x \, dx$

23. $\int \tan^4 x \, dx$

24. $\int \sec^5 x \, dx$

25. $\int \tan^2 x \sec x \, dx$

26. $\int \tan^2 x \sec^3 x \, dx$

In Exercises 27 – 33, evaluate the definite integral. Note: the corresponding indefinite integrals appear in the previous set.

27. $\int_0^\pi \sin x \cos^4 x \, dx$

28. $\int_{-\pi}^\pi \sin^3 x \cos x \, dx$

29. $\int_{-\pi/2}^{\pi/2} \sin^2 x \cos^7 x \, dx$

30. $\int_0^{\pi/2} \sin(5x) \cos(3x) \, dx$

31. $\int_{-\pi/2}^{\pi/2} \cos(x) \cos(2x) \, dx$

32. $\int_0^{\pi/4} \tan^4 x \sec^2 x \, dx$

33. $\int_{-\pi/4}^{\pi/4} \tan^2 x \sec^4 x \, dx$

6.4 Trigonometric Substitution

In Section 5.2 we defined the definite integral as the "signed area under the curve." In that section we had not yet learned the Fundamental Theorem of Calculus, so we evaluated special definite integrals which described nice, geometric shapes. For instance, we were able to evaluate

$$\int_{-3}^{3} \sqrt{9-x^2}\, dx = \frac{9\pi}{2} \tag{6.1}$$

as we recognized that $f(x) = \sqrt{9-x^2}$ described the upper half of a circle with radius 3.

We have since learned a number of integration techniques, including Substitution and Integration by Parts, yet we are still unable to evaluate the above integral without resorting to a geometric interpretation. This section introduces Trigonometric Substitution, a method of integration that fills this gap in our integration skill. This technique works on the same principle as Substitution as found in Section 6.1, though it can feel "backward." In Section 6.1, we set $u = f(x)$, for some function f, and replaced $f(x)$ with u. In this section, we will set $x = f(\theta)$, where f is a trigonometric function, then replace x with $f(\theta)$.

We start by demonstrating this method in evaluating the integral in (6.1). After the example, we will generalize the method and give more examples.

Example 174   **Using Trigonometric Substitution**
Evaluate $\int_{-3}^{3} \sqrt{9-x^2}\, dx$.

SOLUTION   We begin by noting that $9\sin^2\theta + 9\cos^2\theta = 9$, and hence $9\cos^2\theta = 9 - 9\sin^2\theta$. If we let $x = 3\sin\theta$, then $9 - x^2 = 9 - 9\sin^2\theta = 9\cos^2\theta$.

Setting $x = 3\sin\theta$ gives $dx = 3\cos\theta\, d\theta$. We are almost ready to substitute. We also wish to change our bounds of integration. The bound $x = -3$ corresponds to $\theta = -\pi/2$ (for when $\theta = -\pi/2$, $x = 3\sin\theta = -3$). Likewise, the bound of $x = 3$ is replaced by the bound $\theta = \pi/2$. Thus

$$\int_{-3}^{3}\sqrt{9-x^2}\,dx = \int_{-\pi/2}^{\pi/2} \sqrt{9 - 9\sin^2\theta}(3\cos\theta)\,d\theta$$
$$= \int_{-\pi/2}^{\pi/2} 3\sqrt{9\cos^2\theta}\cos\theta\,d\theta$$
$$= \int_{-\pi/2}^{\pi/2} 3|3\cos\theta|\cos\theta\,d\theta.$$

On $[-\pi/2, \pi/2]$, $\cos\theta$ is always positive, so we can drop the absolute value bars, then employ a power-reducing formula:

Notes:

$$= \int_{-\pi/2}^{\pi/2} 9\cos^2\theta\,d\theta$$

$$= \int_{-\pi/2}^{\pi/2} \frac{9}{2}\big(1+\cos(2\theta)\big)\,d\theta$$

$$= \frac{9}{2}\Big(\theta + \frac{1}{2}\sin(2\theta)\Big)\Big|_{-\pi/2}^{\pi/2} = \frac{9}{2}\pi.$$

This matches our answer from before.

We now describe in detail Trigonometric Substitution. This method excels when dealing with integrands that contain $\sqrt{a^2-x^2}$, $\sqrt{x^2-a^2}$ and $\sqrt{x^2+a^2}$. The following Key Idea outlines the procedure for each case, followed by more examples. Each right triangle acts as a reference to help us understand the relationships between x and θ.

Key Idea 13   Trigonometric Substitution

(a) For integrands containing $\sqrt{a^2-x^2}$:

Let $x = a\sin\theta$, $dx = a\cos\theta\,d\theta$

Thus $\theta = \sin^{-1}(x/a)$, for $-\pi/2 \le \theta \le \pi/2$.

On this interval, $\cos\theta \ge 0$, so

$\sqrt{a^2-x^2} = a\cos\theta$

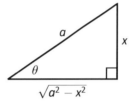

(b) For integrands containing $\sqrt{x^2+a^2}$:

Let $x = a\tan\theta$, $dx = a\sec^2\theta\,d\theta$

Thus $\theta = \tan^{-1}(x/a)$, for $-\pi/2 < \theta < \pi/2$.

On this interval, $\sec\theta > 0$, so

$\sqrt{x^2+a^2} = a\sec\theta$

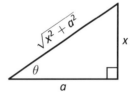

(c) For integrands containing $\sqrt{x^2-a^2}$:

Let $x = a\sec\theta$, $dx = a\sec\theta\tan\theta\,d\theta$

Thus $\theta = \sec^{-1}(x/a)$. If $x/a \ge 1$, then $0 \le \theta < \pi/2$; if $x/a \le -1$, then $\pi/2 < \theta \le \pi$.

We restrict our work to where $x \ge a$, so $x/a \ge 1$, and $0 \le \theta < \pi/2$. On this interval, $\tan\theta \ge 0$, so

$\sqrt{x^2-a^2} = a\tan\theta$

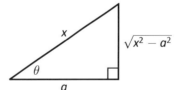

Notes:

Chapter 6 Techniques of Antidifferentiation

Example 175   **Using Trigonometric Substitution**

Evaluate $\int \dfrac{1}{\sqrt{5+x^2}}\, dx$.

SOLUTION   Using Key Idea 13(b), we recognize $a = \sqrt{5}$ and set $x = \sqrt{5}\tan\theta$. This makes $dx = \sqrt{5}\sec^2\theta\, d\theta$. We will use the fact that $\sqrt{5+x^2} = \sqrt{5+5\tan^2\theta} = \sqrt{5\sec^2\theta} = \sqrt{5}\sec\theta$. Substituting, we have:

$$\int \dfrac{1}{\sqrt{5+x^2}}\, dx = \int \dfrac{1}{\sqrt{5+5\tan^2\theta}}\, \sqrt{5}\sec^2\theta\, d\theta$$

$$= \int \dfrac{\sqrt{5}\sec^2\theta}{\sqrt{5}\sec\theta}\, d\theta$$

$$= \int \sec\theta\, d\theta$$

$$= \ln\left|\sec\theta + \tan\theta\right| + C.$$

While the integration steps are over, we are not yet done. The original problem was stated in terms of x, whereas our answer is given in terms of θ. We must convert back to x.

The reference triangle given in Key Idea 13(b) helps. With $x = \sqrt{5}\tan\theta$, we have

$$\tan\theta = \dfrac{x}{\sqrt{5}} \quad \text{and} \quad \sec\theta = \dfrac{\sqrt{x^2+5}}{\sqrt{5}}.$$

This gives

$$\int \dfrac{1}{\sqrt{5+x^2}}\, dx = \ln\left|\sec\theta + \tan\theta\right| + C$$

$$= \ln\left|\dfrac{\sqrt{x^2+5}}{\sqrt{5}} + \dfrac{x}{\sqrt{5}}\right| + C.$$

We can leave this answer as is, or we can use a logarithmic identity to simplify it. Note:

$$\ln\left|\dfrac{\sqrt{x^2+5}}{\sqrt{5}} + \dfrac{x}{\sqrt{5}}\right| + C = \ln\left|\dfrac{1}{\sqrt{5}}\left(\sqrt{x^2+5}+x\right)\right| + C$$

$$= \ln\left|\dfrac{1}{\sqrt{5}}\right| + \ln\left|\sqrt{x^2+5}+x\right| + C$$

$$= \ln\left|\sqrt{x^2+5}+x\right| + C,$$

where the $\ln\left(1/\sqrt{5}\right)$ term is absorbed into the constant C. (In Section 6.6 we will learn another way of approaching this problem.)

Notes:

6.4 Trigonometric Substitution

Example 176   **Using Trigonometric Substitution**
Evaluate $\int \sqrt{4x^2 - 1}\, dx$.

Solution   We start by rewriting the integrand so that it looks like $\sqrt{x^2 - a^2}$ for some value of a:

$$\sqrt{4x^2 - 1} = \sqrt{4\left(x^2 - \frac{1}{4}\right)}$$
$$= 2\sqrt{x^2 - \left(\frac{1}{2}\right)^2}.$$

So we have $a = 1/2$, and following Key Idea 13(c), we set $x = \frac{1}{2}\sec\theta$, and hence $dx = \frac{1}{2}\sec\theta\tan\theta\, d\theta$. We now rewrite the integral with these substitutions:

$$\int \sqrt{4x^2 - 1}\, dx = \int 2\sqrt{x^2 - \left(\frac{1}{2}\right)^2}\, dx$$
$$= \int 2\sqrt{\frac{1}{4}\sec^2\theta - \frac{1}{4}}\left(\frac{1}{2}\sec\theta\tan\theta\right) d\theta$$
$$= \int \sqrt{\frac{1}{4}(\sec^2\theta - 1)}\left(\sec\theta\tan\theta\right) d\theta$$
$$= \int \sqrt{\frac{1}{4}\tan^2\theta}\left(\sec\theta\tan\theta\right) d\theta$$
$$= \int \frac{1}{2}\tan^2\theta \sec\theta\, d\theta$$
$$= \frac{1}{2}\int \left(\sec^2\theta - 1\right)\sec\theta\, d\theta$$
$$= \frac{1}{2}\int \left(\sec^3\theta - \sec\theta\right) d\theta.$$

We integrated $\sec^3\theta$ in Example 172, finding its antiderivatives to be

$$\int \sec^3\theta\, d\theta = \frac{1}{2}\left(\sec\theta\tan\theta + \ln|\sec\theta + \tan\theta|\right) + C.$$

Thus

$$\int \sqrt{4x^2 - 1}\, dx = \frac{1}{2}\int \left(\sec^3\theta - \sec\theta\right) d\theta$$
$$= \frac{1}{2}\left(\frac{1}{2}\left(\sec\theta\tan\theta + \ln|\sec\theta + \tan\theta|\right) - \ln|\sec\theta + \tan\theta|\right) + C$$
$$= \frac{1}{4}\left(\sec\theta\tan\theta - \ln|\sec\theta + \tan\theta|\right) + C.$$

Notes:

We are not yet done. Our original integral is given in terms of x, whereas our final answer, as given, is in terms of θ. We need to rewrite our answer in terms of x. With $a = 1/2$, and $x = \frac{1}{2}\sec\theta$, the reference triangle in Key Idea 13(c) shows that

$$\tan\theta = \sqrt{x^2 - 1/4}\Big/(1/2) = 2\sqrt{x^2 - 1/4} \quad \text{and} \quad \sec\theta = 2x.$$

Thus

$$\frac{1}{4}\Big(\sec\theta\tan\theta - \ln|\sec\theta + \tan\theta|\Big) + C = \frac{1}{4}\Big(2x \cdot 2\sqrt{x^2 - 1/4} - \ln\big|2x + 2\sqrt{x^2 - 1/4}\big|\Big) + C$$

$$= \frac{1}{4}\Big(4x\sqrt{x^2 - 1/4} - \ln\big|2x + 2\sqrt{x^2 - 1/4}\big|\Big) + C.$$

The final answer is given in the last line above, repeated here:

$$\int \sqrt{4x^2 - 1}\, dx = \frac{1}{4}\Big(4x\sqrt{x^2 - 1/4} - \ln\big|2x + 2\sqrt{x^2 - 1/4}\big|\Big) + C.$$

Example 177 **Using Trigonometric Substitution**

Evaluate $\displaystyle\int \frac{\sqrt{4 - x^2}}{x^2}\, dx$.

SOLUTION We use Key Idea 13(a) with $a = 2$, $x = 2\sin\theta$, $dx = 2\cos\theta$ and hence $\sqrt{4 - x^2} = 2\cos\theta$. This gives

$$\int \frac{\sqrt{4 - x^2}}{x^2}\, dx = \int \frac{2\cos\theta}{4\sin^2\theta}(2\cos\theta)\, d\theta$$

$$= \int \cot^2\theta\, d\theta$$

$$= \int (\csc^2\theta - 1)\, d\theta$$

$$= -\cot\theta - \theta + C.$$

We need to rewrite our answer in terms of x. Using the reference triangle found in Key Idea 13(a), we have $\cot\theta = \sqrt{4 - x^2}/x$ and $\theta = \sin^{-1}(x/2)$. Thus

$$\int \frac{\sqrt{4 - x^2}}{x^2}\, dx = -\frac{\sqrt{4 - x^2}}{x} - \sin^{-1}\left(\frac{x}{2}\right) + C.$$

Trigonometric Substitution can be applied in many situations, even those not of the form $\sqrt{a^2 - x^2}$, $\sqrt{x^2 - a^2}$ or $\sqrt{x^2 + a^2}$. In the following example, we apply it to an integral we already know how to handle.

Notes:

6.4 Trigonometric Substitution

Example 178 **Using Trigonometric Substitution**
Evaluate $\int \frac{1}{x^2+1}\,dx$.

SOLUTION We know the answer already as $\tan^{-1} x + C$. We apply Trigonometric Substitution here to show that we get the same answer without inherently relying on knowledge of the derivative of the arctangent function.

Using Key Idea 13(b), let $x = \tan\theta$, $dx = \sec^2\theta\,d\theta$ and note that $x^2 + 1 = \tan^2\theta + 1 = \sec^2\theta$. Thus

$$\int \frac{1}{x^2+1}\,dx = \int \frac{1}{\sec^2\theta}\sec^2\theta\,d\theta$$
$$= \int 1\,d\theta$$
$$= \theta + C.$$

Since $x = \tan\theta$, $\theta = \tan^{-1} x$, and we conclude that $\int \frac{1}{x^2+1}\,dx = \tan^{-1} x + C$.

The next example is similar to the previous one in that it does not involve a square–root. It shows how several techniques and identities can be combined to obtain a solution.

Example 179 **Using Trigonometric Substitution**
Evaluate $\int \frac{1}{(x^2+6x+10)^2}\,dx$.

SOLUTION We start by completing the square, then make the substitution $u = x + 3$, followed by the trigonometric substitution of $u = \tan\theta$:

$$\int \frac{1}{(x^2+6x+10)^2}\,dx = \int \frac{1}{\left((x+3)^2+1\right)^2}\,dx = \int \frac{1}{(u^2+1)^2}\,du.$$

Now make the substitution $u = \tan\theta$, $du = \sec^2\theta\,d\theta$:

$$= \int \frac{1}{(\tan^2\theta+1)^2}\sec^2\theta\,d\theta$$
$$= \int \frac{1}{(\sec^2\theta)^2}\sec^2\theta\,d\theta$$
$$= \int \cos^2\theta\,d\theta.$$

Notes:

Applying a power reducing formula, we have

$$= \int \left(\frac{1}{2} + \frac{1}{2}\cos(2\theta)\right) d\theta$$
$$= \frac{1}{2}\theta + \frac{1}{4}\sin(2\theta) + C. \quad (6.2)$$

We need to return to the variable x. As $u = \tan\theta$, $\theta = \tan^{-1}u$. Using the identity $\sin(2\theta) = 2\sin\theta\cos\theta$ and using the reference triangle found in Key Idea 13(b), we have

$$\frac{1}{4}\sin(2\theta) = \frac{1}{2}\frac{u}{\sqrt{u^2+1}} \cdot \frac{1}{\sqrt{u^2+1}} = \frac{1}{2}\frac{u}{u^2+1}.$$

Finally, we return to x with the substitution $u = x+3$. We start with the expression in Equation (6.2):

$$\frac{1}{2}\theta + \frac{1}{4}\sin(2\theta) + C = \frac{1}{2}\tan^{-1}u + \frac{1}{2}\frac{u}{u^2+1} + C$$
$$= \frac{1}{2}\tan^{-1}(x+3) + \frac{x+3}{2(x^2+6x+10)} + C.$$

Stating our final result in one line,

$$\int \frac{1}{(x^2+6x+10)^2}\,dx = \frac{1}{2}\tan^{-1}(x+3) + \frac{x+3}{2(x^2+6x+10)} + C.$$

Our last example returns us to definite integrals, as seen in our first example. Given a definite integral that can be evaluated using Trigonometric Substitution, we could first evaluate the corresponding indefinite integral (by changing from an integral in terms of x to one in terms of θ, then converting back to x) and then evaluate using the original bounds. It is much more straightforward, though, to change the bounds as we substitute.

Example 180 Definite integration and Trigonometric Substitution
Evaluate $\int_0^5 \frac{x^2}{\sqrt{x^2+25}}\,dx.$

SOLUTION Using Key Idea 13(b), we set $x = 5\tan\theta$, $dx = 5\sec^2\theta\,d\theta$, and note that $\sqrt{x^2+25} = 5\sec\theta$. As we substitute, we can also change the bounds of integration.

The lower bound of the original integral is $x = 0$. As $x = 5\tan\theta$, we solve for θ and find $\theta = \tan^{-1}(x/5)$. Thus the new lower bound is $\theta = \tan^{-1}(0) = 0$. The original upper bound is $x = 5$, thus the new upper bound is $\theta = \tan^{-1}(5/5) = \pi/4$.

Thus we have

$$\int_0^5 \frac{x^2}{\sqrt{x^2+25}}\, dx = \int_0^{\pi/4} \frac{25\tan^2\theta}{5\sec\theta} 5\sec^2\theta\, d\theta$$

$$= 25 \int_0^{\pi/4} \tan^2\theta \sec\theta\, d\theta.$$

We encountered this indefinite integral in Example 176 where we found

$$\int \tan^2\theta \sec\theta\, d\theta = \frac{1}{2}\big(\sec\theta\tan\theta - \ln|\sec\theta + \tan\theta|\big).$$

So

$$25\int_0^{\pi/4} \tan^2\theta \sec\theta\, d\theta = \frac{25}{2}\big(\sec\theta\tan\theta - \ln|\sec\theta + \tan\theta|\big)\Big|_0^{\pi/4}$$

$$= \frac{25}{2}\big(\sqrt{2} - \ln(\sqrt{2}+1)\big)$$

$$\approx 6.661.$$

The following equalities are very useful when evaluating integrals using Trigonometric Substitution.

Key Idea 14   Useful Equalities with Trigonometric Substitution

1. $\sin(2\theta) = 2\sin\theta\cos\theta$
2. $\cos(2\theta) = \cos^2\theta - \sin^2\theta = 2\cos^2\theta - 1 = 1 - 2\sin^2\theta$
3. $\displaystyle\int \sec^3\theta\, d\theta = \frac{1}{2}\Big(\sec\theta\tan\theta + \ln\big|\sec\theta + \tan\theta\big|\Big) + C$
4. $\displaystyle\int \cos^2\theta\, d\theta = \int \frac{1}{2}\big(1 + \cos(2\theta)\big)\, d\theta = \frac{1}{2}\big(\theta + \sin\theta\cos\theta\big) + C.$

The next section introduces Partial Fraction Decomposition, which is an algebraic technique that turns "complicated" fractions into sums of "simpler" fractions, making integration easier.

Notes:

Exercises 6.4

Terms and Concepts

1. Trigonometric Substitution works on the same principles as Integration by Substitution, though it can feel "_____".

2. If one uses Trigonometric Substitution on an integrand containing $\sqrt{25-x^2}$, then one should set $x =$ _____.

3. Consider the Pythagorean Identity $\sin^2\theta + \cos^2\theta = 1$.
   (a) What identity is obtained when both sides are divided by $\cos^2\theta$?
   (b) Use the new identity to simplify $9\tan^2\theta + 9$.

4. Why does Key Idea 13(a) state that $\sqrt{a^2 - x^2} = a\cos\theta$, and not $|a\cos\theta|$?

Problems

In Exercises 5 – 16, apply Trigonometric Substitution to evaluate the indefinite integrals.

5. $\displaystyle\int \sqrt{x^2+1}\, dx$

6. $\displaystyle\int \sqrt{x^2+4}\, dx$

7. $\displaystyle\int \sqrt{1-x^2}\, dx$

8. $\displaystyle\int \sqrt{9-x^2}\, dx$

9. $\displaystyle\int \sqrt{x^2-1}\, dx$

10. $\displaystyle\int \sqrt{x^2-16}\, dx$

11. $\displaystyle\int \sqrt{4x^2+1}\, dx$

12. $\displaystyle\int \sqrt{1-9x^2}\, dx$

13. $\displaystyle\int \sqrt{16x^2-1}\, dx$

14. $\displaystyle\int \frac{8}{\sqrt{x^2+2}}\, dx$

15. $\displaystyle\int \frac{3}{\sqrt{7-x^2}}\, dx$

16. $\displaystyle\int \frac{5}{\sqrt{x^2-8}}\, dx$

In Exercises 17 – 26, evaluate the indefinite integrals. Some may be evaluated without Trigonometric Substitution.

17. $\displaystyle\int \frac{\sqrt{x^2-11}}{x}\, dx$

18. $\displaystyle\int \frac{1}{(x^2+1)^2}\, dx$

19. $\displaystyle\int \frac{x}{\sqrt{x^2-3}}\, dx$

20. $\displaystyle\int x^2\sqrt{1-x^2}\, dx$

21. $\displaystyle\int \frac{x}{(x^2+9)^{3/2}}\, dx$

22. $\displaystyle\int \frac{5x^2}{\sqrt{x^2-10}}\, dx$

23. $\displaystyle\int \frac{1}{(x^2+4x+13)^2}\, dx$

24. $\displaystyle\int x^2(1-x^2)^{-3/2}\, dx$

25. $\displaystyle\int \frac{\sqrt{5-x^2}}{7x^2}\, dx$

26. $\displaystyle\int \frac{x^2}{\sqrt{x^2+3}}\, dx$

In Exercises 27 – 32, evaluate the definite integrals by making the proper trigonometric substitution *and* changing the bounds of integration. (Note: each of the corresponding indefinite integrals has appeared previously in this Exercise set.)

27. $\displaystyle\int_{-1}^{1} \sqrt{1-x^2}\, dx$

28. $\displaystyle\int_{4}^{8} \sqrt{x^2-16}\, dx$

29. $\displaystyle\int_{0}^{2} \sqrt{x^2+4}\, dx$

30. $\displaystyle\int_{-1}^{1} \frac{1}{(x^2+1)^2}\, dx$

31. $\displaystyle\int_{-1}^{1} \sqrt{9-x^2}\, dx$

32. $\displaystyle\int_{-1}^{1} x^2\sqrt{1-x^2}\, dx$

6.5 Partial Fraction Decomposition

In this section we investigate the antiderivatives of rational functions. Recall that rational functions are functions of the form $f(x) = \frac{p(x)}{q(x)}$, where $p(x)$ and $q(x)$ are polynomials and $q(x) \neq 0$. Such functions arise in many contexts, one of which is the solving of certain fundamental differential equations.

We begin with an example that demonstrates the motivation behind this section. Consider the integral $\int \frac{1}{x^2 - 1}\, dx$. We do not have a simple formula for this (if the denominator were $x^2 + 1$, we would recognize the antiderivative as being the arctangent function). It can be solved using Trigonometric Substitution, but note how the integral is easy to evaluate once we realize:

$$\frac{1}{x^2 - 1} = \frac{1/2}{x - 1} - \frac{1/2}{x + 1}.$$

Thus

$$\int \frac{1}{x^2 - 1}\, dx = \int \frac{1/2}{x - 1}\, dx - \int \frac{1/2}{x + 1}\, dx$$
$$= \frac{1}{2} \ln|x - 1| - \frac{1}{2} \ln|x + 1| + C.$$

This section teaches how to *decompose*

$$\frac{1}{x^2 - 1} \quad \text{into} \quad \frac{1/2}{x - 1} - \frac{1/2}{x + 1}.$$

We start with a rational function $f(x) = \frac{p(x)}{q(x)}$, where p and q do not have any common factors and the degree of p is less than the degree of q. It can be shown that any polynomial, and hence q, can be factored into a product of linear and irreducible quadratic terms. The following Key Idea states how to decompose a rational function into a sum of rational functions whose denominators are all of lower degree than q.

Notes:

Chapter 6 Techniques of Antidifferentiation

> **Key Idea 15** **Partial Fraction Decomposition**
>
> Let $\dfrac{p(x)}{q(x)}$ be a rational function, where the degree of p is less than the degree of q.
>
> 1. **Linear Terms:** Let $(x-a)$ divide $q(x)$, where $(x-a)^n$ is the highest power of $(x-a)$ that divides $q(x)$. Then the decomposition of $\dfrac{p(x)}{q(x)}$ will contain the sum
>
> $$\frac{A_1}{(x-a)} + \frac{A_2}{(x-a)^2} + \cdots + \frac{A_n}{(x-a)^n}.$$
>
> 2. **Quadratic Terms:** Let $x^2 + bx + c$ divide $q(x)$, where $(x^2 + bx + c)^n$ is the highest power of $x^2 + bx + c$ that divides $q(x)$. Then the decomposition of $\dfrac{p(x)}{q(x)}$ will contain the sum
>
> $$\frac{B_1 x + C_1}{x^2 + bx + c} + \frac{B_2 x + C_2}{(x^2 + bx + c)^2} + \cdots + \frac{B_n x + C_n}{(x^2 + bx + c)^n}.$$
>
> To find the coefficients A_i, B_i and C_i:
>
> 1. Multiply all fractions by $q(x)$, clearing the denominators. Collect like terms.
>
> 2. Equate the resulting coefficients of the powers of x and solve the resulting system of linear equations.

The following examples will demonstrate how to put this Key Idea into practice. Example 181 stresses the decomposition aspect of the Key Idea.

Example 181 **Decomposing into partial fractions**
Decompose $f(x) = \dfrac{1}{(x+5)(x-2)^3(x^2+x+2)(x^2+x+7)^2}$ without solving for the resulting coefficients.

Solution The denominator is already factored, as both $x^2 + x + 2$ and $x^2 + x + 7$ cannot be factored further. We need to decompose $f(x)$ properly. Since $(x+5)$ is a linear term that divides the denominator, there will be a

$$\frac{A}{x+5}$$

Notes:

term in the decomposition.

As $(x-2)^3$ divides the denominator, we will have the following terms in the decomposition:
$$\frac{B}{x-2}, \quad \frac{C}{(x-2)^2} \quad \text{and} \quad \frac{D}{(x-2)^3}.$$

The x^2+x+2 term in the denominator results in a $\frac{Ex+F}{x^2+x+2}$ term.

Finally, the $(x^2+x+7)^2$ term results in the terms
$$\frac{Gx+H}{x^2+x+7} \quad \text{and} \quad \frac{Ix+J}{(x^2+x+7)^2}.$$

All together, we have
$$\frac{1}{(x+5)(x-2)^3(x^2+x+2)(x^2+x+7)^2} = \frac{A}{x+5} + \frac{B}{x-2} + \frac{C}{(x-2)^2} + \frac{D}{(x-2)^3} + \frac{Ex+F}{x^2+x+2} + \frac{Gx+H}{x^2+x+7} + \frac{Ix+J}{(x^2+x+7)^2}$$

Solving for the coefficients $A, B \ldots J$ would be a bit tedious but not "hard."

Example 182   **Decomposing into partial fractions**

Perform the partial fraction decomposition of $\frac{1}{x^2-1}$.

SOLUTION   The denominator factors into two linear terms: $x^2 - 1 = (x-1)(x+1)$. Thus
$$\frac{1}{x^2-1} = \frac{A}{x-1} + \frac{B}{x+1}.$$

To solve for A and B, first multiply through by $x^2 - 1 = (x-1)(x+1)$:
$$1 = \frac{A(x-1)(x+1)}{x-1} + \frac{B(x-1)(x+1)}{x+1}$$
$$= A(x+1) + B(x-1)$$
$$= Ax + A + Bx - B$$

Now collect like terms.
$$= (A+B)x + (A-B).$$

The next step is key. Note the equality we have:
$$1 = (A+B)x + (A-B).$$

Notes:

Chapter 6 Techniques of Antidifferentiation

For clarity's sake, rewrite the left hand side as

$$0x + 1 = (A + B)x + (A - B).$$

On the left, the coefficient of the x term is 0; on the right, it is $(A + B)$. Since both sides are equal, we must have that $0 = A + B$.

Likewise, on the left, we have a constant term of 1; on the right, the constant term is $(A - B)$. Therefore we have $1 = A - B$.

We have two linear equations with two unknowns. This one is easy to solve by hand, leading to

$$\begin{matrix} A + B = 0 \\ A - B = 1 \end{matrix} \Rightarrow \begin{matrix} A = 1/2 \\ B = -1/2 \end{matrix}.$$

Thus

$$\frac{1}{x^2 - 1} = \frac{1/2}{x - 1} - \frac{1/2}{x + 1}.$$

Example 183   **Integrating using partial fractions**

Use partial fraction decomposition to integrate $\displaystyle\int \frac{1}{(x - 1)(x + 2)^2} \, dx.$

SOLUTION   We decompose the integrand as follows, as described by Key Idea 15:

$$\frac{1}{(x - 1)(x + 2)^2} = \frac{A}{x - 1} + \frac{B}{x + 2} + \frac{C}{(x + 2)^2}.$$

To solve for A, B and C, we multiply both sides by $(x - 1)(x + 2)^2$ and collect like terms:

$$1 = A(x + 2)^2 + B(x - 1)(x + 2) + C(x - 1) \tag{6.3}$$
$$= Ax^2 + 4Ax + 4A + Bx^2 + Bx - 2B + Cx - C$$
$$= (A + B)x^2 + (4A + B + C)x + (4A - 2B - C)$$

We have

$$0x^2 + 0x + 1 = (A + B)x^2 + (4A + B + C)x + (4A - 2B - C)$$

leading to the equations

$$A + B = 0, \quad 4A + B + C = 0 \quad \text{and} \quad 4A - 2B - C = 1.$$

These three equations of three unknowns lead to a unique solution:

$$A = 1/9, \quad B = -1/9 \quad \text{and} \quad C = -1/3.$$

Note: Equation 6.3 offers a direct route to finding the values of A, B and C. Since the equation holds for all values of x, it holds in particular when $x = 1$. However, when $x = 1$, the right hand side simplifies to $A(1 + 2)^2 = 9A$. Since the left hand side is still 1, we have $1 = 9A$. Hence $A = 1/9$. Likewise, the equality holds when $x = -2$; this leads to the equation $1 = -3C$. Thus $C = -1/3$.
Knowing A and C, we can find the value of B by choosing yet another value of x, such as $x = 0$, and solving for B.

Notes:

Thus

$$\int \frac{1}{(x-1)(x+2)^2}\,dx = \int \frac{1/9}{x-1}\,dx + \int \frac{-1/9}{x+2}\,dx + \int \frac{-1/3}{(x+2)^2}\,dx.$$

Each can be integrated with a simple substitution with $u = x-1$ or $u = x+2$ (or by directly applying Key Idea 10 as the denominators are linear functions). The end result is

$$\int \frac{1}{(x-1)(x+2)^2}\,dx = \frac{1}{9}\ln|x-1| - \frac{1}{9}\ln|x+2| + \frac{1}{3(x+2)} + C.$$

Example 184   Integrating using partial fractions

Use partial fraction decomposition to integrate $\displaystyle\int \frac{x^3}{(x-5)(x+3)}\,dx$.

SOLUTION   Key Idea 15 presumes that the degree of the numerator is less than the degree of the denominator. Since this is not the case here, we begin by using polynomial division to reduce the degree of the numerator. We omit the steps, but encourage the reader to verify that

$$\frac{x^3}{(x-5)(x+3)} = x + 2 + \frac{19x + 30}{(x-5)(x+3)}.$$

Note: The values of A and B can be quickly found using the technique described in the margin of Example 183.

Using Key Idea 15, we can rewrite the new rational function as:

$$\frac{19x+30}{(x-5)(x+3)} = \frac{A}{x-5} + \frac{B}{x+3}$$

for appropriate values of A and B. Clearing denominators, we have

$$\begin{aligned}19x + 30 &= A(x+3) + B(x-5) \\ &= (A+B)x + (3A - 5B).\end{aligned}$$

This implies that:

$$\begin{aligned}19 &= A + B \\ 30 &= 3A - 5B.\end{aligned}$$

Solving this system of linear equations gives

$$\begin{aligned}125/8 &= A \\ 27/8 &= B.\end{aligned}$$

Notes:

We can now integrate.

$$\int \frac{x^3}{(x-5)(x+3)}\,dx = \int \left(x+2+\frac{125/8}{x-5}+\frac{27/8}{x+3}\right)dx$$
$$= \frac{x^2}{2}+2x+\frac{125}{8}\ln|x-5|+\frac{27}{8}\ln|x+3|+C.$$

Example 185 **Integrating using partial fractions**

Use partial fraction decomposition to evaluate $\displaystyle\int \frac{7x^2+31x+54}{(x+1)(x^2+6x+11)}\,dx$.

SOLUTION The degree of the numerator is less than the degree of the denominator so we begin by applying Key Idea 15. We have:

$$\frac{7x^2+31x+54}{(x+1)(x^2+6x+11)} = \frac{A}{x+1}+\frac{Bx+C}{x^2+6x+11}.$$

Now clear the denominators.

$$7x^2+31x+54 = A(x^2+6x+11)+(Bx+C)(x+1)$$
$$= (A+B)x^2+(6A+B+C)x+(11A+C).$$

This implies that:

$$7 = A+B$$
$$31 = 6A+B+C$$
$$54 = 11A+C.$$

Solving this system of linear equations gives the nice result of $A = 5$, $B = 2$ and $C = -1$. Thus

$$\int \frac{7x^2+31x+54}{(x+1)(x^2+6x+11)}\,dx = \int \left(\frac{5}{x+1}+\frac{2x-1}{x^2+6x+11}\right)dx.$$

The first term of this new integrand is easy to evaluate; it leads to a $5\ln|x+1|$ term. The second term is not hard, but takes several steps and uses substitution techniques.

The integrand $\dfrac{2x-1}{x^2+6x+11}$ has a quadratic in the denominator and a linear term in the numerator. This leads us to try substitution. Let $u = x^2+6x+11$, so $du = (2x+6)\,dx$. The numerator is $2x-1$, not $2x+6$, but we can get a $2x+6$

term in the numerator by adding 0 in the form of "7 − 7."

$$\frac{2x-1}{x^2+6x+11} = \frac{2x-1+7-7}{x^2+6x+11}$$
$$= \frac{2x+6}{x^2+6x+11} - \frac{7}{x^2+6x+11}.$$

We can now integrate the first term with substitution, leading to a $\ln|x^2+6x+11|$ term. The final term can be integrated using arctangent. First, complete the square in the denominator:

$$\frac{7}{x^2+6x+11} = \frac{7}{(x+3)^2+2}.$$

An antiderivative of the latter term can be found using Theorem 46 and substitution:

$$\int \frac{7}{x^2+6x+11}\,dx = \frac{7}{\sqrt{2}}\tan^{-1}\left(\frac{x+3}{\sqrt{2}}\right)+C.$$

Let's start at the beginning and put all of the steps together.

$$\int \frac{7x^2+31x+54}{(x+1)(x^2+6x+11)}\,dx = \int \left(\frac{5}{x+1}+\frac{2x-1}{x^2+6x+11}\right)dx$$
$$= \int \frac{5}{x+1}\,dx + \int \frac{2x+6}{x^2+6x+11}\,dx - \int \frac{7}{x^2+6x+11}\,dx$$
$$= 5\ln|x+1| + \ln|x^2+6x+11| - \frac{7}{\sqrt{2}}\tan^{-1}\left(\frac{x+3}{\sqrt{2}}\right)+C.$$

As with many other problems in calculus, it is important to remember that one is not expected to "see" the final answer immediately after seeing the problem. Rather, given the initial problem, we break it down into smaller problems that are easier to solve. The final answer is a combination of the answers of the smaller problems.

Partial Fraction Decomposition is an important tool when dealing with rational functions. Note that at its heart, it is a technique of algebra, not calculus, as we are rewriting a fraction in a new form. Regardless, it is very useful in the realm of calculus as it lets us evaluate a certain set of "complicated" integrals.

The next section introduces new functions, called the Hyperbolic Functions. They will allow us to make substitutions similar to those found when studying Trigonometric Substitution, allowing us to approach even more integration problems.

Notes:

Exercises 6.5

Terms and Concepts

1. Fill in the blank: Partial Fraction Decomposition is a method of rewriting _____ functions.

2. T/F: It is sometimes necessary to use polynomial division before using Partial Fraction Decomposition.

3. Decompose $\dfrac{1}{x^2 - 3x}$ without solving for the coefficients, as done in Example 181.

4. Decompose $\dfrac{7-x}{x^2 - 9}$ without solving for the coefficients, as done in Example 181.

5. Decompose $\dfrac{x-3}{x^2 - 7}$ without solving for the coefficients, as done in Example 181.

6. Decompose $\dfrac{2x+5}{x^3 + 7x}$ without solving for the coefficients, as done in Example 181.

Problems

In Exercises 7 – 25, evaluate the indefinite integral.

7. $\displaystyle\int \dfrac{7x+7}{x^2 + 3x - 10}\, dx$

8. $\displaystyle\int \dfrac{7x-2}{x^2 + x}\, dx$

9. $\displaystyle\int \dfrac{-4}{3x^2 - 12}\, dx$

10. $\displaystyle\int \dfrac{x+7}{(x+5)^2}\, dx$

11. $\displaystyle\int \dfrac{-3x - 20}{(x+8)^2}\, dx$

12. $\displaystyle\int \dfrac{9x^2 + 11x + 7}{x(x+1)^2}\, dx$

13. $\displaystyle\int \dfrac{-12x^2 - x + 33}{(x-1)(x+3)(3-2x)}\, dx$

14. $\displaystyle\int \dfrac{94x^2 - 10x}{(7x+3)(5x-1)(3x-1)}\, dx$

15. $\displaystyle\int \dfrac{x^2 + x + 1}{x^2 + x - 2}\, dx$

16. $\displaystyle\int \dfrac{x^3}{x^2 - x - 20}\, dx$

17. $\displaystyle\int \dfrac{2x^2 - 4x + 6}{x^2 - 2x + 3}\, dx$

18. $\displaystyle\int \dfrac{1}{x^3 + 2x^2 + 3x}\, dx$

19. $\displaystyle\int \dfrac{x^2 + x + 5}{x^2 + 4x + 10}\, dx$

20. $\displaystyle\int \dfrac{12x^2 + 21x + 3}{(x+1)(3x^2 + 5x - 1)}\, dx$

21. $\displaystyle\int \dfrac{6x^2 + 8x - 4}{(x-3)(x^2 + 6x + 10)}\, dx$

22. $\displaystyle\int \dfrac{2x^2 + x + 1}{(x+1)(x^2 + 9)}\, dx$

23. $\displaystyle\int \dfrac{x^2 - 20x - 69}{(x-7)(x^2 + 2x + 17)}\, dx$

24. $\displaystyle\int \dfrac{9x^2 - 60x + 33}{(x-9)(x^2 - 2x + 11)}\, dx$

25. $\displaystyle\int \dfrac{6x^2 + 45x + 121}{(x+2)(x^2 + 10x + 27)}\, dx$

In Exercises 26 – 29, evaluate the definite integral.

26. $\displaystyle\int_1^2 \dfrac{8x + 21}{(x+2)(x+3)}\, dx$

27. $\displaystyle\int_0^5 \dfrac{14x + 6}{(3x+2)(x+4)}\, dx$

28. $\displaystyle\int_{-1}^1 \dfrac{x^2 + 5x - 5}{(x-10)(x^2 + 4x + 5)}\, dx$

29. $\displaystyle\int_0^1 \dfrac{x}{(x+1)(x^2 + 2x + 1)}\, dx$

6.6 Hyperbolic Functions

The **hyperbolic functions** are a set of functions that have many applications to mathematics, physics, and engineering. Among many other applications, they are used to describe the formation of satellite rings around planets, to describe the shape of a rope hanging from two points, and have application to the theory of special relativity. This section defines the hyperbolic functions and describes many of their properties, especially their usefulness to calculus.

These functions are sometimes referred to as the "hyperbolic trigonometric functions" as there are many, many connections between them and the standard trigonometric functions. Figure 6.13 demonstrates one such connection. Just as cosine and sine are used to define points on the circle defined by $x^2 + y^2 = 1$, the functions **hyperbolic cosine** and **hyperbolic sine** are used to define points on the hyperbola $x^2 - y^2 = 1$.

We begin with their definition.

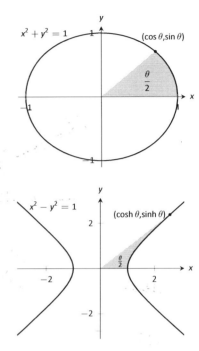

Figure 6.13: Using trigonometric functions to define points on a circle and hyperbolic functions to define points on a hyperbola. The area of the shaded regions are included in them.

Definition 23 Hyperbolic Functions

1. $\cosh x = \dfrac{e^x + e^{-x}}{2}$
2. $\sinh x = \dfrac{e^x - e^{-x}}{2}$
3. $\tanh x = \dfrac{\sinh x}{\cosh x}$
4. $\text{sech}\, x = \dfrac{1}{\cosh x}$
5. $\text{csch}\, x = \dfrac{1}{\sinh x}$
6. $\coth x = \dfrac{\cosh x}{\sinh x}$

These hyperbolic functions are graphed in Figure 6.14. In the graphs of $\cosh x$ and $\sinh x$, graphs of $e^x/2$ and $e^{-x}/2$ are included with dashed lines. As x gets "large," $\cosh x$ and $\sinh x$ each act like $e^x/2$; when x is a large negative number, $\cosh x$ acts like $e^{-x}/2$ whereas $\sinh x$ acts like $-e^{-x}/2$.

Notice the domains of $\tanh x$ and $\text{sech}\, x$ are $(-\infty, \infty)$, whereas both $\coth x$ and $\text{csch}\, x$ have vertical asymptotes at $x = 0$. Also note the ranges of these functions, especially $\tanh x$: as $x \to \infty$, both $\sinh x$ and $\cosh x$ approach $e^{-x}/2$, hence $\tanh x$ approaches 1.

The following example explores some of the properties of these functions that bear remarkable resemblance to the properties of their trigonometric counterparts.

Pronunciation Note:
"cosh" rhymes with "gosh,"
"sinh" rhymes with "pinch," and
"tanh" rhymes with "ranch."

Notes:

Chapter 6 Techniques of Antidifferentiation

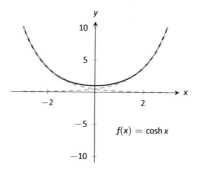
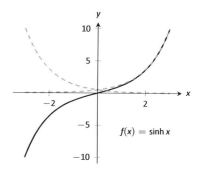
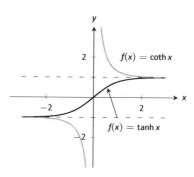
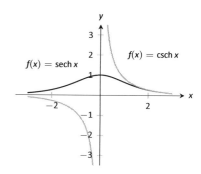

Figure 6.14: Graphs of the hyperbolic functions.

Example 186 **Exploring properties of hyperbolic functions**

Use Definition 23 to rewrite the following expressions.

1. $\cosh^2 x - \sinh^2 x$
2. $\tanh^2 x + \text{sech}^2 x$
3. $2\cosh x \sinh x$
4. $\frac{d}{dx}(\cosh x)$
5. $\frac{d}{dx}(\sinh x)$
6. $\frac{d}{dx}(\tanh x)$

SOLUTION

Notes:

314

1.
$$\cosh^2 x - \sinh^2 x = \left(\frac{e^x + e^{-x}}{2}\right)^2 - \left(\frac{e^x - e^{-x}}{2}\right)^2$$
$$= \frac{e^{2x} + 2e^x e^{-x} + e^{-2x}}{4} - \frac{e^{2x} - 2e^x e^{-x} + e^{-2x}}{4}$$
$$= \frac{4}{4} = 1.$$

So $\cosh^2 x - \sinh^2 x = 1$.

2.
$$\tanh^2 x + \text{sech}^2 x = \frac{\sinh^2 x}{\cosh^2 x} + \frac{1}{\cosh^2 x}$$
$$= \frac{\sinh^2 x + 1}{\cosh^2 x} \quad \text{Now use identity from \#1.}$$
$$= \frac{\cosh^2 x}{\cosh^2 x} = 1.$$

So $\tanh^2 x + \text{sech}^2 x = 1$.

3.
$$2\cosh x \sinh x = 2\left(\frac{e^x + e^{-x}}{2}\right)\left(\frac{e^x - e^{-x}}{2}\right)$$
$$= 2 \cdot \frac{e^{2x} - e^{-2x}}{4}$$
$$= \frac{e^{2x} - e^{-2x}}{2} = \sinh(2x).$$

Thus $2\cosh x \sinh x = \sinh(2x)$.

4.
$$\frac{d}{dx}(\cosh x) = \frac{d}{dx}\left(\frac{e^x + e^{-x}}{2}\right)$$
$$= \frac{e^x - e^{-x}}{2}$$
$$= \sinh x.$$

So $\frac{d}{dx}(\cosh x) = \sinh x$.

5.
$$\frac{d}{dx}(\sinh x) = \frac{d}{dx}\left(\frac{e^x - e^{-x}}{2}\right)$$
$$= \frac{e^x + e^{-x}}{2}$$
$$= \cosh x.$$

So $\frac{d}{dx}(\sinh x) = \cosh x$.

Notes:

Chapter 6 Techniques of Antidifferentiation

6. $\dfrac{d}{dx}\left(\tanh x\right) = \dfrac{d}{dx}\left(\dfrac{\sinh x}{\cosh x}\right)$

$= \dfrac{\cosh x \cosh x - \sinh x \sinh x}{\cosh^2 x}$

$= \dfrac{1}{\cosh^2 x}$

$= \text{sech}^2 x.$

So $\dfrac{d}{dx}\left(\tanh x\right) = \text{sech}^2 x.$

The following Key Idea summarizes many of the important identities relating to hyperbolic functions. Each can be verified by referring back to Definition 23.

Key Idea 16 Useful Hyperbolic Function Properties

Basic Identities

1. $\cosh^2 x - \sinh^2 x = 1$
2. $\tanh^2 x + \text{sech}^2 x = 1$
3. $\coth^2 x - \text{csch}^2 x = 1$
4. $\cosh 2x = \cosh^2 x + \sinh^2 x$
5. $\sinh 2x = 2 \sinh x \cosh x$
6. $\cosh^2 x = \dfrac{\cosh 2x + 1}{2}$
7. $\sinh^2 x = \dfrac{\cosh 2x - 1}{2}$

Derivatives

1. $\dfrac{d}{dx}\left(\cosh x\right) = \sinh x$
2. $\dfrac{d}{dx}\left(\sinh x\right) = \cosh x$
3. $\dfrac{d}{dx}\left(\tanh x\right) = \text{sech}^2 x$
4. $\dfrac{d}{dx}\left(\text{sech } x\right) = -\text{sech } x \tanh x$
5. $\dfrac{d}{dx}\left(\text{csch } x\right) = -\text{csch } x \coth x$
6. $\dfrac{d}{dx}\left(\coth x\right) = -\text{csch}^2 x$

Integrals

1. $\displaystyle\int \cosh x\, dx = \sinh x + C$
2. $\displaystyle\int \sinh x\, dx = \cosh x + C$
3. $\displaystyle\int \tanh x\, dx = \ln(\cosh x) + C$
4. $\displaystyle\int \coth x\, dx = \ln|\sinh x| + C$

We practice using Key Idea 16.

Example 187 Derivatives and integrals of hyperbolic functions
Evaluate the following derivatives and integrals.

1. $\dfrac{d}{dx}\left(\cosh 2x\right)$

2. $\displaystyle\int \text{sech}^2(7t - 3)\, dt$

3. $\displaystyle\int_0^{\ln 2} \cosh x\, dx$

Notes:

SOLUTION

1. Using the Chain Rule directly, we have $\frac{d}{dx}\big(\cosh 2x\big) = 2\sinh 2x$.

    Just to demonstrate that it works, let's also use the Basic Identity found in Key Idea 16: $\cosh 2x = \cosh^2 x + \sinh^2 x$.

    $$\frac{d}{dx}\big(\cosh 2x\big) = \frac{d}{dx}\big(\cosh^2 x + \sinh^2 x\big) = 2\cosh x \sinh x + 2 \sinh x \cosh x$$
    $$= 4 \cosh x \sinh x.$$

    Using another Basic Identity, we can see that $4 \cosh x \sinh x = 2 \sinh 2x$. We get the same answer either way.

2. We employ substitution, with $u = 7t - 3$ and $du = 7dt$. Applying Key Ideas 10 and 16 we have:

    $$\int \text{sech}^2(7t - 3)\, dt = \frac{1}{7}\tanh(7t - 3) + C.$$

3.
    $$\int_0^{\ln 2} \cosh x\, dx = \sinh x \Big|_0^{\ln 2} = \sinh(\ln 2) - \sinh 0 = \sinh(\ln 2).$$

    We can simplify this last expression as $\sinh x$ is based on exponentials:

    $$\sinh(\ln 2) = \frac{e^{\ln 2} - e^{-\ln 2}}{2} = \frac{2 - 1/2}{2} = \frac{3}{4}.$$

Inverse Hyperbolic Functions

Just as the inverse trigonometric functions are useful in certain integrations, the inverse hyperbolic functions are useful with others. Figure 6.15 shows the restrictions on the domains to make each function one-to-one and the resulting domains and ranges of their inverse functions. Their graphs are shown in Figure 6.16.

Because the hyperbolic functions are defined in terms of exponential functions, their inverses can be expressed in terms of logarithms as shown in Key Idea 17. It is often more convenient to refer to $\sinh^{-1} x$ than to $\ln\big(x + \sqrt{x^2 + 1}\big)$, especially when one is working on theory and does not need to compute actual values. On the other hand, when computations are needed, technology is often helpful but many hand-held calculators lack a *convenient* $\sinh^{-1} x$ button. (Often it can be accessed under a menu system, but not conveniently.) In such a situation, the logarithmic representation is useful. The reader is not encouraged to memorize these, but rather know they exist and know how to use them when needed.

Notes:

Chapter 6 Techniques of Antidifferentiation

Function	Domain	Range
$\cosh x$	$[0, \infty)$	$[1, \infty)$
$\sinh x$	$(-\infty, \infty)$	$(-\infty, \infty)$
$\tanh x$	$(-\infty, \infty)$	$(-1, 1)$
$\text{sech } x$	$[0, \infty)$	$(0, 1]$
$\text{csch } x$	$(-\infty, 0) \cup (0, \infty)$	$(-\infty, 0) \cup (0, \infty)$
$\coth x$	$(-\infty, 0) \cup (0, \infty)$	$(-\infty, -1) \cup (1, \infty)$

Function	Domain	Range
$\cosh^{-1} x$	$[1, \infty)$	$[0, \infty)$
$\sinh^{-1} x$	$(-\infty, \infty)$	$(-\infty, \infty)$
$\tanh^{-1} x$	$(-1, 1)$	$(-\infty, \infty)$
$\text{sech}^{-1} x$	$(0, 1]$	$[0, \infty)$
$\text{csch}^{-1} x$	$(-\infty, 0) \cup (0, \infty)$	$(-\infty, 0) \cup (0, \infty)$
$\coth^{-1} x$	$(-\infty, -1) \cup (1, \infty)$	$(-\infty, 0) \cup (0, \infty)$

Figure 6.15: Domains and ranges of the hyperbolic and inverse hyperbolic functions.

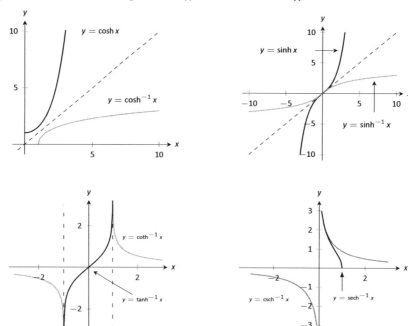

Figure 6.16: Graphs of the hyperbolic functions and their inverses.

Key Idea 17 **Logarithmic definitions of Inverse Hyperbolic Functions**

1. $\cosh^{-1} x = \ln\left(x + \sqrt{x^2 - 1}\right);\ x \geq 1$

2. $\tanh^{-1} x = \dfrac{1}{2} \ln\left(\dfrac{1+x}{1-x}\right);\ |x| < 1$

3. $\text{sech}^{-1} x = \ln\left(\dfrac{1 + \sqrt{1 - x^2}}{x}\right);\ 0 < x \leq 1$

4. $\sinh^{-1} x = \ln\left(x + \sqrt{x^2 + 1}\right)$

5. $\coth^{-1} x = \dfrac{1}{2} \ln\left(\dfrac{x+1}{x-1}\right);\ |x| > 1$

6. $\text{csch}^{-1} x = \ln\left(\dfrac{1}{x} + \dfrac{\sqrt{1 + x^2}}{|x|}\right);\ x \neq 0$

Notes:

The following Key Ideas give the derivatives and integrals relating to the inverse hyperbolic functions. In Key Idea 19, both the inverse hyperbolic and logarithmic function representations of the antiderivative are given, based on Key Idea 17. Again, these latter functions are often more useful than the former. Note how inverse hyperbolic functions can be used to solve integrals we used Trigonometric Substitution to solve in Section 6.4.

Key Idea 18   Derivatives Involving Inverse Hyperbolic Functions

1. $\dfrac{d}{dx}\left(\cosh^{-1} x\right) = \dfrac{1}{\sqrt{x^2-1}}; \; x > 1$

2. $\dfrac{d}{dx}\left(\sinh^{-1} x\right) = \dfrac{1}{\sqrt{x^2+1}}$

3. $\dfrac{d}{dx}\left(\tanh^{-1} x\right) = \dfrac{1}{1-x^2}; \; |x| < 1$

4. $\dfrac{d}{dx}\left(\text{sech}^{-1} x\right) = \dfrac{-1}{x\sqrt{1-x^2}}; \; 0 < x < 1$

5. $\dfrac{d}{dx}\left(\text{csch}^{-1} x\right) = \dfrac{-1}{|x|\sqrt{1+x^2}}; \; x \neq 0$

6. $\dfrac{d}{dx}\left(\coth^{-1} x\right) = \dfrac{1}{1-x^2}; \; |x| > 1$

Key Idea 19   Integrals Involving Inverse Hyperbolic Functions

1. $\displaystyle\int \dfrac{1}{\sqrt{x^2-a^2}}\,dx \;=\; \cosh^{-1}\left(\dfrac{x}{a}\right) + C; \; 0 < a < x \;=\; \ln\left|x + \sqrt{x^2-a^2}\right| + C$

2. $\displaystyle\int \dfrac{1}{\sqrt{x^2+a^2}}\,dx \;=\; \sinh^{-1}\left(\dfrac{x}{a}\right) + C; \; a > 0 \;=\; \ln\left|x + \sqrt{x^2+a^2}\right| + C$

3. $\displaystyle\int \dfrac{1}{a^2-x^2}\,dx \;=\; \begin{cases} \dfrac{1}{a}\tanh^{-1}\left(\dfrac{x}{a}\right) + C & x^2 < a^2 \\[4pt] \dfrac{1}{a}\coth^{-1}\left(\dfrac{x}{a}\right) + C & a^2 < x^2 \end{cases} \;=\; \dfrac{1}{2}\ln\left|\dfrac{a+x}{a-x}\right| + C$

4. $\displaystyle\int \dfrac{1}{x\sqrt{a^2-x^2}}\,dx \;=\; -\dfrac{1}{a}\text{sech}^{-1}\left(\dfrac{x}{a}\right) + C; \; 0 < x < a \;=\; \dfrac{1}{a}\ln\left(\dfrac{x}{a+\sqrt{a^2-x^2}}\right) + C$

5. $\displaystyle\int \dfrac{1}{x\sqrt{x^2+a^2}}\,dx \;=\; -\dfrac{1}{a}\text{csch}^{-1}\left|\dfrac{x}{a}\right| + C; \; x \neq 0, \, a > 0 \;=\; \dfrac{1}{a}\ln\left|\dfrac{x}{a+\sqrt{a^2+x^2}}\right| + C$

We practice using the derivative and integral formulas in the following example.

Notes:

Chapter 6 Techniques of Antidifferentiation

Example 188   **Derivatives and integrals involving inverse hyperbolic functions**

Evaluate the following.

1. $\dfrac{d}{dx}\left[\cosh^{-1}\left(\dfrac{3x-2}{5}\right)\right]$

2. $\displaystyle\int \dfrac{1}{x^2-1}\,dx$

3. $\displaystyle\int \dfrac{1}{\sqrt{9x^2+10}}\,dx$

SOLUTION

1. Applying Key Idea 18 with the Chain Rule gives:

$$\dfrac{d}{dx}\left[\cosh^{-1}\left(\dfrac{3x-2}{5}\right)\right] = \dfrac{1}{\sqrt{\left(\dfrac{3x-2}{5}\right)^2 - 1}} \cdot \dfrac{3}{5}.$$

2. Multiplying the numerator and denominator by (-1) gives: $\displaystyle\int \dfrac{1}{x^2-1}\,dx = \int \dfrac{-1}{1-x^2}\,dx$. The second integral can be solved with a direct application of item #3 from Key Idea 19, with $a = 1$. Thus

$$\begin{aligned}
\int \dfrac{1}{x^2-1}\,dx &= -\int \dfrac{1}{1-x^2}\,dx \\
&= \begin{cases} -\tanh^{-1}(x) + C & x^2 < 1 \\ -\coth^{-1}(x) + C & 1 < x^2 \end{cases} \\
&= -\dfrac{1}{2}\ln\left|\dfrac{x+1}{x-1}\right| + C \\
&= \dfrac{1}{2}\ln\left|\dfrac{x-1}{x+1}\right| + C. \qquad (6.4)
\end{aligned}$$

We should note that this exact problem was solved at the beginning of Section 6.5. In that example the answer was given as $\frac{1}{2}\ln|x-1| - \frac{1}{2}\ln|x+1| + C$. Note that this is equivalent to the answer given in Equation 6.4, as $\ln(a/b) = \ln a - \ln b$.

3. This requires a substitution, then item #2 of Key Idea 19 can be applied. Let $u = 3x$, hence $du = 3dx$. We have

$$\int \dfrac{1}{\sqrt{9x^2+10}}\,dx = \dfrac{1}{3}\int \dfrac{1}{\sqrt{u^2+10}}\,du.$$

Notes:

Note $a^2 = 10$, hence $a = \sqrt{10}$. Now apply the integral rule.

$$= \frac{1}{3} \sinh^{-1}\left(\frac{3x}{\sqrt{10}}\right) + C$$
$$= \frac{1}{3} \ln\left|3x + \sqrt{9x^2 + 10}\right| + C.$$

This section covers a lot of ground. New functions were introduced, along with some of their fundamental identities, their derivatives and antiderivatives, their inverses, and the derivatives and antiderivatives of these inverses. Four Key Ideas were presented, each including quite a bit of information.

Do not view this section as containing a source of information to be memorized, but rather as a reference for future problem solving. Key Idea 19 contains perhaps the most useful information. Know the integration forms it helps evaluate and understand how to use the inverse hyperbolic answer and the logarithmic answer.

The next section takes a brief break from demonstrating new integration techniques. It instead demonstrates a technique of evaluating limits that return indeterminate forms. This technique will be useful in Section 6.8, where limits will arise in the evaluation of certain definite integrals.

Notes:

Exercises 6.6

Terms and Concepts

1. In Key Idea 16, the equation $\int \tanh x \, dx = \ln(\cosh x) + C$ is given. Why is "$\ln|\cosh x|$" not used – i.e., why are absolute values not necessary?

2. The hyperbolic functions are used to define points on the right hand portion of the hyperbola $x^2 - y^2 = 1$, as shown in Figure 6.13. How can we use the hyperbolic functions to define points on the left hand portion of the hyperbola?

Problems

In Exercises 3 – 10, verify the given identity using Definition 23, as done in Example 186.

3. $\coth^2 x - \operatorname{csch}^2 x = 1$

4. $\cosh 2x = \cosh^2 x + \sinh^2 x$

5. $\cosh^2 x = \dfrac{\cosh 2x + 1}{2}$

6. $\sinh^2 x = \dfrac{\cosh 2x - 1}{2}$

7. $\dfrac{d}{dx}[\operatorname{sech} x] = -\operatorname{sech} x \tanh x$

8. $\dfrac{d}{dx}[\coth x] = -\operatorname{csch}^2 x$

9. $\int \tanh x \, dx = \ln(\cosh x) + C$

10. $\int \coth x \, dx = \ln|\sinh x| + C$

In Exercises 11 – 21, find the derivative of the given function.

11. $f(x) = \cosh 2x$

12. $f(x) = \tanh(x^2)$

13. $f(x) = \ln(\sinh x)$

14. $f(x) = \sinh x \cosh x$

15. $f(x) = x \sinh x - \cosh x$

16. $f(x) = \operatorname{sech}^{-1}(x^2)$

17. $f(x) = \sinh^{-1}(3x)$

18. $f(x) = \cosh^{-1}(2x^2)$

19. $f(x) = \tanh^{-1}(x + 5)$

20. $f(x) = \tanh^{-1}(\cos x)$

21. $f(x) = \cosh^{-1}(\sec x)$

In Exercises 22 – 26, find the equation of the line tangent to the function at the given x-value.

22. $f(x) = \sinh x$ at $x = 0$

23. $f(x) = \cosh x$ at $x = \ln 2$

24. $f(x) = \operatorname{sech}^2 x$ at $x = \ln 3$

25. $f(x) = \sinh^{-1} x$ at $x = 0$

26. $f(x) = \cosh^{-1} x$ at $x = \sqrt{2}$

In Exercises 27 – 40, evaluate the given indefinite integral.

27. $\int \tanh(2x) \, dx$

28. $\int \cosh(3x - 7) \, dx$

29. $\int \sinh x \cosh x \, dx$

30. $\int x \cosh x \, dx$

31. $\int x \sinh x \, dx$

32. $\int \dfrac{1}{9 - x^2} \, dx$

33. $\int \dfrac{2x}{\sqrt{x^4 - 4}} \, dx$

34. $\int \dfrac{\sqrt{x}}{\sqrt{1 + x^3}} \, dx$

35. $\int \dfrac{1}{x^4 - 16} \, dx$

36. $\int \dfrac{1}{x^2 + x} \, dx$

37. $\int \dfrac{e^x}{e^{2x} + 1} \, dx$

38. $\int \sinh^{-1} x \, dx$

39. $\int \tanh^{-1} x \, dx$

40. $\int \text{sech}\, x\, dx$ (Hint: mutiply by $\frac{\cosh x}{\cosh x}$; set $u = \sinh x$.)

42. $\int_{-\ln 2}^{\ln 2} \cosh x\, dx$

In Exercises 41 – 43, evaluate the given definite integral.

43. $\int_0^1 \tanh^{-1} x\, dx$

41. $\int_{-1}^{1} \sinh x\, dx$

6.7 L'Hôpital's Rule

While this chapter is devoted to learning techniques of integration, this section is not about integration. Rather, it is concerned with a technique of evaluating certain limits that will be useful in the following section, where integration is once more discussed.

Our treatment of limits exposed us to "0/0", an indeterminate form. If $\lim\limits_{x \to c} f(x) = 0$ and $\lim\limits_{x \to c} g(x) = 0$, we do not conclude that $\lim\limits_{x \to c} f(x)/g(x)$ is 0/0; rather, we use 0/0 as notation to describe the fact that both the numerator and denominator approach 0. The expression 0/0 has no numeric value; other work must be done to evaluate the limit.

Other indeterminate forms exist; they are: ∞/∞, $0 \cdot \infty$, $\infty - \infty$, 0^0, 1^∞ and ∞^0. Just as "0/0" does not mean "divide 0 by 0," the expression "∞/∞" does not mean "divide infinity by infinity." Instead, it means "a quantity is growing without bound and is being divided by another quantity that is growing without bound." We cannot determine from such a statement what value, if any, results in the limit. Likewise, "$0 \cdot \infty$" does not mean "multiply zero by infinity." Instead, it means "one quantity is shrinking to zero, and is being multiplied by a quantity that is growing without bound." We cannot determine from such a description what the result of such a limit will be.

This section introduces l'Hôpital's Rule, a method of resolving limits that produce the indeterminate forms 0/0 and ∞/∞. We'll also show how algebraic manipulation can be used to convert other indeterminate expressions into one of these two forms so that our new rule can be applied.

Theorem 49 **L'Hôpital's Rule, Part 1**

Let $\lim\limits_{x \to c} f(x) = 0$ and $\lim\limits_{x \to c} g(x) = 0$, where f and g are differentiable functions on an open interval I containing c, and $g'(x) \neq 0$ on I except possibly at c. Then

$$\lim_{x \to c} \frac{f(x)}{g(x)} = \lim_{x \to c} \frac{f'(x)}{g'(x)}.$$

We demonstrate the use of l'Hôpital's Rule in the following examples; we will often use "LHR" as an abbreviation of "l'Hôpital's Rule."

Notes:

6.7 L'Hôpital's Rule

Example 189   **Using l'Hôpital's Rule**

Evaluate the following limits, using l'Hôpital's Rule as needed.

1. $\lim\limits_{x\to 0} \dfrac{\sin x}{x}$

2. $\lim\limits_{x\to 1} \dfrac{\sqrt{x+3}-2}{1-x}$

3. $\lim\limits_{x\to 0} \dfrac{x^2}{1-\cos x}$

4. $\lim\limits_{x\to 2} \dfrac{x^2+x-6}{x^2-3x+2}$

SOLUTION

1. We proved this limit is 1 in Example 13 using the Squeeze Theorem. Here we use l'Hôpital's Rule to show its power.

$$\lim_{x\to 0} \frac{\sin x}{x} \stackrel{\text{by LHR}}{=} \lim_{x\to 0} \frac{\cos x}{1} = 1.$$

2. $\lim\limits_{x\to 1} \dfrac{\sqrt{x+3}-2}{1-x} \stackrel{\text{by LHR}}{=} \lim\limits_{x\to 1} \dfrac{\frac{1}{2}(x+3)^{-1/2}}{-1} = -\dfrac{1}{4}.$

3. $\lim\limits_{x\to 0} \dfrac{x^2}{1-\cos x} \stackrel{\text{by LHR}}{=} \lim\limits_{x\to 0} \dfrac{2x}{\sin x}.$

This latter limit also evaluates to the 0/0 indeterminate form. To evaluate it, we apply l'Hôpital's Rule again.

$$\lim_{x\to 0} \frac{2x}{\sin x} \stackrel{\text{by LHR}}{=} \frac{2}{\cos x} = 2.$$

Thus $\lim\limits_{x\to 0} \dfrac{x^2}{1-\cos x} = 2.$

4. We already know how to evaluate this limit; first factor the numerator and denominator. We then have:

$$\lim_{x\to 2} \frac{x^2+x-6}{x^2-3x+2} = \lim_{x\to 2} \frac{(x-2)(x+3)}{(x-2)(x-1)} = \lim_{x\to 2} \frac{x+3}{x-1} = 5.$$

We now show how to solve this using l'Hôpital's Rule.

$$\lim_{x\to 2} \frac{x^2+x-6}{x^2-3x+2} \stackrel{\text{by LHR}}{=} \lim_{x\to 2} \frac{2x+1}{2x-3} = 5.$$

Note that at each step where l'Hôpital's Rule was applied, it was *needed*: the initial limit returned the indeterminate form of "0/0." If the initial limit returns, for example, 1/2, then l'Hôpital's Rule does not apply.

Notes:

The following theorem extends our initial version of l'Hôpital's Rule in two ways. It allows the technique to be applied to the indeterminate form ∞/∞ and to limits where x approaches $\pm\infty$.

Theorem 50   **L'Hôpital's Rule, Part 2**

1. Let $\lim\limits_{x\to a} f(x) = \pm\infty$ and $\lim\limits_{x\to a} g(x) = \pm\infty$, where f and g are differentiable on an open interval I containing a. Then
$$\lim_{x\to a}\frac{f(x)}{g(x)} = \lim_{x\to a}\frac{f'(x)}{g'(x)}.$$

2. Let f and g be differentiable functions on the open interval (a,∞) for some value a, where $g'(x) \neq 0$ on (a,∞) and $\lim\limits_{x\to\infty} f(x)/g(x)$ returns either $0/0$ or ∞/∞. Then
$$\lim_{x\to\infty}\frac{f(x)}{g(x)} = \lim_{x\to\infty}\frac{f'(x)}{g'(x)}.$$

A similar statement can be made for limits where x approaches $-\infty$.

Example 190   **Using l'Hôpital's Rule with limits involving ∞**
Evaluate the following limits.

1. $\lim\limits_{x\to\infty} \dfrac{3x^2 - 100x + 2}{4x^2 + 5x - 1000}$

2. $\lim\limits_{x\to\infty} \dfrac{e^x}{x^3}$.

Solution

1. We can evaluate this limit already using Theorem 11; the answer is 3/4. We apply l'Hôpital's Rule to demonstrate its applicability.
$$\lim_{x\to\infty}\frac{3x^2 - 100x + 2}{4x^2 + 5x - 1000} \overset{\text{by LHR}}{=} \lim_{x\to\infty}\frac{6x - 100}{8x + 5} \overset{\text{by LHR}}{=} \lim_{x\to\infty}\frac{6}{8} = \frac{3}{4}.$$

2. $\lim\limits_{x\to\infty}\dfrac{e^x}{x^3} \overset{\text{by LHR}}{=} \lim\limits_{x\to\infty}\dfrac{e^x}{3x^2} \overset{\text{by LHR}}{=} \lim\limits_{x\to\infty}\dfrac{e^x}{6x} \overset{\text{by LHR}}{=} \lim\limits_{x\to\infty}\dfrac{e^x}{6} = \infty.$

Recall that this means that the limit does not exist; as x approaches ∞, the expression e^x/x^3 grows without bound. We can infer from this that e^x grows "faster" than x^3; as x gets large, e^x is far larger than x^3. (This

---

Notes:

has important implications in computing when considering efficiency of algorithms.)

Indeterminate Forms $0 \cdot \infty$ and $\infty - \infty$

L'Hôpital's Rule can only be applied to ratios of functions. When faced with an indeterminate form such as $0 \cdot \infty$ or $\infty - \infty$, we can sometimes apply algebra to rewrite the limit so that l'Hôpital's Rule can be applied. We demonstrate the general idea in the next example.

Example 191   **Applying l'Hôpital's Rule to other indeterminate forms**
Evaluate the following limits.

1. $\lim\limits_{x \to 0^+} x \cdot e^{1/x}$

2. $\lim\limits_{x \to 0^-} x \cdot e^{1/x}$

3. $\lim\limits_{x \to \infty} \ln(x+1) - \ln x$

4. $\lim\limits_{x \to \infty} x^2 - e^x$

SOLUTION

1. As $x \to 0^+$, $x \to 0$ and $e^{1/x} \to \infty$. Thus we have the indeterminate form $0 \cdot \infty$. We rewrite the expression $x \cdot e^{1/x}$ as $\dfrac{e^{1/x}}{1/x}$; now, as $x \to 0^+$, we get the indeterminate form ∞/∞ to which l'Hôpital's Rule can be applied.

$$\lim_{x \to 0^+} x \cdot e^{1/x} = \lim_{x \to 0^+} \frac{e^{1/x}}{1/x} \stackrel{\text{by LHR}}{=} \lim_{x \to 0^+} \frac{(-1/x^2)e^{1/x}}{-1/x^2} = \lim_{x \to 0^+} e^{1/x} = \infty.$$

Interpretation: $e^{1/x}$ grows "faster" than x shrinks to zero, meaning their product grows without bound.

2. As $x \to 0^-$, $x \to 0$ and $e^{1/x} \to e^{-\infty} \to 0$. The the limit evaluates to $0 \cdot 0$ which is not an indeterminate form. We conclude then that

$$\lim_{x \to 0^-} x \cdot e^{1/x} = 0.$$

3. This limit initially evaluates to the indeterminate form $\infty - \infty$. By applying a logarithmic rule, we can rewrite the limit as

$$\lim_{x \to \infty} \ln(x+1) - \ln x = \lim_{x \to \infty} \ln\left(\frac{x+1}{x}\right).$$

As $x \to \infty$, the argument of the ln term approaches ∞/∞, to which we can apply l'Hôpital's Rule.

$$\lim_{x \to \infty} \frac{x+1}{x} \stackrel{\text{by LHR}}{=} \frac{1}{1} = 1.$$

Notes:

Since $x \to \infty$ implies $\dfrac{x+1}{x} \to 1$, it follows that

$$x \to \infty \quad \text{implies} \quad \ln\left(\frac{x+1}{x}\right) \to \ln 1 = 0.$$

Thus

$$\lim_{x \to \infty} \ln(x+1) - \ln x = \lim_{x \to \infty} \ln\left(\frac{x+1}{x}\right) = 0.$$

Interpretation: since this limit evaluates to 0, it means that for large x, there is essentially no difference between $\ln(x+1)$ and $\ln x$; their difference is essentially 0.

4. The limit $\lim\limits_{x \to \infty} x^2 - e^x$ initially returns the indeterminate form $\infty - \infty$. We can rewrite the expression by factoring out x^2; $x^2 - e^x = x^2\left(1 - \dfrac{e^x}{x^2}\right)$. We need to evaluate how e^x/x^2 behaves as $x \to \infty$:

$$\lim_{x \to \infty} \frac{e^x}{x^2} \stackrel{\text{by LHR}}{=} \lim_{x \to \infty} \frac{e^x}{2x} \stackrel{\text{by LHR}}{=} \lim_{x \to \infty} \frac{e^x}{2} = \infty.$$

Thus $\lim_{x \to \infty} x^2(1 - e^x/x^2)$ evaluates to $\infty \cdot (-\infty)$, which is not an indeterminate form; rather, $\infty \cdot (-\infty)$ evaluates to $-\infty$. We conclude that $\lim\limits_{x \to \infty} x^2 - e^x = -\infty$.

Interpretation: as x gets large, the difference between x^2 and e^x grows very large.

Indeterminate Forms 0^0, 1^∞ and ∞^0

When faced with an indeterminate form that involves a power, it often helps to employ the natural logarithmic function. The following Key Idea expresses the concept, which is followed by an example that demonstrates its use.

Key Idea 20 **Evaluating Limits Involving Indeterminate Forms 0^0, 1^∞ and ∞^0**

If $\lim\limits_{x \to c} \ln\bigl(f(x)\bigr) = L$, then $\lim\limits_{x \to c} f(x) = \lim\limits_{x \to c} e^{\ln(f(x))} = e^L$.

6.7 L'Hôpital's Rule

Example 192 **Using l'Hôpital's Rule with indeterminate forms involving exponents**

Evaluate the following limits.

1. $\lim\limits_{x\to\infty} \left(1+\dfrac{1}{x}\right)^x$
2. $\lim\limits_{x\to 0^+} x^x.$

SOLUTION

1. This equivalent to a special limit given in Theorem 3; these limits have important applications within mathematics and finance. Note that the exponent approaches ∞ while the base approaches 1, leading to the indeterminate form 1^∞. Let $f(x) = (1+1/x)^x$; the problem asks to evaluate $\lim\limits_{x\to\infty} f(x)$. Let's first evaluate $\lim\limits_{x\to\infty} \ln\bigl(f(x)\bigr)$.

$$\begin{aligned}
\lim_{x\to\infty} \ln\bigl(f(x)\bigr) &= \lim_{x\to\infty} \ln\left(1+\frac{1}{x}\right)^x \\
&= \lim_{x\to\infty} x \ln\left(1+\frac{1}{x}\right) \\
&= \lim_{x\to\infty} \frac{\ln\left(1+\frac{1}{x}\right)}{1/x}
\end{aligned}$$

This produces the indeterminate form 0/0, so we apply l'Hôpital's Rule.

$$\begin{aligned}
&= \lim_{x\to\infty} \frac{\frac{1}{1+1/x} \cdot (-1/x^2)}{(-1/x^2)} \\
&= \lim_{x\to\infty} \frac{1}{1+1/x} \\
&= 1.
\end{aligned}$$

Thus $\lim\limits_{x\to\infty} \ln\bigl(f(x)\bigr) = 1$. We return to the original limit and apply Key Idea 20.

$$\lim_{x\to\infty} \left(1+\frac{1}{x}\right)^x = \lim_{x\to\infty} f(x) = \lim_{x\to\infty} e^{\ln(f(x))} = e^1 = e.$$

2. This limit leads to the indeterminate form 0^0. Let $f(x) = x^x$ and consider

Notes:

first $\lim_{x\to 0^+} \ln(f(x))$.

$$\lim_{x\to 0^+} \ln(f(x)) = \lim_{x\to 0^+} \ln(x^x)$$
$$= \lim_{x\to 0^+} x \ln x$$
$$= \lim_{x\to 0^+} \frac{\ln x}{1/x}.$$

This produces the indeterminate form $-\infty/\infty$ so we apply l'Hôpital's Rule.

$$= \lim_{x\to 0^+} \frac{1/x}{-1/x^2}$$
$$= \lim_{x\to 0^+} -x$$
$$= 0.$$

Thus $\lim_{x\to 0^+} \ln(f(x)) = 0$. We return to the original limit and apply Key Idea 20.

$$\lim_{x\to 0^+} x^x = \lim_{x\to 0^+} f(x) = \lim_{x\to 0^+} e^{\ln(f(x))} = e^0 = 1.$$

This result is supported by the graph of $f(x) = x^x$ given in Figure 6.17.

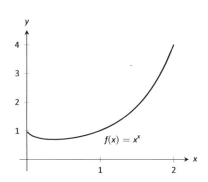

Figure 6.17: A graph of $f(x) = x^x$ supporting the fact that as $x \to 0^+$, $f(x) \to 1$.

Our brief revisit of limits will be rewarded in the next section where we consider *improper integration*. So far, we have only considered definite integrals where the bounds are finite numbers, such as $\int_0^1 f(x)\, dx$. Improper integration considers integrals where one, or both, of the bounds are "infinity." Such integrals have many uses and applications, in addition to generating ideas that are enlightening.

Notes:

Exercises 6.7

Terms and Concepts

1. List the different indeterminate forms described in this section.

2. T/F: l'Hôpital's Rule provides a faster method of computing derivatives.

3. T/F: l'Hôpital's Rule states that $\dfrac{d}{dx}\left[\dfrac{f(x)}{g(x)}\right] = \dfrac{f'(x)}{g'(x)}$.

4. Explain what the indeterminate form "1^∞" means.

5. Fill in the blanks: The Quotient Rule is applied to $\dfrac{f(x)}{g(x)}$ when taking _____; l'Hôpital's Rule is applied when taking certain _____.

6. Create (but do not evaluate!) a limit that returns "∞^0".

7. Create a function $f(x)$ such that $\lim_{x\to 1} f(x)$ returns "0^0".

Problems

In Exercises 8 – 52, evaluate the given limit.

8. $\lim\limits_{x\to 1} \dfrac{x^2+x-2}{x-1}$

9. $\lim\limits_{x\to 2} \dfrac{x^2+x-6}{x^2-7x+10}$

10. $\lim\limits_{x\to \pi} \dfrac{\sin x}{x-\pi}$

11. $\lim\limits_{x\to \pi/4} \dfrac{\sin x - \cos x}{\cos(2x)}$

12. $\lim\limits_{x\to 0} \dfrac{\sin(5x)}{x}$

13. $\lim\limits_{x\to 0} \dfrac{\sin(2x)}{x+2}$

14. $\lim\limits_{x\to 0} \dfrac{\sin(2x)}{\sin(3x)}$

15. $\lim\limits_{x\to 0} \dfrac{\sin(ax)}{\sin(bx)}$

16. $\lim\limits_{x\to 0^+} \dfrac{e^x-1}{x^2}$

17. $\lim\limits_{x\to 0^+} \dfrac{e^x-x-1}{x^2}$

18. $\lim\limits_{x\to 0^+} \dfrac{x-\sin x}{x^3-x^2}$

19. $\lim\limits_{x\to \infty} \dfrac{x^4}{e^x}$

20. $\lim\limits_{x\to \infty} \dfrac{\sqrt{x}}{e^x}$

21. $\lim\limits_{x\to \infty} \dfrac{e^x}{\sqrt{x}}$

22. $\lim\limits_{x\to \infty} \dfrac{e^x}{2^x}$

23. $\lim\limits_{x\to \infty} \dfrac{e^x}{3^x}$

24. $\lim\limits_{x\to 3} \dfrac{x^3-5x^2+3x+9}{x^3-7x^2+15x-9}$

25. $\lim\limits_{x\to -2} \dfrac{x^3+4x^2+4x}{x^3+7x^2+16x+12}$

26. $\lim\limits_{x\to \infty} \dfrac{\ln x}{x}$

27. $\lim\limits_{x\to \infty} \dfrac{\ln(x^2)}{x}$

28. $\lim\limits_{x\to \infty} \dfrac{(\ln x)^2}{x}$

29. $\lim\limits_{x\to 0^+} x \cdot \ln x$

30. $\lim\limits_{x\to 0^+} \sqrt{x} \cdot \ln x$

31. $\lim\limits_{x\to 0^+} x e^{1/x}$

32. $\lim\limits_{x\to \infty} x^3 - x^2$

33. $\lim\limits_{x\to \infty} \sqrt{x} - \ln x$

34. $\lim\limits_{x\to -\infty} x e^x$

35. $\lim\limits_{x\to 0^+} \dfrac{1}{x^2} e^{-1/x}$

36. $\lim\limits_{x\to 0^+} (1+x)^{1/x}$

37. $\lim\limits_{x\to 0^+} (2x)^x$

38. $\lim\limits_{x\to 0^+} (2/x)^x$

39. $\lim\limits_{x\to 0^+} (\sin x)^x$ Hint: use the Squeeze Theorem.

40. $\lim_{x \to 1^+} (1-x)^{1-x}$

41. $\lim_{x \to \infty} (x)^{1/x}$

42. $\lim_{x \to \infty} (1/x)^x$

43. $\lim_{x \to 1^+} (\ln x)^{1-x}$

44. $\lim_{x \to \infty} (1+x)^{1/x}$

45. $\lim_{x \to \infty} (1+x^2)^{1/x}$

46. $\lim_{x \to \pi/2} \tan x \cos x$

47. $\lim_{x \to \pi/2} \tan x \sin(2x)$

48. $\lim_{x \to 1^+} \dfrac{1}{\ln x} - \dfrac{1}{x-1}$

49. $\lim_{x \to 3^+} \dfrac{5}{x^2-9} - \dfrac{x}{x-3}$

50. $\lim_{x \to \infty} x \tan(1/x)$

51. $\lim_{x \to \infty} \dfrac{(\ln x)^3}{x}$

52. $\lim_{x \to 1} \dfrac{x^2+x-2}{\ln x}$

6.8 Improper Integration

We begin this section by considering the following definite integrals:

- $\int_0^{100} \dfrac{1}{1+x^2}\, dx \approx 1.5608,$

- $\int_0^{1000} \dfrac{1}{1+x^2}\, dx \approx 1.5698,$

- $\int_0^{10,000} \dfrac{1}{1+x^2}\, dx \approx 1.5707.$

Notice how the integrand is $1/(1+x^2)$ in each integral (which is sketched in Figure 6.18). As the upper bound gets larger, one would expect the "area under the curve" would also grow. While the definite integrals do increase in value as the upper bound grows, they are not increasing by much. In fact, consider:

$$\int_0^b \dfrac{1}{1+x^2}\, dx = \tan^{-1} x \Big|_0^b = \tan^{-1} b - \tan^{-1} 0 = \tan^{-1} b.$$

As $b \to \infty$, $\tan^{-1} b \to \pi/2$. Therefore it seems that as the upper bound b grows, the value of the definite integral $\int_0^b \dfrac{1}{1+x^2}\, dx$ approaches $\pi/2 \approx 1.5708$. This should strike the reader as being a bit amazing: even though the curve extends "to infinity," it has a finite amount of area underneath it.

When we defined the definite integral $\int_a^b f(x)\, dx$, we made two stipulations:

1. The interval over which we integrated, $[a, b]$, was a finite interval, and

2. The function $f(x)$ was continuous on $[a, b]$ (ensuring that the range of f was finite).

In this section we consider integrals where one or both of the above conditions do not hold. Such integrals are called **improper integrals.**

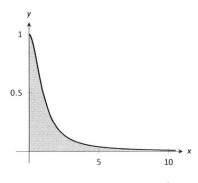

Figure 6.18: Graphing $f(x) = \dfrac{1}{1+x^2}$.

Improper Integrals with Infinite Bounds

> **Definition 24**   **Improper Integrals with Infinite Bounds; Converge, Diverge**
>
> 1. Let f be a continuous function on $[a, \infty)$. Define
> $$\int_a^\infty f(x)\,dx \quad \text{to be} \quad \lim_{b\to\infty} \int_a^b f(x)\,dx.$$
>
> 2. Let f be a continuous function on $(-\infty, b]$. Define
> $$\int_{-\infty}^b f(x)\,dx \quad \text{to be} \quad \lim_{a\to-\infty} \int_a^b f(x)\,dx.$$
>
> 3. Let f be a continuous function on $(-\infty, \infty)$. Let c be any real number; define
> $$\int_{-\infty}^\infty f(x)\,dx \quad \text{to be} \quad \lim_{a\to-\infty}\int_a^c f(x)\,dx + \lim_{b\to\infty}\int_c^b f(x)\,dx.$$
>
> An improper integral is said to **converge** if its corresponding limit exists; otherwise, it **diverges**. The improper integral in part 3 converges if and only if both of its limits exist.

Example 193   **Evaluating improper integrals**

Evaluate the following improper integrals.

1. $\displaystyle\int_1^\infty \frac{1}{x^2}\,dx$
2. $\displaystyle\int_1^\infty \frac{1}{x}\,dx$
3. $\displaystyle\int_{-\infty}^0 e^x\,dx$
4. $\displaystyle\int_{-\infty}^\infty \frac{1}{1+x^2}\,dx$

Solution

1.
$$\int_1^\infty \frac{1}{x^2}\,dx = \lim_{b\to\infty}\int_1^b \frac{1}{x^2}\,dx = \lim_{b\to\infty}\frac{-1}{x}\bigg|_1^b$$
$$= \lim_{b\to\infty}\frac{-1}{b}+1$$
$$= 1.$$

A graph of the area defined by this integral is given in Figure 6.19.

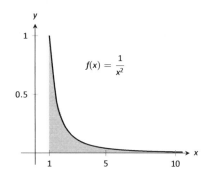

Figure 6.19: A graph of $f(x) = \frac{1}{x^2}$ in Example 193.

Notes:

2.
$$\int_1^\infty \frac{1}{x}\,dx = \lim_{b\to\infty}\int_1^b \frac{1}{x}\,dx$$
$$= \lim_{b\to\infty} \ln|x|\Big|_1^b$$
$$= \lim_{b\to\infty} \ln(b)$$
$$= \infty.$$

The limit does not exist, hence the improper integral $\int_1^\infty \frac{1}{x}\,dx$ diverges.

Compare the graphs in Figures 6.19 and 6.20; notice how the graph of $f(x) = 1/x$ is noticeably larger. This difference is enough to cause the improper integral to diverge.

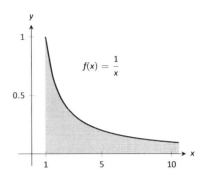

Figure 6.20: A graph of $f(x) = \frac{1}{x}$ in Example 193.

3.
$$\int_{-\infty}^0 e^x\,dx = \lim_{a\to-\infty}\int_a^0 e^x\,dx$$
$$= \lim_{a\to-\infty} e^x\Big|_a^0$$
$$= \lim_{a\to-\infty} e^0 - e^a$$
$$= 1.$$

A graph of the area defined by this integral is given in Figure 6.21.

4. We will need to break this into two improper integrals and choose a value of c as in part 3 of Definition 24. Any value of c is fine; we choose $c = 0$.

$$\int_{-\infty}^\infty \frac{1}{1+x^2}\,dx = \lim_{a\to-\infty}\int_a^0 \frac{1}{1+x^2}\,dx + \lim_{b\to\infty}\int_0^b \frac{1}{1+x^2}\,dx$$
$$= \lim_{a\to-\infty} \tan^{-1}x\Big|_a^0 + \lim_{b\to\infty} \tan^{-1}x\Big|_0^b$$
$$= \lim_{a\to-\infty}\left(\tan^{-1}0 - \tan^{-1}a\right) + \lim_{b\to\infty}\left(\tan^{-1}b - \tan^{-1}0\right)$$
$$= \left(0 - \frac{-\pi}{2}\right) + \left(\frac{\pi}{2} - 0\right).$$

Each limit exists, hence the original integral converges and has value:

$$= \pi.$$

A graph of the area defined by this integral is given in Figure 6.22.

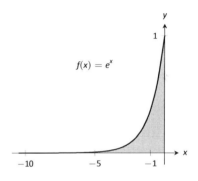

Figure 6.21: A graph of $f(x) = e^x$ in Example 193.

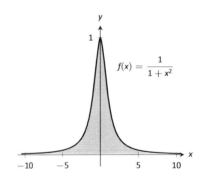

Figure 6.22: A graph of $f(x) = \frac{1}{1+x^2}$ in Example 193.

Notes:

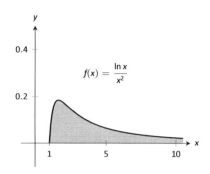

Figure 6.23: A graph of $f(x) = \frac{\ln x}{x^2}$ in Example 194.

The previous section introduced l'Hôpital's Rule, a method of evaluating limits that return indeterminate forms. It is not uncommon for the limits resulting from improper integrals to need this rule as demonstrated next.

Example 194   **Improper integration and l'Hôpital's Rule**
Evaluate the improper integral $\displaystyle\int_1^\infty \frac{\ln x}{x^2}\, dx$.

SOLUTION   This integral will require the use of Integration by Parts. Let $u = \ln x$ and $dv = 1/x^2\, dx$. Then

$$\int_1^\infty \frac{\ln x}{x^2}\, dx = \lim_{b\to\infty} \int_1^b \frac{\ln x}{x^2}\, dx$$

$$= \lim_{b\to\infty} \left(-\frac{\ln x}{x}\Big|_1^b + \int_1^b \frac{1}{x^2}\, dx \right)$$

$$= \lim_{b\to\infty} \left(-\frac{\ln x}{x} - \frac{1}{x} \right)\Big|_1^b$$

$$= \lim_{b\to\infty} \left(-\frac{\ln b}{b} - \frac{1}{b} - (-\ln 1 - 1) \right).$$

The $1/b$ and $\ln 1$ terms go to 0, leaving $\displaystyle\lim_{b\to\infty} -\frac{\ln b}{b} + 1$. We need to evaluate $\displaystyle\lim_{b\to\infty} \frac{\ln b}{b}$ with l'Hôpital's Rule. We have:

$$\lim_{b\to\infty} \frac{\ln b}{b} \stackrel{\text{by LHR}}{=} \lim_{b\to\infty} \frac{1/b}{1}$$
$$= 0.$$

Thus the improper integral evaluates as:

$$\int_1^\infty \frac{\ln x}{x^2}\, dx = 1.$$

Improper Integrals with Infinite Range

We have just considered definite integrals where the interval of integration was infinite. We now consider another type of improper integration, where the range of the integrand is infinite.

Notes:

6.8 Improper Integration

Definition 25   **Improper Integration with Infinite Range**

Let $f(x)$ be a continuous function on $[a, b]$ except at c, $a \leq c \leq b$, where $x = c$ is a vertical asymptote of f. Define

$$\int_a^b f(x)\, dx = \lim_{t \to c^-} \int_a^t f(x)\, dx + \lim_{t \to c^+} \int_t^b f(x)\, dx.$$

Note: In Definition 25, c can be one of the endpoints (a or b). In that case, there is only one limit to consider as part of the definition.

Example 195   **Improper integration of functions with infinite range**
Evaluate the following improper integrals:

1. $\displaystyle\int_0^1 \frac{1}{\sqrt{x}}\, dx$ 2. $\displaystyle\int_{-1}^1 \frac{1}{x^2}\, dx.$

SOLUTION

1. A graph of $f(x) = 1/\sqrt{x}$ is given in Figure 6.24. Notice that f has a vertical asymptote at $x = 0$; in some sense, we are trying to compute the area of a region that has no "top." Could this have a finite value?

$$\int_0^1 \frac{1}{\sqrt{x}}\, dx = \lim_{a \to 0^+} \int_a^1 \frac{1}{\sqrt{x}}\, dx$$
$$= \lim_{a \to 0^+} 2\sqrt{x}\Big|_a^1$$
$$= \lim_{a \to 0^+} 2\left(\sqrt{1} - \sqrt{a}\right)$$
$$= 2.$$

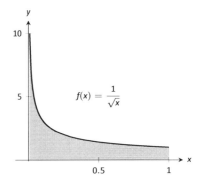

Figure 6.24: A graph of $f(x) = \frac{1}{\sqrt{x}}$ in Example 195.

It turns out that the region does have a finite area even though it has no upper bound (strange things can occur in mathematics when considering the infinite).

2. The function $f(x) = 1/x^2$ has a vertical asymptote at $x = 0$, as shown in Figure 6.25, so this integral is an improper integral. Let's eschew using limits for a moment and proceed without recognizing the improper nature of the integral. This leads to:

$$\int_{-1}^1 \frac{1}{x^2}\, dx = -\frac{1}{x}\Big|_{-1}^1$$
$$= -1 - (1)$$
$$= -2!$$

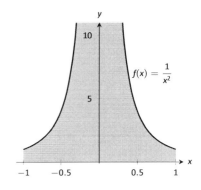

Figure 6.25: A graph of $f(x) = \frac{1}{x^2}$ in Example 195.

Notes:

Chapter 6 Techniques of Antidifferentiation

Clearly the area in question is above the x-axis, yet the area is supposedly negative! Why does our answer not match our intuition? To answer this, evaluate the integral using Definition 25.

$$\int_{-1}^{1} \frac{1}{x^2}\, dx = \lim_{t \to 0^-} \int_{-1}^{t} \frac{1}{x^2}\, dx + \lim_{t \to 0^+} \int_{t}^{1} \frac{1}{x^2}\, dx$$

$$= \lim_{t \to 0^-} -\frac{1}{x}\Big|_{-1}^{t} + \lim_{t \to 0^+} -\frac{1}{x}\Big|_{t}^{1}$$

$$= \lim_{t \to 0^-} -\frac{1}{t} - 1 + \lim_{t \to 0^+} -1 + \frac{1}{t}$$

$$\Rightarrow \big(\infty - 1\big) + \big(-1 + \infty\big).$$

Neither limit converges hence the original improper integral diverges. The nonsensical answer we obtained by ignoring the improper nature of the integral is just that: nonsensical.

Understanding Convergence and Divergence

Oftentimes we are interested in knowing simply whether or not an improper integral converges, and not necessarily the value of a convergent integral. We provide here several tools that help determine the convergence or divergence of improper integrals without integrating.

Our first tool is to understand the behavior of functions of the form $\frac{1}{x^p}$.

Example 196 **Improper integration of** $1/x^p$

Determine the values of p for which $\displaystyle\int_{1}^{\infty} \frac{1}{x^p}\, dx$ converges.

SOLUTION We begin by integrating and then evaluating the limit.

$$\int_{1}^{\infty} \frac{1}{x^p}\, dx = \lim_{b \to \infty} \int_{1}^{b} \frac{1}{x^p}\, dx$$

$$= \lim_{b \to \infty} \int_{1}^{b} x^{-p}\, dx \qquad \text{(assume } p \neq 1\text{)}$$

$$= \lim_{b \to \infty} \frac{1}{-p+1} x^{-p+1}\Big|_{1}^{b}$$

$$= \lim_{b \to \infty} \frac{1}{1-p}\big(b^{1-p} - 1^{1-p}\big).$$

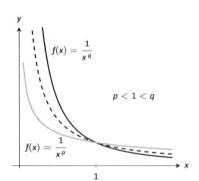

Figure 6.26: Plotting functions of the form $1/x^p$ in Example 196.

When does this limit converge – i.e., when is this limit *not* ∞? This limit converges precisely when the power of b is less than 0: when $1 - p < 0 \Rightarrow 1 < p$.

Notes:

Our analysis shows that if $p > 1$, then $\int_1^\infty \frac{1}{x^p}\,dx$ converges. When $p < 1$ the improper integral diverges; we showed in Example 193 that when $p = 1$ the integral also diverges.

Figure 6.26 graphs $y = 1/x$ with a dashed line, along with graphs of $y = 1/x^p$, $p < 1$, and $y = 1/x^q$, $q > 1$. Somehow the dashed line forms a dividing line between convergence and divergence.

The result of Example 196 provides an important tool in determining the convergence of other integrals. A similar result is proved in the exercises about improper integrals of the form $\int_0^1 \frac{1}{x^p}\,dx$. These results are summarized in the following Key Idea.

Key Idea 21   **Convergence of Improper Integrals** $\int_1^\infty \frac{1}{x^p}\,dx$ **and** $\int_0^1 \frac{1}{x^p}\,dx$.

1. The improper integral $\int_1^\infty \frac{1}{x^p}\,dx$ converges when $p > 1$ and diverges when $p \leq 1$.

2. The improper integral $\int_0^1 \frac{1}{x^p}\,dx$ converges when $p < 1$ and diverges when $p \geq 1$.

A basic technique in determining convergence of improper integrals is to compare an integrand whose convergence is unknown to an integrand whose convergence is known. We often use integrands of the form $1/x^p$ to compare to as their convergence on certain intervals is known. This is described in the following theorem.

Note: We used the upper and lower bound of "1" in Key Idea 21 for convenience. It can be replaced by any a where $a > 0$.

Theorem 51   **Direct Comparison Test for Improper Integrals**

Let f and g be continuous on $[a, \infty)$ where $0 \leq f(x) \leq g(x)$ for all x in $[a, \infty)$.

1. If $\int_a^\infty g(x)\,dx$ converges, then $\int_a^\infty f(x)\,dx$ converges.

2. If $\int_a^\infty f(x)\,dx$ diverges, then $\int_a^\infty g(x)\,dx$ diverges.

Notes:

Chapter 6 Techniques of Antidifferentiation

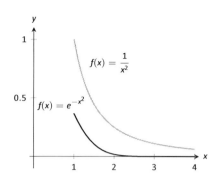

Figure 6.27: Graphs of $f(x) = e^{-x^2}$ and $f(x) = 1/x^2$ in Example 197.

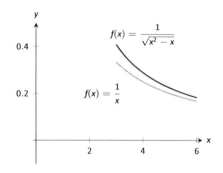

Figure 6.28: Graphs of $f(x) = 1/\sqrt{x^2 - x}$ and $f(x) = 1/x$ in Example 197.

Example 197   **Determining convergence of improper integrals**

Determine the convergence of the following improper integrals.

1. $\displaystyle\int_1^\infty e^{-x^2}\,dx$ 2. $\displaystyle\int_3^\infty \frac{1}{\sqrt{x^2 - x}}\,dx$

SOLUTION

1. The function $f(x) = e^{-x^2}$ does not have an antiderivative expressible in terms of elementary functions, so we cannot integrate directly. It is comparable to $g(x) = 1/x^2$, and as demonstrated in Figure 6.27, $e^{-x^2} < 1/x^2$ on $[1, \infty)$. We know from Key Idea 21 that $\displaystyle\int_1^\infty \frac{1}{x^2}\,dx$ converges, hence $\displaystyle\int_1^\infty e^{-x^2}\,dx$ also converges.

2. Note that for large values of x, $\dfrac{1}{\sqrt{x^2 - x}} \approx \dfrac{1}{\sqrt{x^2}} = \dfrac{1}{x}$. We know from Key Idea 21 and the subsequent note that $\displaystyle\int_3^\infty \frac{1}{x}\,dx$ diverges, so we seek to compare the original integrand to $1/x$.

   It is easy to see that when $x > 0$, we have $x = \sqrt{x^2} > \sqrt{x^2 - x}$. Taking reciprocals reverses the inequality, giving

   $$\frac{1}{x} < \frac{1}{\sqrt{x^2 - x}}.$$

   Using Theorem 51, we conclude that since $\displaystyle\int_3^\infty \frac{1}{x}\,dx$ diverges, $\displaystyle\int_3^\infty \frac{1}{\sqrt{x^2 - x}}\,dx$ diverges as well. Figure 6.28 illustrates this.

Being able to compare "unknown" integrals to "known" integrals is very useful in determining convergence. However, some of our examples were a little "too nice." For instance, it was convenient that $\dfrac{1}{x} < \dfrac{1}{\sqrt{x^2 - x}}$, but what if the "$-x$" were replaced with a "$+2x+5$"? That is, what can we say about the convergence of $\displaystyle\int_3^\infty \frac{1}{\sqrt{x^2 + 2x + 5}}\,dx$? We have $\dfrac{1}{x} > \dfrac{1}{\sqrt{x^2 + 2x + 5}}$, so we cannot use Theorem 51.

In cases like this (and many more) it is useful to employ the following theorem.

Notes:

6.8 Improper Integration

Theorem 52   **Limit Comparison Test for Improper Integrals**

Let f and g be continuous functions on $[a, \infty)$ where $f(x) > 0$ and $g(x) > 0$ for all x. If
$$\lim_{x \to \infty} \frac{f(x)}{g(x)} = L, \qquad 0 < L < \infty,$$
then
$$\int_a^\infty f(x)\, dx \quad \text{and} \quad \int_a^\infty g(x)\, dx$$
either both converge or both diverge.

Example 198   **Determining convergence of improper integrals**

Determine the convergence of $\displaystyle\int_3^\infty \frac{1}{\sqrt{x^2 + 2x + 5}}\, dx$.

SOLUTION   As x gets large, the quadratic inside the square root function will begin to behave much like $y = x$. So we compare $\dfrac{1}{\sqrt{x^2 + 2x + 5}}$ to $\dfrac{1}{x}$ with the Limit Comparison Test:
$$\lim_{x \to \infty} \frac{1/\sqrt{x^2 + 2x + 5}}{1/x} = \lim_{x \to \infty} \frac{x}{\sqrt{x^2 + 2x + 5}}.$$

The immediate evaluation of this limit returns ∞/∞, an indeterminate form. Using l'Hôpital's Rule seems appropriate, but in this situation, it does not lead to useful results. (We encourage the reader to employ l'Hôpital's Rule at least once to verify this.)

The trouble is the square root function. To get rid of it, we employ the following fact: If $\lim_{x \to c} f(x) = L$, then $\lim_{x \to c} f(x)^2 = L^2$. (This is true when either c or L is ∞.) So we consider now the limit
$$\lim_{x \to \infty} \frac{x^2}{x^2 + 2x + 5}.$$

This converges to 1, meaning the original limit also converged to 1. As x gets very large, the function $\dfrac{1}{\sqrt{x^2 + 2x + 5}}$ looks very much like $\dfrac{1}{x}$. Since we know that $\displaystyle\int_3^\infty \frac{1}{x}\, dx$ diverges, by the Limit Comparison Test we know that $\displaystyle\int_3^\infty \frac{1}{\sqrt{x^2 + 2x + 5}}\, dx$ also diverges. Figure 6.29 graphs $f(x) = 1/\sqrt{x^2 + 2x + 5}$ and $f(x) = 1/x$, illustrating that as x gets large, the functions become indistinguishable.

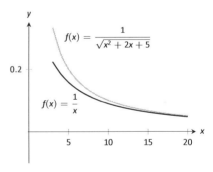

Figure 6.29: Graphing $f(x) = \dfrac{1}{\sqrt{x^2 + 2x + 5}}$ and $f(x) = \dfrac{1}{x}$ in Example 198.

Notes:

Both the Direct and Limit Comparison Tests were given in terms of integrals over an infinite interval. There are versions that apply to improper integrals with an infinite range, but as they are a bit wordy and a little more difficult to employ, they are omitted from this text.

This chapter has explored many integration techniques. We learned Substitution, which "undoes" the Chain Rule of differentiation, as well as Integration by Parts, which "undoes" the Product Rule. We learned specialized techniques for handling trigonometric functions and introduced the hyperbolic functions, which are closely related to the trigonometric functions. All techniques effectively have this goal in common: rewrite the integrand in a new way so that the integration step is easier to see and implement.

As stated before, integration is, in general, hard. It is easy to write a function whose antiderivative is impossible to write in terms of elementary functions, and even when a function does have an antiderivative expressible by elementary functions, it may be really hard to discover what it is. The powerful computer algebra system *Mathematica*® has approximately 1,000 pages of code dedicated to integration.

Do not let this difficulty discourage you. There is great value in learning integration techniques, as they allow one to manipulate an integral in ways that can illuminate a concept for greater understanding. There is also great value in understanding the need for good numerical techniques: the Trapezoidal and Simpson's Rules are just the beginning of powerful techniques for approximating the value of integration.

The next chapter stresses the uses of integration. We generally do not find antiderivatives for antiderivative's sake, but rather because they provide the solution to some type of problem. The following chapter introduces us to a number of different problems whose solution is provided by integration.

Notes:

Exercises 6.8

Terms and Concepts

1. The definite integral was defined with what two stipulations?

2. If $\lim\limits_{b \to \infty} \int_0^b f(x)\, dx$ exists, then the integral $\int_0^\infty f(x)\, dx$ is said to _____.

3. If $\int_1^\infty f(x)\, dx = 10$, and $0 \leq g(x) \leq f(x)$ for all x, then we know that $\int_1^\infty g(x)\, dx$ _____.

4. For what values of p will $\int_1^\infty \frac{1}{x^p}\, dx$ converge?

5. For what values of p will $\int_{10}^\infty \frac{1}{x^p}\, dx$ converge?

6. For what values of p will $\int_0^1 \frac{1}{x^p}\, dx$ converge?

Problems

In Exercises 7 – 33, evaluate the given improper integral.

7. $\int_0^\infty e^{5-2x}\, dx$

8. $\int_1^\infty \frac{1}{x^3}\, dx$

9. $\int_1^\infty x^{-4}\, dx$

10. $\int_{-\infty}^\infty \frac{1}{x^2+9}\, dx$

11. $\int_{-\infty}^0 2^x\, dx$

12. $\int_{-\infty}^0 \left(\frac{1}{2}\right)^x\, dx$

13. $\int_{-\infty}^\infty \frac{x}{x^2+1}\, dx$

14. $\int_{-\infty}^\infty \frac{x}{x^2+4}\, dx$

15. $\int_2^\infty \frac{1}{(x-1)^2}\, dx$

16. $\int_1^2 \frac{1}{(x-1)^2}\, dx$

17. $\int_2^\infty \frac{1}{x-1}\, dx$

18. $\int_1^2 \frac{1}{x-1}\, dx$

19. $\int_{-1}^1 \frac{1}{x}\, dx$

20. $\int_1^3 \frac{1}{x-2}\, dx$

21. $\int_0^\pi \sec^2 x\, dx$

22. $\int_{-2}^1 \frac{1}{\sqrt{|x|}}\, dx$

23. $\int_0^\infty xe^{-x}\, dx$

24. $\int_0^\infty xe^{-x^2}\, dx$

25. $\int_{-\infty}^\infty xe^{-x^2}\, dx$

26. $\int_{-\infty}^\infty \frac{1}{e^x + e^{-x}}\, dx$

27. $\int_0^1 x \ln x\, dx$

28. $\int_1^\infty \frac{\ln x}{x}\, dx$

29. $\int_0^1 \ln x\, dx$

30. $\int_1^\infty \frac{\ln x}{x^2}\, dx$

31. $\int_1^\infty \frac{\ln x}{\sqrt{x}}\, dx$

32. $\int_0^\infty e^{-x} \sin x\, dx$

33. $\int_0^\infty e^{-x} \cos x\, dx$

In Exercises 34 – 43, use the Direct Comparison Test or the Limit Comparison Test to determine whether the given definite integral converges or diverges. Clearly state what test is being used and what function the integrand is being compared to.

34. $\displaystyle\int_{10}^{\infty} \frac{3}{\sqrt{3x^2 + 2x - 5}}\, dx$

35. $\displaystyle\int_{2}^{\infty} \frac{4}{\sqrt{7x^3 - x}}\, dx$

36. $\displaystyle\int_{0}^{\infty} \frac{\sqrt{x+3}}{\sqrt{x^3 - x^2 + x + 1}}\, dx$

37. $\displaystyle\int_{1}^{\infty} e^{-x} \ln x\, dx$

38. $\displaystyle\int_{5}^{\infty} e^{-x^2 + 3x + 1}\, dx$

39. $\displaystyle\int_{0}^{\infty} \frac{\sqrt{x}}{e^x}\, dx$

40. $\displaystyle\int_{2}^{\infty} \frac{1}{x^2 + \sin x}\, dx$

41. $\displaystyle\int_{0}^{\infty} \frac{x}{x^2 + \cos x}\, dx$

42. $\displaystyle\int_{0}^{\infty} \frac{1}{x + e^x}\, dx$

43. $\displaystyle\int_{0}^{\infty} \frac{1}{e^x - x}\, dx$

7: Applications of Integration

We begin this chapter with a reminder of a few key concepts from Chapter 5. Let f be a continuous function on $[a, b]$ which is partitioned into n equally spaced subintervals as

$$a < x_1 < x_2 < \cdots < x_n < x_{n+1} = b.$$

Let $\Delta x = (b - a)/n$ denote the length of the subintervals, and let c_i be any x-value in the ith subinterval. Definition 21 states that the sum

$$\sum_{i=1}^{n} f(c_i) \Delta x$$

is a *Riemann Sum*. Riemann Sums are often used to approximate some quantity (area, volume, work, pressure, etc.). The *approximation* becomes *exact* by taking the limit

$$\lim_{n \to \infty} \sum_{i=1}^{n} f(c_i) \Delta x.$$

Theorem 38 connects limits of Riemann Sums to definite integrals:

$$\lim_{n \to \infty} \sum_{i=1}^{n} f(c_i) \Delta x = \int_{a}^{b} f(x)\, dx.$$

Finally, the Fundamental Theorem of Calculus states how definite integrals can be evaluated using antiderivatives.

This chapter employs the following technique to a variety of applications. Suppose the value Q of a quantity is to be calculated. We first approximate the value of Q using a Riemann Sum, then find the exact value via a definite integral. We spell out this technique in the following Key Idea.

Key Idea 22   Application of Definite Integrals Strategy

Let a quantity be given whose value Q is to be computed.

1. Divide the quantity into n smaller "subquantities" of value Q_i.

2. Identify a variable x and function $f(x)$ such that each subquantity can be approximated with the product $f(c_i)\Delta x$, where Δx represents a small change in x. Thus $Q_i \approx f(c_i)\Delta x$. A sample approximation $f(c_i)\Delta x$ of Q_i is called a *differential element*.

3. Recognize that $Q = \sum_{i=1}^{n} Q_i \approx \sum_{i=1}^{n} f(c_i)\Delta x$, which is a Riemann Sum.

4. Taking the appropriate limit gives $Q = \int_{a}^{b} f(x)\, dx$

This Key Idea will make more sense after we have had a chance to use it several times. We begin with Area Between Curves, which we addressed briefly in Section 5.5.4.

Chapter 7 Applications of Integration

7.1 Area Between Curves

We are often interested in knowing the area of a region. Forget momentarily that we addressed this already in Section 5.5.4 and approach it instead using the technique described in Key Idea 22.

Let Q be the area of a region bounded by continuous functions f and g. If we break the region into many subregions, we have an obvious equation:

Total Area = sum of the areas of the subregions.

The issue to address next is how to systematically break a region into subregions. A graph will help. Consider Figure 7.1 (a) where a region between two curves is shaded. While there are many ways to break this into subregions, one particularly efficient way is to "slice" it vertically, as shown in Figure 7.1 (b), into n equally spaced slices.

We now approximate the area of a slice. Again, we have many options, but using a rectangle seems simplest. Picking any x-value c_i in the ith slice, we set the height of the rectangle to be $f(c_i) - g(c_i)$, the difference of the corresponding y-values. The width of the rectangle is a small difference in x-values, which we represent with Δx. Figure 7.1 (c) shows sample points c_i chosen in each subinterval and appropriate rectangles drawn. (Each of these rectangles represents a differential element.) Each slice has an area approximately equal to $\bigl(f(c_i) - g(c_i)\bigr)\Delta x$; hence, the total area is approximately the Riemann Sum

$$Q = \sum_{i=1}^{n} \bigl(f(c_i) - g(c_i)\bigr)\Delta x.$$

Taking the limit as $n \to \infty$ gives the exact area as $\int_a^b \bigl(f(x) - g(x)\bigr)\,dx$.

> **Theorem 53**   **Area Between Curves**   (restatement of Theorem 41)
>
> Let $f(x)$ and $g(x)$ be continuous functions defined on $[a, b]$ where $f(x) \geq g(x)$ for all x in $[a, b]$. The area of the region bounded by the curves $y = f(x)$, $y = g(x)$ and the lines $x = a$ and $x = b$ is
>
> $$\int_a^b \bigl(f(x) - g(x)\bigr)\,dx.$$

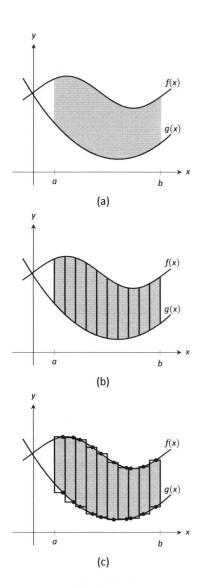

Figure 7.1: Subdividing a region into vertical slices and approximating the areas with rectangles.

Example 199   **Finding area enclosed by curves**
Find the area of the region bounded by $f(x) = \sin x + 2$, $g(x) = \frac{1}{2}\cos(2x) - 1$, $x = 0$ and $x = 4\pi$, as shown in Figure 7.2.

SOLUTION   The graph verifies that the upper boundary of the region is

Notes:

given by f and the lower bound is given by g. Therefore the area of the region is the value of the integral

$$\int_0^{4\pi} \big(f(x) - g(x)\big)\, dx = \int_0^{4\pi} \left(\sin x + 2 - \left(\frac{1}{2}\cos(2x) - 1\right)\right) dx$$
$$= -\cos x - \frac{1}{4}\sin(2x) + 3x \Big|_0^{4\pi}$$
$$= 12\pi \approx 37.7 \text{ units}^2.$$

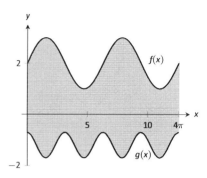

Figure 7.2: Graphing an enclosed region in Example 199.

Example 200   **Finding total area enclosed by curves**
Find the total area of the region enclosed by the functions $f(x) = -2x + 5$ and $g(x) = x^3 - 7x^2 + 12x - 3$ as shown in Figure 7.3.

SOLUTION   A quick calculation shows that $f = g$ at $x = 1, 2$ and 4. One can proceed thoughtlessly by computing $\int_1^4 \big(f(x) - g(x)\big)\, dx$, but this ignores the fact that on $[1, 2]$, $g(x) > f(x)$. (In fact, the thoughtless integration returns $-9/4$, hardly the expected value of an *area*.) Thus we compute the total area by breaking the interval $[1, 4]$ into two subintervals, $[1, 2]$ and $[2, 4]$ and using the proper integrand in each.

$$\text{Total Area} = \int_1^2 \big(g(x) - f(x)\big)\, dx + \int_2^4 \big(f(x) - g(x)\big)\, dx$$
$$= \int_1^2 \big(x^3 - 7x^2 + 14x - 8\big)\, dx + \int_2^4 \big(-x^3 + 7x^2 - 14x + 8\big)\, dx$$
$$= 5/12 + 8/3$$
$$= 37/12 = 3.083 \text{ units}^2.$$

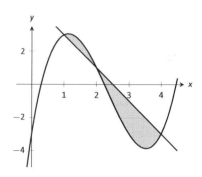

Figure 7.3: Graphing a region enclosed by two functions in Example 200.

The previous example makes note that we are expecting area to be *positive*. When first learning about the definite integral, we interpreted it as "signed area under the curve," allowing for "negative area." That doesn't apply here; area is to be positive.

The previous example also demonstrates that we often have to break a given region into subregions before applying Theorem 53. The following example shows another situation where this is applicable, along with an alternate view of applying the Theorem.

Example 201   **Finding area: integrating with respect to y**
Find the area of the region enclosed by the functions $y = \sqrt{x} + 2$, $y = -(x - 1)^2 + 3$ and $y = 2$, as shown in Figure 7.4.

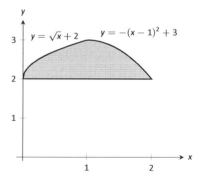

Figure 7.4: Graphing a region for Example 201.

Notes:

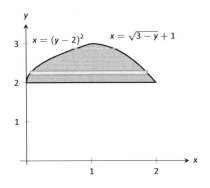

Figure 7.5: The region used in Example 201 with boundaries relabeled as functions of y.

Solution We give two approaches to this problem. In the first approach, we notice that the region's "top" is defined by two different curves. On $[0, 1]$, the top function is $y = \sqrt{x} + 2$; on $[1, 2]$, the top function is $y = -(x-1)^2 + 3$. Thus we compute the area as the sum of two integrals:

$$\text{Total Area} = \int_0^1 \left((\sqrt{x} + 2) - 2\right) dx + \int_1^2 \left((-(x-1)^2 + 3) - 2\right) dx$$
$$= 2/3 + 2/3$$
$$= 4/3.$$

The second approach is clever and very useful in certain situations. We are used to viewing curves as functions of x; we input an x-value and a y-value is returned. Some curves can also be described as functions of y: input a y-value and an x-value is returned. We can rewrite the equations describing the boundary by solving for x:

$$y = \sqrt{x} + 2 \quad \Rightarrow \quad x = (y-2)^2$$
$$y = -(x-1)^2 + 3 \quad \Rightarrow \quad x = \sqrt{3-y} + 1.$$

Figure 7.5 shows the region with the boundaries relabeled. A differential element, a horizontal rectangle, is also pictured. The width of the rectangle is a small change in y: Δy. The height of the rectangle is a difference in x-values. The "top" x-value is the largest value, i.e., the rightmost. The "bottom" x-value is the smaller, i.e., the leftmost. Therefore the height of the rectangle is

$$(\sqrt{3-y} + 1) - (y-2)^2.$$

The area is found by integrating the above function with respect to y with the appropriate bounds. We determine these by considering the y-values the region occupies. It is bounded below by $y = 2$, and bounded above by $y = 3$. That is, both the "top" and "bottom" functions exist on the y interval $[2, 3]$. Thus

$$\text{Total Area} = \int_2^3 \left(\sqrt{3-y} + 1 - (y-2)^2\right) dy$$
$$= \left(-\frac{2}{3}(3-y)^{3/2} + y - \frac{1}{3}(y-2)^3\right)\Big|_2^3$$
$$= 4/3.$$

This calculus–based technique of finding area can be useful even with shapes that we normally think of as "easy." Example 202 computes the area of a triangle. While the formula "$\frac{1}{2} \times$ base $\times$ height" is well known, in arbitrary triangles it can be nontrivial to compute the height. Calculus makes the problem simple.

Notes:

7.1 Area Between Curves

Example 202   **Finding the area of a triangle**
Compute the area of the regions bounded by the lines
$y = x + 1$, $y = -2x + 7$ and $y = -\frac{1}{2}x + \frac{5}{2}$, as shown in Figure 7.6.

SOLUTION   Recognize that there are two "top" functions to this region, causing us to use two definite integrals.

$$\text{Total Area} = \int_1^2 \left((x+1) - \left(-\frac{1}{2}x + \frac{5}{2}\right)\right) dx + \int_2^3 \left((-2x+7) - \left(-\frac{1}{2}x + \frac{5}{2}\right)\right) dx$$
$$= 3/4 + 3/4$$
$$= 3/2.$$

We can also approach this by converting each function into a function of y. This also requires 2 integrals, so there isn't really any advantage to doing so. We do it here for demonstration purposes.

The "top" function is always $x = \frac{7-y}{2}$ while there are two "bottom" functions. Being mindful of the proper integration bounds, we have

$$\text{Total Area} = \int_1^2 \left(\frac{7-y}{2} - (5 - 2y)\right) dy + \int_2^3 \left(\frac{7-y}{2} - (y - 1)\right) dy$$
$$= 3/4 + 3/4$$
$$= 3/2.$$

Of course, the final answer is the same. (It is interesting to note that the area of all 4 subregions used is 3/4. This is coincidental.)

While we have focused on producing exact answers, we are also able to make approximations using the principle of Theorem 53. The integrand in the theorem is a distance ("top minus bottom"); integrating this distance function gives an area. By taking discrete measurements of distance, we can approximate an area using numerical integration techniques developed in Section 5.5. The following example demonstrates this.

Example 203   **Numerically approximating area**
To approximate the area of a lake, shown in Figure 7.7 (a), the "length" of the lake is measured at 200-foot increments as shown in Figure 7.7 (b), where the lengths are given in hundreds of feet. Approximate the area of the lake.

SOLUTION   The measurements of length can be viewed as measuring "top minus bottom" of two functions. The exact answer is found by integrating $\int_0^{12} \bigl(f(x) - g(x)\bigr) dx$, but of course we don't know the functions f and g. Our discrete measurements instead allow us to approximate.

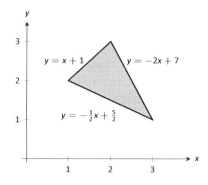

Figure 7.6: Graphing a triangular region in Example 202.

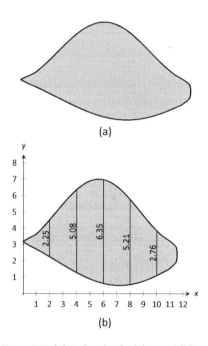

Figure 7.7: (a) A sketch of a lake, and (b) the lake with length measurements.

We have the following data points:

$$(0, 0),\ (2, 2.25),\ (4, 5.08),\ (6, 6.35),\ (8, 5.21),\ (10, 2.76),\ (12, 0).$$

We also have that $\Delta x = \frac{b-a}{n} = 2$, so Simpson's Rule gives

$$\text{Area} \approx \frac{2}{3}\Big(1 \cdot 0 + 4 \cdot 2.25 + 2 \cdot 5.08 + 4 \cdot 6.35 + 2 \cdot 5.21 + 4 \cdot 2.76 + 1 \cdot 0\Big)$$
$$= 44.01\overline{3} \text{ units}^2.$$

Since the measurements are in hundreds of feet, units^2 = (100 ft)^2 = 10,000 ft^2, giving a total area of 440,133 ft^2. (Since we are approximating, we'd likely say the area was about 440,000 ft^2, which is a little more than 10 acres.)

In the next section we apply our applications–of–integration techniques to finding the volumes of certain solids.

Notes:

Exercises 7.1

Terms and Concepts

1. T/F: The area between curves is always positive.

2. T/F: Calculus can be used to find the area of basic geometric shapes.

3. In your own words, describe how to find the total area enclosed by $y = f(x)$ and $y = g(x)$.

Problems

In Exercises 4 – 10, find the area of the shaded region in the given graph.

4.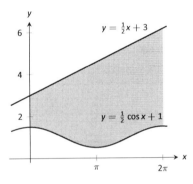

5.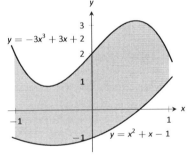

6.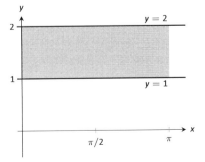

7.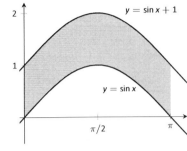

8.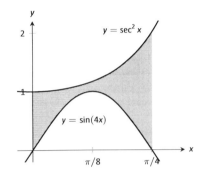

9.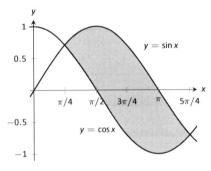

10.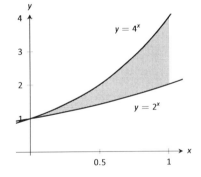

In Exercises 11 – 16, find the total area enclosed by the functions f and g.

11. $f(x) = 2x^2 + 5x - 3, g(x) = x^2 + 4x - 1$

12. $f(x) = x^2 - 3x + 2, g(x) = -3x + 3$

13. $f(x) = \sin x, g(x) = 2x/\pi$

14. $f(x) = x^3 - 4x^2 + x - 1, g(x) = -x^2 + 2x - 4$

15. $f(x) = x$, $g(x) = \sqrt{x}$

16. $f(x) = -x^3 + 5x^2 + 2x + 1$, $g(x) = 3x^2 + x + 3$

17. The functions $f(x) = \cos(2x)$ and $g(x) = \sin x$ intersect infinitely many times, forming an infinite number of repeated, enclosed regions. Find the areas of these regions.

In Exercises 18 – 22, find the area of the enclosed region in two ways:
1. **by treating the boundaries as functions of x, and**
2. **by treating the boundaries as functions of y.**

18.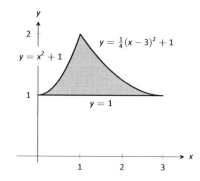

19.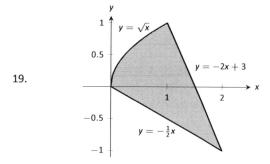

20.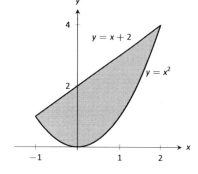

21.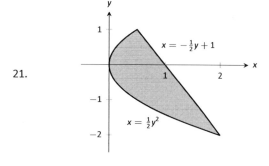

22.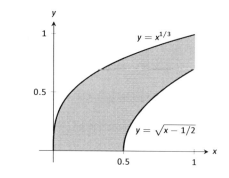

In Exercises 23 – 26, find the area triangle formed by the given three points.

23. $(1, 1)$, $(2, 3)$, and $(3, 3)$

24. $(-1, 1)$, $(1, 3)$, and $(2, -1)$

25. $(1, 1)$, $(3, 3)$, and $(3, 3)$

26. $(0, 0)$, $(2, 5)$, and $(5, 2)$

27. Use the Trapezoidal Rule to approximate the area of the pictured lake whose lengths, in hundreds of feet, are measured in 100-foot increments.

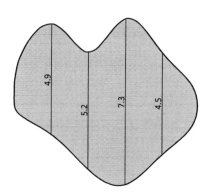

28. Use Simpson's Rule to approximate the area of the pictured lake whose lengths, in hundreds of feet, are measured in 200-foot increments.

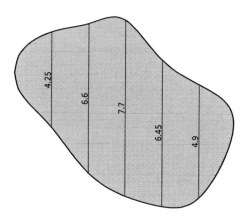

7.2 Volume by Cross-Sectional Area; Disk and Washer Methods

The volume of a general right cylinder, as shown in Figure 7.8, is
$$\text{Area of the base} \times \text{height}.$$
We can use this fact as the building block in finding volumes of a variety of shapes.

Given an arbitrary solid, we can *approximate* its volume by cutting it into n thin slices. When the slices are thin, each slice can be approximated well by a general right cylinder. Thus the volume of each slice is approximately its cross-sectional area × thickness. (These slices are the differential elements.)

By orienting a solid along the x-axis, we can let $A(x_i)$ represent the cross-sectional area of the ith slice, and let Δx_i represent the thickness of this slice (the thickness is a small change in x). The total volume of the solid is approximately:

$$\text{Volume} \approx \sum_{i=1}^{n} \Big[\text{Area} \times \text{thickness}\Big]$$
$$= \sum_{i=1}^{n} A(x_i)\Delta x_i.$$

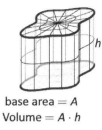

base area = A
Volume = $A \cdot h$

Figure 7.8: The volume of a general right cylinder

Recognize that this is a Riemann Sum. By taking a limit (as the thickness of the slices goes to 0) we can find the volume exactly.

Theorem 54   **Volume By Cross-Sectional Area**

The volume V of a solid, oriented along the x-axis with cross-sectional area $A(x)$ from $x = a$ to $x = b$, is

$$V = \int_a^b A(x)\, dx.$$

Example 204   **Finding the volume of a solid**
Find the volume of a pyramid with a square base of side length 10 in and a height of 5 in.

SOLUTION   There are many ways to "orient" the pyramid along the x-axis; Figure 7.9 gives one such way, with the pointed top of the pyramid at the origin and the x-axis going through the center of the base.

Each cross section of the pyramid is a square; this is a sample differential element. To determine its area $A(x)$, we need to determine the side lengths of

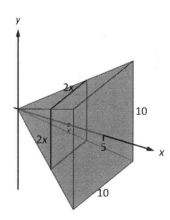

Figure 7.9: Orienting a pyramid along the x-axis in Example 204.

Notes:

the square.

When $x = 5$, the square has side length 10; when $x = 0$, the square has side length 0. Since the edges of the pyramid are lines, it is easy to figure that each cross-sectional square has side length $2x$, giving $A(x) = (2x)^2 = 4x^2$.

If one were to cut a slice out of the pyramid at $x = 3$, as shown in Figure 7.10, one would have a shape with square bottom and top with sloped sides. If the slice were thin, both the bottom and top squares would have sides lengths of about 6, and thus the cross–sectional area of the bottom and top would be about 36in^2. Letting Δx_i represent the thickness of the slice, the volume of this slice would then be about $36\Delta x_i$in^3.

Cutting the pyramid into n slices divides the total volume into n equally–spaced smaller pieces, each with volume $(2x_i)^2 \Delta x$, where x_i is the approximate location of the slice along the x-axis and Δx represents the thickness of each slice. One can approximate total volume of the pyramid by summing up the volumes of these slices:

$$\text{Approximate volume} = \sum_{i=1}^{n}(2x_i)^2 \Delta x.$$

Taking the limit as $n \to \infty$ gives the actual volume of the pyramid; recoginizing this sum as a Riemann Sum allows us to find the exact answer using a definite integral, matching the definite integral given by Theorem 54.

We have

$$V = \lim_{n\to\infty} \sum_{i=1}^{n}(2x_i)^2 \Delta x$$
$$= \int_0^5 4x^2\, dx$$
$$= \frac{4}{3}x^3 \Big|_0^5$$
$$= \frac{500}{3} \text{ in}^3 \approx 166.67 \text{ in}^3.$$

We can check our work by consulting the general equation for the volume of a pyramid (see the back cover under "Volume of A General Cone"):
$\frac{1}{3} \times$ area of base $\times$ height.
Certainly, using this formula from geometry is faster than our new method, but the calculus–based method can be applied to much more than just cones.

An important special case of Theorem 54 is when the solid is a **solid of revolution**, that is, when the solid is formed by rotating a shape around an axis.

Start with a function $y = f(x)$ from $x = a$ to $x = b$. Revolving this curve about a horizontal axis creates a three-dimensional solid whose cross sections

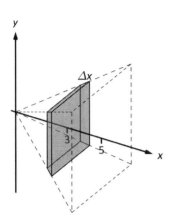

Figure 7.10: Cutting a slice in they pyramid in Example 204 at $x = 3$.

Notes:

7.2 Volume by Cross-Sectional Area; Disk and Washer Methods

are disks (thin circles). Let $R(x)$ represent the radius of the cross-sectional disk at x; the area of this disk is $\pi R(x)^2$. Applying Theorem 54 gives the Disk Method.

Key Idea 23   **The Disk Method**

Let a solid be formed by revolving the curve $y = f(x)$ from $x = a$ to $x = b$ around a horizontal axis, and let $R(x)$ be the radius of the cross-sectional disk at x. The volume of the solid is

$$V = \pi \int_a^b R(x)^2 \, dx.$$

Example 205   **Finding volume using the Disk Method**
Find the volume of the solid formed by revolving the curve $y = 1/x$, from $x = 1$ to $x = 2$, around the x-axis.

Solution   A sketch can help us understand this problem. In Figure 7.11(a) the curve $y = 1/x$ is sketched along with the differential element – a disk – at x with radius $R(x) = 1/x$. In Figure 7.11 (b) the whole solid is pictured, along with the differential element.

The volume of the differential element shown in part (a) of the figure is approximately $\pi R(x_i)^2 \Delta x$, where $R(x_i)$ is the radius of the disk shown and Δx is the thickness of that slice. The radius $R(x_i)$ is the distance from the x-axis to the curve, hence $R(x_i) = 1/x_i$.

Slicing the solid into n equally–spaced slices, we can approximate the total volume by adding up the approximate volume of each slice:

$$\text{Approximate volume} = \sum_{i=1}^{n} \pi \left(\frac{1}{x_i}\right)^2 \Delta x.$$

Taking the limit of the above sum as $n \to \infty$ gives the actual volume; recognizing this sum as a Riemann sum allows us to evaluate the limit with a definite integral, which matches the formula given in Key Idea 23:

$$V = \lim_{n \to \infty} \sum_{i=1}^{n} \pi \left(\frac{1}{x_i}\right)^2 \Delta x$$

$$= \pi \int_1^2 \left(\frac{1}{x}\right)^2 dx$$

$$= \pi \int_1^2 \frac{1}{x^2} dx$$

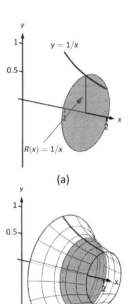

Figure 7.11: Sketching a solid in Example 205.

Notes:

Chapter 7 Applications of Integration

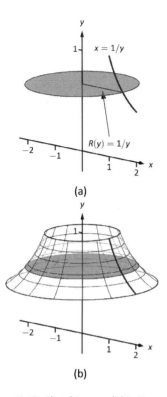

(a)

(b)

Figure 7.12: Sketching a solid in Example 206.

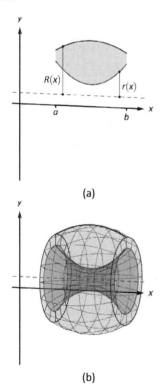

(a)

(b)

Figure 7.13: Establishing the Washer Method; see also Figure 7.14.

$$= \pi \left[-\frac{1}{x}\right]\Big|_1^2$$
$$= \pi \left[-\frac{1}{2} - (-1)\right]$$
$$= \frac{\pi}{2} \text{ units}^3.$$

While Key Idea 23 is given in terms of functions of x, the principle involved can be applied to functions of y when the axis of rotation is vertical, not horizontal. We demonstrate this in the next example.

Example 206 Finding volume using the Disk Method
Find the volume of the solid formed by revolving the curve $y = 1/x$, from $x = 1$ to $x = 2$, about the y-axis.

SOLUTION Since the axis of rotation is vertical, we need to convert the function into a function of y and convert the x-bounds to y-bounds. Since $y = 1/x$ defines the curve, we rewrite it as $x = 1/y$. The bound $x = 1$ corresponds to the y-bound $y = 1$, and the bound $x = 2$ corresponds to the y-bound $y = 1/2$.

Thus we are rotating the curve $x = 1/y$, from $y = 1/2$ to $y = 1$ about the y-axis to form a solid. The curve and sample differential element are sketched in Figure 7.12 (a), with a full sketch of the solid in Figure 7.12 (b). We integrate to find the volume:

$$V = \pi \int_{1/2}^{1} \frac{1}{y^2}\, dy$$
$$= -\frac{\pi}{y}\Big|_{1/2}^{1}$$
$$= \pi \text{ units}^3.$$

We can also compute the volume of solids of revolution that have a hole in the center. The general principle is simple: compute the volume of the solid irrespective of the hole, then subtract the volume of the hole. If the outside radius of the solid is $R(x)$ and the inside radius (defining the hole) is $r(x)$, then the volume is

$$V = \pi \int_a^b R(x)^2\, dx - \pi \int_a^b r(x)^2\, dx = \pi \int_a^b \left(R(x)^2 - r(x)^2\right)\, dx.$$

One can generate a solid of revolution with a hole in the middle by revolving a region about an axis. Consider Figure 7.13(a), where a region is sketched along

Notes:

7.2 Volume by Cross-Sectional Area; Disk and Washer Methods

with a dashed, horizontal axis of rotation. By rotating the region about the axis, a solid is formed as sketched in Figure 7.13(b). The outside of the solid has radius $R(x)$, whereas the inside has radius $r(x)$. Each cross section of this solid will be a washer (a disk with a hole in the center) as sketched in Figure 7.14(c). This leads us to the Washer Method.

Key Idea 24   **The Washer Method**

Let a region bounded by $y = f(x)$, $y = g(x)$, $x = a$ and $x = b$ be rotated about a horizontal axis that does not intersect the region, forming a solid. Each cross section at x will be a washer with outside radius $R(x)$ and inside radius $r(x)$. The volume of the solid is

$$V = \pi \int_a^b \left(R(x)^2 - r(x)^2 \right) dx.$$

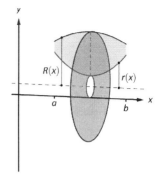

Figure 7.14: Establishing the Washer Method; see also Figure 7.13.

Even though we introduced it first, the Disk Method is just a special case of the Washer Method with an inside radius of $r(x) = 0$.

Example 207   **Finding volume with the Washer Method**
Find the volume of the solid formed by rotating the region bounded by $y = x^2 - 2x + 2$ and $y = 2x - 1$ about the x-axis.

SOLUTION   A sketch of the region will help, as given in Figure 7.15(a). Rotating about the x-axis will produce cross sections in the shape of washers, as shown in Figure 7.15(b); the complete solid is shown in part (c). The outside radius of this washer is $R(x) = 2x + 1$; the inside radius is $r(x) = x^2 - 2x + 2$. As the region is bounded from $x = 1$ to $x = 3$, we integrate as follows to compute the volume.

$$V = \pi \int_1^3 \left((2x-1)^2 - (x^2 - 2x + 2)^2 \right) dx$$

$$= \pi \int_1^3 \left(-x^4 + 4x^3 - 4x^2 + 4x - 3 \right) dx$$

$$= \pi \left[-\frac{1}{5}x^5 + x^4 - \frac{4}{3}x^3 + 2x^2 - 3x \right]\Big|_1^3$$

$$= \frac{104}{15}\pi \approx 21.78 \text{ units}^3.$$

When rotating about a vertical axis, the outside and inside radius functions must be functions of y.

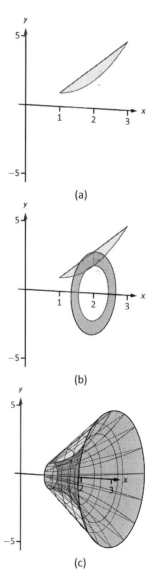

Figure 7.15: Sketching the differential element and solid in Example 207.

Notes:

Chapter 7 Applications of Integration

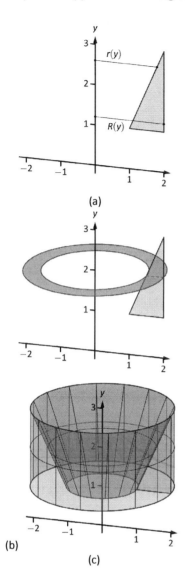

Figure 7.16: Sketching the solid in Example 208.

Example 208   **Finding volume with the Washer Method**
Find the volume of the solid formed by rotating the triangular region with vertices at $(1,1)$, $(2,1)$ and $(2,3)$ about the y-axis.

SOLUTION   The triangular region is sketched in Figure 7.16(a); the differential element is sketched in (b) and the full solid is drawn in (c). They help us establish the outside and inside radii. Since the axis of rotation is vertical, each radius is a function of y.

The outside radius $R(y)$ is formed by the line connecting $(2,1)$ and $(2,3)$; it is a constant function, as regardless of the y-value the distance from the line to the axis of rotation is 2. Thus $R(y) = 2$.

The inside radius is formed by the line connecting $(1,1)$ and $(2,3)$. The equation of this line is $y = 2x - 1$, but we need to refer to it as a function of y. Solving for x gives $r(y) = \frac{1}{2}(y+1)$.

We integrate over the y-bounds of $y = 1$ to $y = 3$. Thus the volume is

$$V = \pi \int_1^3 \left(2^2 - \left(\frac{1}{2}(y+1)\right)^2\right) dy$$
$$= \pi \int_1^3 \left(-\frac{1}{4}y^2 - \frac{1}{2}y + \frac{15}{4}\right) dy$$
$$= \pi \left[-\frac{1}{12}y^3 - \frac{1}{4}y^2 + \frac{15}{4}y\right]\Big|_1^3$$
$$= \frac{10}{3}\pi \approx 10.47 \text{ units}^3.$$

This section introduced a new application of the definite integral. Our default view of the definite integral is that it gives "the area under the curve." However, we can establish definite integrals that represent other quantities; in this section, we computed volume.

The ultimate goal of this section is not to compute volumes of solids. That can be useful, but what is more useful is the understanding of this basic principle of integral calculus, outlined in Key Idea 22: to find the exact value of some quantity,

- we start with an approximation (in this section, slice the solid and approximate the volume of each slice),
- then make the approximation better by refining our original approximation (i.e., use more slices),
- then use limits to establish a definite integral which gives the exact value.

We practice this principle in the next section where we find volumes by slicing solids in a different way.

Notes:

Exercises 7.2

Terms and Concepts

1. T/F: A solid of revolution is formed by revolving a shape around an axis.

2. In your own words, explain how the Disk and Washer Methods are related.

3. Explain the how the units of volume are found in the integral of Theorem 54: if $A(x)$ has units of in^2, how does $\int A(x)\,dx$ have units of in^3?

Problems

In Exercises 4 – 7, a region of the Cartesian plane is shaded. Use the Disk/Washer Method to find the volume of the solid of revolution formed by revolving the region about the x-axis.

4.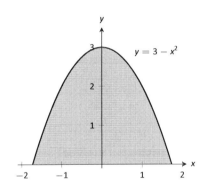

5.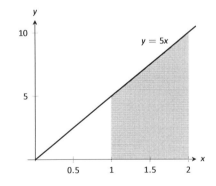

6.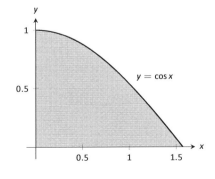

7.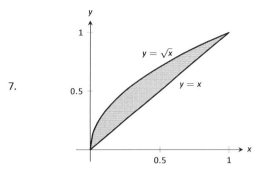

In Exercises 8 – 11, a region of the Cartesian plane is shaded. Use the Disk/Washer Method to find the volume of the solid of revolution formed by revolving the region about the y-axis.

8.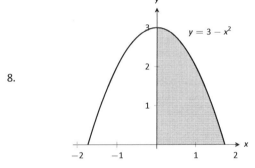

9.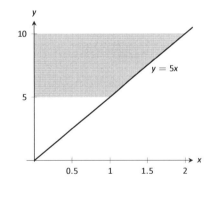

10.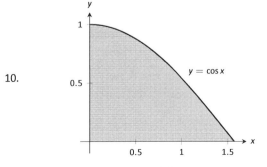

(Hint: Integration By Parts will be necessary, twice. First let $u = \arccos^2 x$, then let $u = \arccos x$.)

11.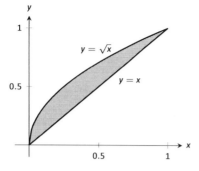

In Exercises 12 – 17, a region of the Cartesian plane is described. Use the Disk/Washer Method to find the volume of the solid of revolution formed by rotating the region about each of the given axes.

12. Region bounded by: $y = \sqrt{x}$, $y = 0$ and $x = 1$.
    Rotate about:

    (a) the x-axis (c) the y-axis
    (b) $y = 1$ (d) $x = 1$

13. Region bounded by: $y = 4 - x^2$ and $y = 0$.
    Rotate about:

    (a) the x-axis (c) $y = -1$
    (b) $y = 4$ (d) $x = 2$

14. The triangle with vertices $(1, 1)$, $(1, 2)$ and $(2, 1)$.
    Rotate about:

    (a) the x-axis (c) the y-axis
    (b) $y = 2$ (d) $x = 1$

15. Region bounded by $y = x^2 - 2x + 2$ and $y = 2x - 1$.
    Rotate about:

    (a) the x-axis (c) $y = 5$
    (b) $y = 1$

16. Region bounded by $y = 1/\sqrt{x^2 + 1}$, $x = -1$, $x = 1$ and the x-axis.
    Rotate about:

    (a) the x-axis (c) $y = -1$
    (b) $y = 1$

17. Region bounded by $y = 2x$, $y = x$ and $x = 2$.
    Rotate about:

    (a) the x-axis (c) the y-axis
    (b) $y = 4$ (d) $x = 2$

In Exercises 18 – 21, a solid is described. Orient the solid along the x-axis such that a cross-sectional area function $A(x)$ can be obtained, then apply Theorem 54 to find the volume of the solid.

18. A right circular cone with height of 10 and base radius of 5.

    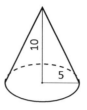

19. A skew right circular cone with height of 10 and base radius of 5. (Hint: all cross-sections are circles.)

    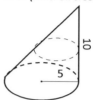

20. A right triangular cone with height of 10 and whose base is a right, isosceles triangle with side length 4.

21. A solid with length 10 with a rectangular base and triangular top, wherein one end is a square with side length 5 and the other end is a triangle with base and height of 5.

    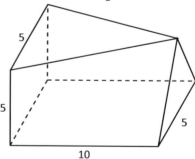

7.3 The Shell Method

Often a given problem can be solved in more than one way. A particular method may be chosen out of convenience, personal preference, or perhaps necessity. Ultimately, it is good to have options.

The previous section introduced the Disk and Washer Methods, which computed the volume of solids of revolution by integrating the cross–sectional area of the solid. This section develops another method of computing volume, the **Shell Method**. Instead of slicing the solid perpendicular to the axis of rotation creating cross-sections, we now slice it parallel to the axis of rotation, creating "shells."

Consider Figure 7.17, where the region shown in (a) is rotated around the y-axis forming the solid shown in (b). A small slice of the region is drawn in (a), parallel to the axis of rotation. When the region is rotated, this thin slice forms a **cylindrical shell**, as pictured in part (c) of the figure. The previous section approximated a solid with lots of thin disks (or washers); we now approximate a solid with many thin cylindrical shells.

To compute the volume of one shell, first consider the paper label on a soup can with radius r and height h. What is the area of this label? A simple way of determining this is to cut the label and lay it out flat, forming a rectangle with height h and length $2\pi r$. Thus the area is $A = 2\pi rh$; see Figure 7.18 (a).

Do a similar process with a cylindrical shell, with height h, thickness Δx, and approximate radius r. Cutting the shell and laying it flat forms a rectangular solid with length $2\pi r$, height h and depth Δx. Thus the volume is $V \approx 2\pi rh\Delta x$; see Figure 7.18 (b). (We say "approximately" since our radius was an approximation.)

By breaking the solid into n cylindrical shells, we can approximate the volume of the solid as

$$V = \sum_{i=1}^{n} 2\pi r_i h_i \Delta x_i,$$

where r_i, h_i and Δx_i are the radius, height and thickness of the ith shell, respectively.

This is a Riemann Sum. Taking a limit as the thickness of the shells approaches 0 leads to a definite integral.

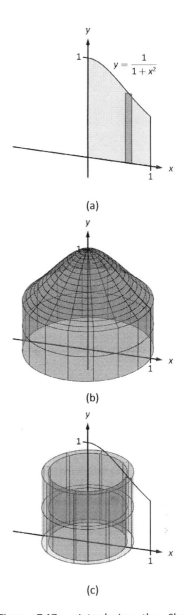

Figure 7.17: Introducing the Shell Method.

Notes:

Chapter 7 Applications of Integration

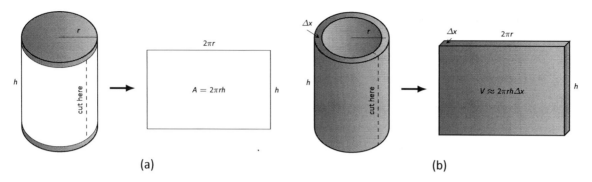

Figure 7.18: Determining the volume of a thin cylindrical shell.

> **Key Idea 25** **The Shell Method**
>
> Let a solid be formed by revolving a region R, bounded by $x = a$ and $x = b$, around a vertical axis. Let $r(x)$ represent the distance from the axis of rotation to x (i.e., the radius of a sample shell) and let $h(x)$ represent the height of the solid at x (i.e., the height of the shell). The volume of the solid is
> $$V = 2\pi \int_a^b r(x)h(x)\,dx.$$

Special Cases:

1. When the region R is bounded above by $y = f(x)$ and below by $y = g(x)$, then $h(x) = f(x) - g(x)$.

2. When the axis of rotation is the y-axis (i.e., $x = 0$) then $r(x) = x$.

Let's practice using the Shell Method.

Example 209 **Finding volume using the Shell Method**
Find the volume of the solid formed by rotating the region bounded by $y = 0$, $y = 1/(1+x^2)$, $x = 0$ and $x = 1$ about the y-axis.

SOLUTION This is the region used to introduce the Shell Method in Figure 7.17, but is sketched again in Figure 7.19 for closer reference. A line is drawn in the region parallel to the axis of rotation representing a shell that will be

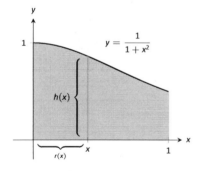

Figure 7.19: Graphing a region in Example 209.

Notes:

carved out as the region is rotated about the y-axis. (This is the differential element.)

The distance this line is from the axis of rotation determines $r(x)$; as the distance from x to the y-axis is x, we have $r(x) = x$. The height of this line determines $h(x)$; the top of the line is at $y = 1/(1+x^2)$, whereas the bottom of the line is at $y = 0$. Thus $h(x) = 1/(1+x^2) - 0 = 1/(1+x^2)$. The region is bounded from $x = 0$ to $x = 1$, so the volume is

$$V = 2\pi \int_0^1 \frac{x}{1+x^2}\, dx.$$

This requires substitution. Let $u = 1 + x^2$, so $du = 2x\, dx$. We also change the bounds: $u(0) = 1$ and $u(1) = 2$. Thus we have:

$$= \pi \int_1^2 \frac{1}{u}\, du$$
$$= \pi \ln u \Big|_1^2$$
$$= \pi \ln 2 \approx 2.178 \text{ units}^3.$$

Note: in order to find this volume using the Disk Method, two integrals would be needed to account for the regions above and below $y = 1/2$.

With the Shell Method, nothing special needs to be accounted for to compute the volume of a solid that has a hole in the middle, as demonstrated next.

Example 210   **Finding volume using the Shell Method**
Find the volume of the solid formed by rotating the triangular region determined by the points $(0,1)$, $(1,1)$ and $(1,3)$ about the line $x = 3$.

SOLUTION   The region is sketched in Figure 7.20(a) along with the differential element, a line within the region parallel to the axis of rotation. In part (b) of the figure, we see the shell traced out by the differential element, and in part (c) the whole solid is shown.

The height of the differential element is the distance from $y = 1$ to $y = 2x + 1$, the line that connects the points $(0,1)$ and $(1,3)$. Thus $h(x) = 2x+1-1 = 2x$. The radius of the shell formed by the differential element is the distance from x to $x = 3$; that is, it is $r(x) = 3 - x$. The x-bounds of the region are $x = 0$ to

Notes:

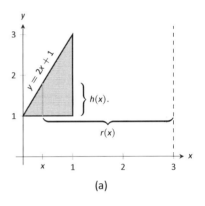

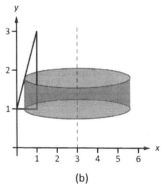

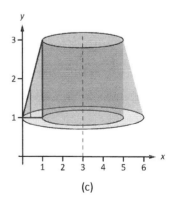

Figure 7.20: Graphing a region in Example 210.

$x = 1$, giving

$$V = 2\pi \int_0^1 (3-x)(2x)\,dx$$
$$= 2\pi \int_0^1 (6x - 2x^2)\,dx$$
$$= 2\pi \left(3x^2 - \frac{2}{3}x^3\right)\Big|_0^1$$
$$= \frac{14}{3}\pi \approx 14.66 \text{ units}^3.$$

When revolving a region around a horizontal axis, we must consider the radius and height functions in terms of y, not x.

Example 211   **Finding volume using the Shell Method**
Find the volume of the solid formed by rotating the region given in Example 210 about the x-axis.

SOLUTION   The region is sketched in Figure 7.21(a) with a sample differential element. In part (b) of the figure the shell formed by the differential element is drawn, and the solid is sketched in (c). (Note that the triangular region looks "short and wide" here, whereas in the previous example the same region looked "tall and narrow." This is because the bounds on the graphs are different.)

The height of the differential element is an x-distance, between $x = \frac{1}{2}y - \frac{1}{2}$ and $x = 1$. Thus $h(y) = 1 - (\frac{1}{2}y - \frac{1}{2}) = -\frac{1}{2}y + \frac{3}{2}$. The radius is the distance from y to the x-axis, so $r(y) = y$. The y bounds of the region are $y = 1$ and $y = 3$, leading to the integral

$$V = 2\pi \int_1^3 \left[y\left(-\frac{1}{2}y + \frac{3}{2}\right)\right] dy$$
$$= 2\pi \int_1^3 \left[-\frac{1}{2}y^2 + \frac{3}{2}y\right] dy$$
$$= 2\pi \left[-\frac{1}{6}y^3 + \frac{3}{4}y^2\right]\Big|_1^3$$
$$= 2\pi \left[\frac{9}{4} - \frac{7}{12}\right]$$
$$= \frac{10}{3}\pi \approx 10.472 \text{ units}^3.$$

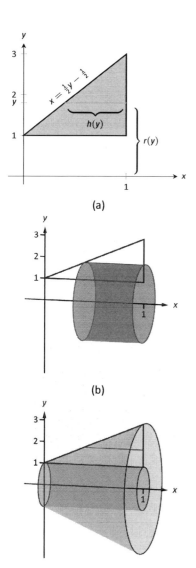

Figure 7.21: Graphing a region in Example 211.

At the beginning of this section it was stated that "it is good to have options." The next example finds the volume of a solid rather easily with the Shell Method, but using the Washer Method would be quite a chore.

Example 212   **Finding volume using the Shell Method**
Find the volume of the solid formed by revolving the region bounded by $y = \sin x$ and the x-axis from $x = 0$ to $x = \pi$ about the y-axis.

SOLUTION   The region and a differential element, the shell formed by this differential element, and the resulting solid are given in Figure 7.22. The radius of a sample shell is $r(x) = x$; the height of a sample shell is $h(x) = \sin x$, each from $x = 0$ to $x = \pi$. Thus the volume of the solid is

$$V = 2\pi \int_0^\pi x \sin x \, dx.$$

This requires Integration By Parts. Set $u = x$ and $dv = \sin x \, dx$; we leave it to the reader to fill in the rest. We have:

$$= 2\pi \left[-x \cos x \Big|_0^\pi + \int_0^\pi \cos x \, dx \right]$$
$$= 2\pi \left[\pi + \sin x \Big|_0^\pi \right]$$
$$= 2\pi \left[\pi + 0 \right]$$
$$= 2\pi^2 \approx 19.74 \text{ units}^3.$$

Note that in order to use the Washer Method, we would need to solve $y = \sin x$ for x, requiring the use of the arcsine function. We leave it to the reader to verify that the outside radius function is $R(y) = \pi - \arcsin y$ and the inside radius function is $r(y) = \arcsin y$. Thus the volume can be computed as

$$\pi \int_0^1 \left[(\pi - \arcsin y)^2 - (\arcsin y)^2 \right] dy.$$

This integral isn't terrible given that the $\arcsin^2 y$ terms cancel, but it is more onerous than the integral created by the Shell Method.

We end this section with a table summarizing the usage of the Washer and Shell Methods.

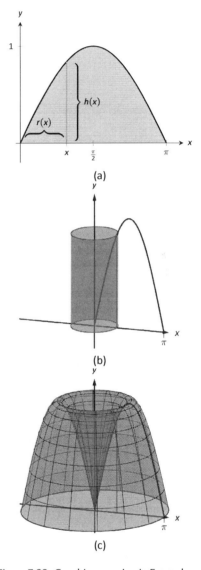

Figure 7.22: Graphing a region in Example 212.

Chapter 7 Applications of Integration

> **Key Idea 26**   **Summary of the Washer and Shell Methods**
>
> Let a region R be given with x-bounds $x = a$ and $x = b$ and y-bounds $y = c$ and $y = d$.
>
> | | Washer Method | Shell Method |
> |---|---|---|
> | Horizontal Axis | $\pi \int_a^b \left(R(x)^2 - r(x)^2\right) dx$ | $2\pi \int_c^d r(y)h(y)\, dy$ |
> | Vertical Axis | $\pi \int_c^d \left(R(y)^2 - r(y)^2\right) dy$ | $2\pi \int_a^b r(x)h(x)\, dx$ |

As in the previous section, the real goal of this section is not to be able to compute volumes of certain solids. Rather, it is to be able to solve a problem by first approximating, then using limits to refine the approximation to give the exact value. In this section, we approximate the volume of a solid by cutting it into thin cylindrical shells. By summing up the volumes of each shell, we get an approximation of the volume. By taking a limit as the number of equally spaced shells goes to infinity, our summation can be evaluated as a definite integral, giving the exact value.

We use this same principle again in the next section, where we find the length of curves in the plane.

Notes:

Exercises 7.3

Terms and Concepts

1. T/F: A solid of revolution is formed by revolving a shape around an axis.

2. T/F: The Shell Method can only be used when the Washer Method fails.

3. T/F: The Shell Method works by integrating cross–sectional areas of a solid.

4. T/F: When finding the volume of a solid of revolution that was revolved around a vertical axis, the Shell Method integrates with respect to x.

Problems

In Exercises 5 – 8, a region of the Cartesian plane is shaded. Use the Shell Method to find the volume of the solid of revolution formed by revolving the region about the y-axis.

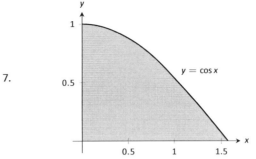

7.

8.

5.

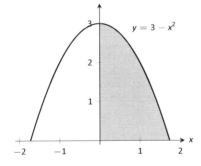

6.

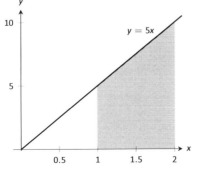

In Exercises 9 – 12, a region of the Cartesian plane is shaded. Use the Shell Method to find the volume of the solid of revolution formed by revolving the region about the x-axis.

9.

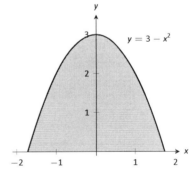

10.

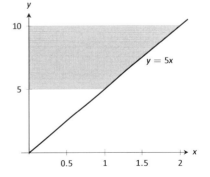

11.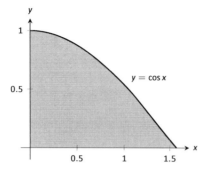

12.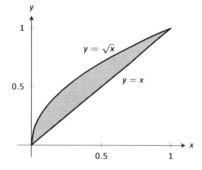

In Exercises 13 – 18, a region of the Cartesian plane is described. Use the Shell Method to find the volume of the solid of revolution formed by rotating the region about each of the given axes.

13. Region bounded by: $y = \sqrt{x}$, $y = 0$ and $x = 1$.
    Rotate about:

    (a) the y-axis (c) the x-axis
    (b) $x = 1$ (d) $y = 1$

14. Region bounded by: $y = 4 - x^2$ and $y = 0$.
    Rotate about:

    (a) $x = 2$ (c) the x-axis
    (b) $x = -2$ (d) $y = 4$

15. The triangle with vertices $(1, 1)$, $(1, 2)$ and $(2, 1)$.
    Rotate about:

    (a) the y-axis (c) the x-axis
    (b) $x = 1$ (d) $y = 2$

16. Region bounded by $y = x^2 - 2x + 2$ and $y = 2x - 1$.
    Rotate about:

    (a) the y-axis (c) $x = -1$
    (b) $x = 1$

17. Region bounded by $y = 1/\sqrt{x^2 + 1}$, $x = 1$ and the x and y-axes.
    Rotate about:

    (a) the y-axis (b) $x = 1$

18. Region bounded by $y = 2x$, $y = x$ and $x = 2$.
    Rotate about:

    (a) the y-axis (c) the x-axis
    (b) $x = 2$ (d) $y = 4$

7.4 Arc Length and Surface Area

In previous sections we have used integration to answer the following questions:

1. Given a region, what is its area?

2. Given a solid, what is its volume?

In this section, we address a related question: Given a curve, what is its length? This is often referred to as **arc length**.

Consider the graph of $y = \sin x$ on $[0, \pi]$ given in Figure 7.23 (a). How long is this curve? That is, if we were to use a piece of string to exactly match the shape of this curve, how long would the string be?

As we have done in the past, we start by approximating; later, we will refine our answer using limits to get an exact solution.

The length of straight–line segments is easy to compute using the Distance Formula. We can approximate the length of the given curve by approximating the curve with straight lines and measuring their lengths.

In Figure 7.23 (b), the curve $y = \sin x$ has been approximated with 4 line segments (the interval $[0, \pi]$ has been divided into 4 equally–lengthed subintervals). It is clear that these four line segments approximate $y = \sin x$ very well on the first and last subinterval, though not so well in the middle. Regardless, the sum of the lengths of the line segments is 3.79, so we approximate the arc length of $y = \sin x$ on $[0, \pi]$ to be 3.79.

In general, we can approximate the arc length of $y = f(x)$ on $[a, b]$ in the following manner. Let $a = x_1 < x_2 < \ldots < x_n < x_{n+1} = b$ be a partition of $[a, b]$ into n subintervals. Let Δx_i represent the length of the ith subinterval $[x_i, x_{i+1}]$.

Figure 7.24 zooms in on the ith subinterval where $y = f(x)$ is approximated by a straight line segment. The dashed lines show that we can view this line segment as they hypotenuse of a right triangle whose sides have length Δx_i and Δy_i. Using the Pythagorean Theorem, the length of this line segment is $\sqrt{\Delta x_i^2 + \Delta y_i^2}$. Summing over all subintervals gives an arc length approximation

$$L \approx \sum_{i=1}^{n} \sqrt{\Delta x_i^2 + \Delta y_i^2}.$$

As shown here, this is *not* a Riemann Sum. While we could conclude that taking a limit as the subinterval length goes to zero gives the exact arc length, we would not be able to compute the answer with a definite integral. We need first to do a little algebra.

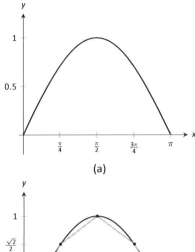

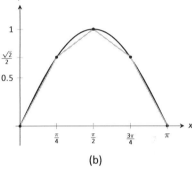

Figure 7.23: Graphing $y = \sin x$ on $[0, \pi]$ and approximating the curve with line segments.

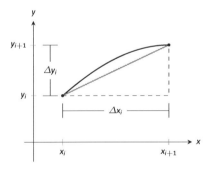

Figure 7.24: Zooming in on the ith subinterval $[x_i, x_{i+1}]$ of a partition of $[a, b]$.

Notes:

In the above expression factor out a Δx_i^2 term:

$$\sum_{i=1}^n \sqrt{\Delta x_i^2 + \Delta y_i^2} = \sum_{i=1}^n \sqrt{\Delta x_i^2 \left(1 + \frac{\Delta y_i^2}{\Delta x_i^2}\right)}.$$

Now pull the Δx_i^2 term out of the square root:

$$= \sum_{i=1}^n \sqrt{1 + \frac{\Delta y_i^2}{\Delta x_i^2}}\, \Delta x_i.$$

This is nearly a Riemann Sum. Consider the $\Delta y_i^2 / \Delta x_i^2$ term. The expression $\Delta y_i / \Delta x_i$ measures the "change in y/change in x," that is, the "rise over run" of f on the ith subinterval. The Mean Value Theorem of Differentiation (Theorem 27) states that there is a c_i in the ith subinterval where $f'(c_i) = \Delta y_i / \Delta x_i$. Thus we can rewrite our above expression as:

$$= \sum_{i=1}^n \sqrt{1 + f'(c_i)^2}\, \Delta x_i.$$

This *is* a Riemann Sum. As long as f' is continuous, we can invoke Theorem 38 and conclude

$$= \int_a^b \sqrt{1 + f'(x)^2}\, dx.$$

Key Idea 27   **Arc Length**

Let f be differentiable on an open interval containing $[a, b]$, where f' is also continuous on $[a, b]$. Then the arc length of f from $x = a$ to $x = b$ is

$$L = \int_a^b \sqrt{1 + f'(x)^2}\, dx.$$

As the integrand contains a square root, it is often difficult to use the formula in Key Idea 27 to find the length exactly. When exact answers are difficult to come by, we resort to using numerical methods of approximating definite integrals. The following examples will demonstrate this.

Notes:

7.4 Arc Length and Surface Area

Example 213   **Finding arc length**
Find the arc length of $f(x) = x^{3/2}$ from $x = 0$ to $x = 4$.

SOLUTION   We begin by finding $f'(x) = \frac{3}{2}x^{1/2}$. Using the formula, we find the arc length L as

$$L = \int_0^4 \sqrt{1 + \left(\frac{3}{2}x^{1/2}\right)^2}\, dx$$

$$= \int_0^4 \sqrt{1 + \frac{9}{4}x}\, dx$$

$$= \int_0^4 \left(1 + \frac{9}{4}x\right)^{1/2} dx$$

$$= \frac{2}{3}\frac{4}{9}\left(1 + \frac{9}{4}x\right)^{3/2}\bigg|_0^4$$

$$= \frac{8}{27}\left(10^{3/2} - 1\right) \approx 9.07 \text{units}.$$

A graph of f is given in Figure 7.25.

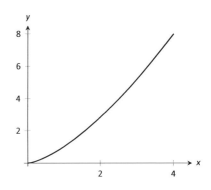

Figure 7.25: A graph of $f(x) = x^{3/2}$ from Example 213.

Example 214   **Finding arc length**
Find the arc length of $f(x) = \frac{1}{8}x^2 - \ln x$ from $x = 1$ to $x = 2$.

SOLUTION   This function was chosen specifically because the resulting integral can be evaluated exactly. We begin by finding $f'(x) = x/4 - 1/x$. The arc length is

$$L = \int_1^2 \sqrt{1 + \left(\frac{x}{4} - \frac{1}{x}\right)^2}\, dx$$

$$= \int_1^2 \sqrt{1 + \frac{x^2}{16} - \frac{1}{2} + \frac{1}{x^2}}\, dx$$

$$= \int_1^2 \sqrt{\frac{x^2}{16} + \frac{1}{2} + \frac{1}{x^2}}\, dx$$

$$= \int_1^2 \sqrt{\left(\frac{x}{4} + \frac{1}{x}\right)^2}\, dx$$

Notes:

Chapter 7 Applications of Integration

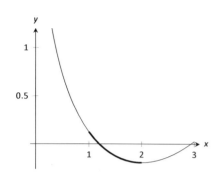

Figure 7.26: A graph of $f(x) = \frac{1}{8}x^2 - \ln x$ from Example 214.

x	$\sqrt{1+\cos^2 x}$
0	$\sqrt{2}$
$\pi/4$	$\sqrt{3/2}$
$\pi/2$	1
$3\pi/4$	$\sqrt{3/2}$
π	$\sqrt{2}$

Figure 7.27: A table of values of $y = \sqrt{1+\cos^2 x}$ to evaluate a definite integral in Example 215.

$$= \int_1^2 \left(\frac{x}{4} + \frac{1}{x}\right) dx$$

$$= \left(\frac{x^2}{8} + \ln x\right)\Big|_1^2$$

$$= \frac{3}{8} + \ln 2 \approx 1.07 \text{ units.}$$

A graph of f is given in Figure 7.26; the portion of the curve measured in this problem is in bold.

The previous examples found the arc length exactly through careful choice of the functions. In general, exact answers are much more difficult to come by and numerical approximations are necessary.

Example 215   **Approximating arc length numerically**
Find the length of the sine curve from $x = 0$ to $x = \pi$.

SOLUTION   This is somewhat of a mathematical curiosity; in Example 127 we found the area under one "hump" of the sine curve is 2 square units; now we are measuring its arc length.

The setup is straightforward: $f(x) = \sin x$ and $f'(x) = \cos x$. Thus

$$L = \int_0^\pi \sqrt{1+\cos^2 x}\, dx.$$

This integral *cannot* be evaluated in terms of elementary functions so we will approximate it with Simpson's Method with $n = 4$. Figure 7.27 gives $\sqrt{1+\cos^2 x}$ evaluated at 5 evenly spaced points in $[0, \pi]$. Simpson's Rule then states that

$$\int_0^\pi \sqrt{1+\cos^2 x}\, dx \approx \frac{\pi - 0}{4 \cdot 3}\left(\sqrt{2} + 4\sqrt{3/2} + 2(1) + 4\sqrt{3/2} + \sqrt{2}\right)$$

$$= 3.82918.$$

Using a computer with $n = 100$ the approximation is $L \approx 3.8202$; our approximation with $n = 4$ is quite good.

Notes:

Surface Area of Solids of Revolution

We have already seen how a curve $y = f(x)$ on $[a, b]$ can be revolved around an axis to form a solid. Instead of computing its volume, we now consider its surface area.

We begin as we have in the previous sections: we partition the interval $[a, b]$ with n subintervals, where the ith subinterval is $[x_i, x_{i+1}]$. On each subinterval, we can approximate the curve $y = f(x)$ with a straight line that connects $f(x_i)$ and $f(x_{i+1})$ as shown in Figure 7.28(a). Revolving this line segment about the x-axis creates part of a cone (called a *frustum* of a cone) as shown in Figure 7.28(b). The surface area of a frustum of a cone is

$$2\pi \cdot \text{length} \cdot \text{average of the two radii } R \text{ and } r.$$

The length is given by L; we use the material just covered by arc length to state that

$$L \approx \sqrt{1 + f'(c_i)}\Delta x_i$$

for some c_i in the ith subinterval. The radii are just the function evaluated at the endpoints of the interval. That is,

$$R = f(x_{i+1}) \quad \text{and} \quad r = f(x_i).$$

Thus the surface area of this sample frustum of the cone is approximately

$$2\pi \frac{f(x_i) + f(x_{i+1})}{2}\sqrt{1 + f'(c_i)^2}\Delta x_i.$$

Since f is a continuous function, the Intermediate Value Theorem states there is some d_i in $[x_i, x_{i+1}]$ such that $f(d_i) = \dfrac{f(x_i) + f(x_{i+1})}{2}$; we can use this to rewrite the above equation as

$$2\pi f(d_i)\sqrt{1 + f'(c_i)^2}\Delta x_i.$$

Summing over all the subintervals we get the total surface area to be approximately

$$\text{Surface Area} \approx \sum_{i=1}^{n} 2\pi f(d_i)\sqrt{1 + f'(c_i)^2}\Delta x_i,$$

which is a Riemann Sum. Taking the limit as the subinterval lengths go to zero gives us the exact surface area, given in the following Key Idea.

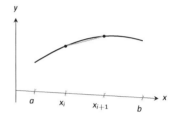

(a)

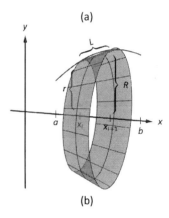

(b)

Figure 7.28: Establishing the formula for surface area.

Notes:

Chapter 7 Applications of Integration

> **Key Idea 28   Surface Area of a Solid of Revolution**
>
> Let f be differentiable on an open interval containing $[a,b]$ where f' is also continuous on $[a,b]$.
>
> 1. The surface area of the solid formed by revolving the graph of $y = f(x)$, where $f(x) \geq 0$, about the x-axis is
>
> $$\text{Surface Area} = 2\pi \int_a^b f(x)\sqrt{1 + f'(x)^2}\, dx.$$
>
> 2. The surface area of the solid formed by revolving the graph of $y = f(x)$ about the y-axis, where $a, b \geq 0$, is
>
> $$\text{Surface Area} = 2\pi \int_a^b x\sqrt{1 + f'(x)^2}\, dx.$$

(When revolving $y = f(x)$ about the y-axis, the radii of the resulting frustum are x_i and x_{i+1}; their average value is simply the midpoint of the interval. In the limit, this midpoint is just x. This gives the second part of Key Idea 28.)

Example 216   Finding surface area of a solid of revolution
Find the surface area of the solid formed by revolving $y = \sin x$ on $[0, \pi]$ around the x-axis, as shown in Figure 7.29.

SOLUTION   The setup is relatively straightforward. Using Key Idea 28, we have the surface area SA is:

$$SA = 2\pi \int_0^\pi \sin x \sqrt{1 + \cos^2 x}\, dx$$
$$= -2\pi \frac{1}{2}\left(\sinh^{-1}(\cos x) + \cos x\sqrt{1 + \cos^2 x}\right)\Big|_0^\pi$$
$$= 2\pi\left(\sqrt{2} + \sinh^{-1} 1\right)$$
$$\approx 14.42 \text{ units}^2.$$

The integration step above is nontrivial, utilizing an integration method called Trigonometric Substitution.

It is interesting to see that the surface area of a solid, whose shape is defined by a trigonometric function, involves both a square root and an inverse hyperbolic trigonometric function.

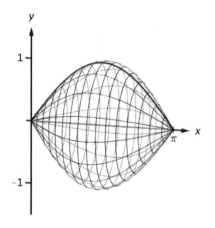

Figure 7.29: Revolving $y = \sin x$ on $[0, \pi]$ about the x-axis.

Notes:

7.4 Arc Length and Surface Area

Example 217   **Finding surface area of a solid of revolution**

Find the surface area of the solid formed by revolving the curve $y = x^2$ on $[0, 1]$ about:

1. the x-axis
2. the y-axis.

Solution

1. The integral is straightforward to setup:

$$SA = 2\pi \int_0^1 x^2 \sqrt{1 + (2x)^2} \, dx.$$

Like the integral in Example 216, this requires Trigonometric Substitution.

$$= \frac{\pi}{32} \left(2(8x^3 + x)\sqrt{1 + 4x^2} - \sinh^{-1}(2x) \right) \Big|_0^1$$

$$= \frac{\pi}{32} \left(18\sqrt{5} - \sinh^{-1} 2 \right)$$

$$\approx 3.81 \text{ units}^2.$$

The solid formed by revolving $y = x^2$ around the x-axis is graphed in Figure 7.30 (a).

2. Since we are revolving around the y-axis, the "radius" of the solid is not $f(x)$ but rather x. Thus the integral to compute the surface area is:

$$SA = 2\pi \int_0^1 x\sqrt{1 + (2x)^2} \, dx.$$

This integral can be solved using substitution. Set $u = 1 + 4x^2$; the new bounds are $u = 1$ to $u = 5$. We then have

$$= \frac{\pi}{4} \int_1^5 \sqrt{u} \, du$$

$$= \frac{\pi}{4} \frac{2}{3} u^{3/2} \Big|_1^5$$

$$= \frac{\pi}{6} \left(5\sqrt{5} - 1 \right)$$

$$\approx 5.33 \text{ units}^2.$$

The solid formed by revolving $y = x^2$ about the y-axis is graphed in Figure 7.30 (b).

Our final example is a famous mathematical "paradox."

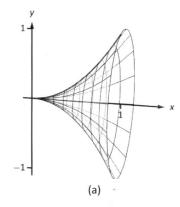

(a)

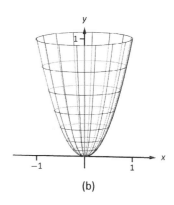

(b)

Figure 7.30: The solids used in Example 217.

Chapter 7 Applications of Integration

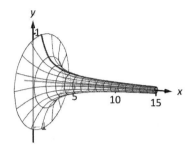

Figure 7.31: A graph of Gabriel's Horn.

Example 218 **The surface area and volume of Gabriel's Horn**

Consider the solid formed by revolving $y = 1/x$ about the x-axis on $[1, \infty)$. Find the volume and surface area of this solid. (This shape, as graphed in Figure 7.31, is known as "Gabriel's Horn" since it looks like a very long horn that only a supernatural person, such as an angel, could play.)

SOLUTION To compute the volume it is natural to use the Disk Method. We have:

$$V = \pi \int_1^\infty \frac{1}{x^2}\, dx$$

$$= \lim_{b \to \infty} \pi \int_1^b \frac{1}{x^2}\, dx$$

$$= \lim_{b \to \infty} \pi \left(\frac{-1}{x}\right)\bigg|_1^b$$

$$= \lim_{b \to \infty} \pi \left(1 - \frac{1}{b}\right)$$

$$= \pi \text{ units}^3.$$

Gabriel's Horn has a finite volume of π cubic units. Since we have already seen that regions with infinite length can have a finite area, this is not too difficult to accept.

We now consider its surface area. The integral is straightforward to setup:

$$SA = 2\pi \int_1^\infty \frac{1}{x}\sqrt{1 + 1/x^4}\, dx.$$

Integrating this expression is not trivial. We can, however, compare it to other improper integrals. Since $1 < \sqrt{1 + 1/x^4}$ on $[1, \infty)$, we can state that

$$2\pi \int_1^\infty \frac{1}{x}\, dx < 2\pi \int_1^\infty \frac{1}{x}\sqrt{1 + 1/x^4}\, dx.$$

By Key Idea 21, the improper integral on the left diverges. Since the integral on the right is larger, we conclude it also diverges, meaning Gabriel's Horn has infinite surface area.

Hence the "paradox": we can fill Gabriel's Horn with a finite amount of paint, but since it has infinite surface area, we can never paint it.

Somehow this paradox is striking when we think about it in terms of volume and area. However, we have seen a similar paradox before, as referenced above. We know that the area under the curve $y = 1/x^2$ on $[1, \infty)$ is finite, yet the shape has an infinite perimeter. Strange things can occur when we deal with the infinite.

A standard equation from physics is "Work = force × distance", when the force applied is constant. In the next section we learn how to compute work when the force applied is variable.

Notes:

Exercises 7.4

Terms and Concepts

1. T/F: The integral formula for computing Arc Length was found by first approximating arc length with straight line segments.

2. T/F: The integral formula for computing Arc Length includes a square–root, meaning the integration is probably easy.

Problems

In Exercises 3 – 12, find the arc length of the function on the given interval.

3. $f(x) = x$ on $[0, 1]$.

4. $f(x) = \sqrt{8x}$ on $[-1, 1]$.

5. $f(x) = \frac{1}{3}x^{3/2} - x^{1/2}$ on $[0, 1]$.

6. $f(x) = \frac{1}{12}x^3 + \frac{1}{x}$ on $[1, 4]$.

7. $f(x) = 2x^{3/2} - \frac{1}{6}\sqrt{x}$ on $[0, 9]$.

8. $f(x) = \cosh x$ on $[-\ln 2, \ln 2]$.

9. $f(x) = \frac{1}{2}\left(e^x + e^{-x}\right)$ on $[0, \ln 5]$.

10. $f(x) = \frac{1}{12}x^5 + \frac{1}{5x^3}$ on $[.1, 1]$.

11. $f(x) = \ln(\sin x)$ on $[\pi/6, \pi/2]$.

12. $f(x) = \ln(\cos x)$ on $[0, \pi/4]$.

In Exercises 13 – 20, set up the integral to compute the arc length of the function on the given interval. Do not evaluate the integral.

13. $f(x) = x^2$ on $[0, 1]$.

14. $f(x) = x^{10}$ on $[0, 1]$.

15. $f(x) = \sqrt{x}$ on $[0, 1]$.

16. $f(x) = \ln x$ on $[1, e]$.

17. $f(x) = \sqrt{1 - x^2}$ on $[-1, 1]$. (Note: this describes the top half of a circle with radius 1.)

18. $f(x) = \sqrt{1 - x^2/9}$ on $[-3, 3]$. (Note: this describes the top half of an ellipse with a major axis of length 6 and a minor axis of length 2.)

19. $f(x) = \frac{1}{x}$ on $[1, 2]$.

20. $f(x) = \sec x$ on $[-\pi/4, \pi/4]$.

In Exercises 21 – 28, use Simpson's Rule, with $n = 4$, to approximate the arc length of the function on the given interval. Note: these are the same problems as in Exercises 13–20.

21. $f(x) = x^2$ on $[0, 1]$.

22. $f(x) = x^{10}$ on $[0, 1]$.

23. $f(x) = \sqrt{x}$ on $[0, 1]$. (Note: $f'(x)$ is not defined at $x = 0$.)

24. $f(x) = \ln x$ on $[1, e]$.

25. $f(x) = \sqrt{1 - x^2}$ on $[-1, 1]$. (Note: $f'(x)$ is not defined at the endpoints.)

26. $f(x) = \sqrt{1 - x^2/9}$ on $[-3, 3]$. (Note: $f'(x)$ is not defined at the endpoints.)

27. $f(x) = \frac{1}{x}$ on $[1, 2]$.

28. $f(x) = \sec x$ on $[-\pi/4, \pi/4]$.

In Exercises 29 – 33, find the surface area of the described solid of revolution.

29. The solid formed by revolving $y = 2x$ on $[0, 1]$ about the x-axis.

30. The solid formed by revolving $y = x^2$ on $[0, 1]$ about the y-axis.

31. The solid formed by revolving $y = x^3$ on $[0, 1]$ about the x-axis.

32. The solid formed by revolving $y = \sqrt{x}$ on $[0, 1]$ about the x-axis.

33. The sphere formed by revolving $y = \sqrt{1 - x^2}$ on $[-1, 1]$ about the x-axis.

Chapter 7 Applications of Integration

7.5 Work

Work is the scientific term used to describe the action of a force which moves an object. When a constant force F is applied to move an object a distance d, the amount of work performed is $W = F \cdot d$.

The SI unit of force is the Newton, (kg·m/s^2), and the SI unit of distance is a meter (m). The fundamental unit of work is one Newton–meter, or a joule (J). That is, applying a force of one Newton for one meter performs one joule of work. In Imperial units (as used in the United States), force is measured in pounds (lb) and distance is measured in feet (ft), hence work is measured in ft–lb.

When force is constant, the measurement of work is straightforward. For instance, lifting a 200 lb object 5 ft performs $200 \cdot 5 = 1000$ ft–lb of work.

What if the force applied is variable? For instance, imagine a climber pulling a 200 ft rope up a vertical face. The rope becomes lighter as more is pulled in, requiring less force and hence the climber performs less work.

In general, let $F(x)$ be a force function on an interval $[a, b]$. We want to measure the amount of work done applying the force F from $x = a$ to $x = b$. We can approximate the amount of work being done by partitioning $[a, b]$ into subintervals $a = x_1 < x_2 < \cdots < x_{n+1} = b$ and assuming that F is constant on each subinterval. Let c_i be a value in the ith subinterval $[x_i, x_{i+1}]$. Then the work done on this interval is approximately $W_i \approx F(c_i) \cdot (x_{i+1} - x_i) = F(c_i)\Delta x_i$, a constant force × the distance over which it is applied. The total work is

$$W = \sum_{i=1}^{n} W_i \approx \sum_{i=1}^{n} F(c_i)\Delta x_i.$$

This, of course, is a Riemann sum. Taking a limit as the subinterval lengths go to zero give an exact value of work which can be evaluated through a definite integral.

Note: *Mass* and *weight* are closely related, yet different, concepts. The mass m of an object is a quantitative measure of that object's resistance to acceleration. The weight w of an object is a measurement of the force applied to the object by the acceleration of gravity g.

Since the two measurements are proportional, $w = m \cdot g$, they are often used interchangeably in everyday conversation. When computing work, one must be careful to note which is being referred to. When mass is given, it must be multiplied by the acceleration of gravity to reference the related force.

Key Idea 29   **Work**

Let $F(x)$ be a continuous function on $[a, b]$ describing the amount of force being applied to an object in the direction of travel from distance $x = a$ to distance $x = b$. The total work W done on $[a, b]$ is

$$W = \int_a^b F(x)\, dx.$$

Notes:

7.5 Work

Example 219   **Computing work performed: applying variable force**
A 60m climbing rope is hanging over the side of a tall cliff. How much work is performed in pulling the rope up to the top, where the rope has a mass of 66g/m?

SOLUTION   We need to create a force function $F(x)$ on the interval $[0, 60]$. To do so, we must first decide what x is measuring: it is the length of the rope still hanging or is it the amount of rope pulled in? As long as we are consistent, either approach is fine. We adopt for this example the convention that x is the amount of rope pulled in. This seems to match intuition better; pulling up the first 10 meters of rope involves $x = 0$ to $x = 10$ instead of $x = 60$ to $x = 50$.

As x is the amount of rope pulled in, the amount of rope still hanging is $60-x$. This length of rope has a mass of 66 g/m, or 0.066 kg/m. The the mass of the rope still hanging is $0.066(60 - x)$ kg; multiplying this mass by the acceleration of gravity, 9.8 m/s^2, gives our variable force function

$$F(x) = (9.8)(0.066)(60 - x) = 0.6468(60 - x).$$

Thus the total work performed in pulling up the rope is

$$W = \int_0^{60} 0.6468(60 - x)\, dx = 1,164.24 \text{ J}.$$

By comparison, consider the work done in lifting the entire rope 60 meters. The rope weights $60 \times 0.066 \times 9.8 = 38.808$ N, so the work applying this force for 60 meters is $60 \times 38.808 = 2,328.48$ J. This is exactly twice the work calculated before (and we leave it to the reader to understand why.)

Example 220   **Computing work performed: applying variable force**
Consider again pulling a 60 m rope up a cliff face, where the rope has a mass of 66 g/m. At what point is exactly half the work performed?

SOLUTION   From Example 219 we know the total work performed is 1,164.24 J. We want to find a height h such that the work in pulling the rope from a height of $x = 0$ to a height of $x = h$ is 582.12, half the total work. Thus we want to solve the equation

$$\int_0^h 0.6468(60 - x)\, dx = 582.12$$

for h.

Notes:

Chapter 7 Applications of Integration

Note: In Example 220, we find that half of the work performed in pulling up a 60 m rope is done in the last 42.43 m. Why is it not coincidental that $60/\sqrt{2} = 42.43$?

$$\int_0^h 0.6468(60 - x)\, dx = 582.12$$

$$\left(38.808x - 0.3234x^2\right)\Big|_0^h = 582.12$$

$$38.808h - 0.3234h^2 = 582.12$$

$$-0.3234h^2 + 38.808h - 582.12 = 0.$$

Apply the Quadratic Formula.

$$h = 17.57 \text{ and } 102.43$$

As the rope is only 60m long, the only sensible answer is $h = 17.57$. Thus about half the work is done pulling up the first 17.5m the other half of the work is done pulling up the remaining 42.43m.

Example 221   **Computing work performed: applying variable force**
A box of 100 lb of sand is being pulled up at a uniform rate a distance of 50 ft over 1 minute. The sand is leaking from the box at a rate of 1 lb/s. The box itself weighs 5 lb and is pulled by a rope weighing .2 lb/ft.

1. How much work is done lifting just the rope?

2. How much work is done lifting just the box and sand?

3. What is the total amount of work performed?

SOLUTION

1. We start by forming the force function $F_r(x)$ for the rope (where the subscript denotes we are considering the rope). As in the previous example, let x denote the amount of rope, in feet, pulled in. (This is the same as saying x denotes the height of the box.) The weight of the rope with x feet pulled in is $F_r(x) = 0.2(50 - x) = 10 - 0.2x$. (Note that we do not have to include the acceleration of gravity here, for the *weight* of the rope per foot is given, not its *mass* per meter as before.) The work performed lifting the rope is

$$W_r = \int_0^{50} (10 - 0.2x)\, dx = 250 \text{ ft–lb}.$$

Notes:

2. The sand is leaving the box at a rate of 1 lb/s. As the vertical trip is to take one minute, we know that 60 lb will have left when the box reaches its final height of 50 ft. Again letting x represent the height of the box, we have two points on the line that describes the weight of the sand: when $x = 0$, the sand weight is 100 lb, producing the point $(0, 100)$; when $x = 50$, the sand in the box weighs 40 lb, producing the point $(50, 40)$. The slope of this line is $\frac{100-40}{0-50} = -1.2$, giving the equation of the weight of the sand at height x as $w(x) = -1.2x + 100$. The box itself weighs a constant 5 lb, so the total force function is $F_b(x) = -1.2x + 105$. Integrating from $x = 0$ to $x = 50$ gives the work performed in lifting box and sand:

$$W_b = \int_0^{50} (-1.2x + 105)\, dx = 3750 \text{ ft–lb}.$$

3. The total work is the sum of W_r and W_b: $250 + 3750 = 4000$ ft–lb. We can also arrive at this via integration:

$$\begin{aligned} W &= \int_0^{50} (F_r(x) + F_b(x))\, dx \\ &= \int_0^{50} (10 - 0.2x - 1.2x + 105)\, dx \\ &= \int_0^{50} (-1.4x + 115)\, dx \\ &= 4000 \text{ ft–lb}. \end{aligned}$$

Hooke's Law and Springs

Hooke's Law states that the force required to compress or stretch a spring x units from its natural length is proportional to x; that is, this force is $F(x) = kx$ for some constant k. For example, if a force of 1 N stretches a given spring 2 cm, then a force of 5 N will stretch the spring 10 cm. Converting the distances to meters, we have that stretching this spring 0.02 m requires a force of $F(0.02) = k(0.02) = 1$ N, hence $k = 1/0.02 = 50$ N/m.

Example 222 **Computing work performed: stretching a spring**
A force of 20 lb stretches a spring from a natural length of 7 inches to a length of 12 inches. How much work was performed in stretching the spring to this length?

SOLUTION In many ways, we are not at all concerned with the actual length of the spring, only with the amount of its change. Hence, we do not care

Notes:

that 20 lb of force stretches the spring to a length of 12 inches, but rather that a force of 20 lb stretches the spring by 5 in. This is illustrated in Figure 7.32; we only measure the change in the spring's length, not the overall length of the spring.

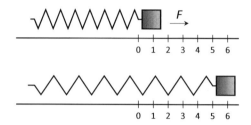

Figure 7.32: Illustrating the important aspects of stretching a spring in computing work in Example 222.

Converting the units of length to feet, we have

$$F(5/12) = 5/12k = 20 \text{ lb}.$$

Thus $k = 48$ lb/ft and $F(x) = 48x$.

We compute the total work performed by integrating $F(x)$ from $x = 0$ to $x = 5/12$:

$$W = \int_0^{5/12} 48x\,dx$$
$$= 24x^2 \Big|_0^{5/12}$$
$$= 25/6 \approx 4.1667 \text{ ft-lb}.$$

Pumping Fluids

Another useful example of the application of integration to compute work comes in the pumping of fluids, often illustrated in the context of emptying a storage tank by pumping the fluid out the top. This situation is different than our previous examples for the forces involved are constant. After all, the force required to move one cubic foot of water (about 62.4 lb) is the same regardless of its location in the tank. What is variable is the distance that cubic foot of water has to travel; water closer to the top travels less distance than water at the bottom, producing less work.

We demonstrate how to compute the total work done in pumping a fluid out of the top of a tank in the next two examples.

Fluid	lb/ft^3	kg/m^3
Concrete	150	2400
Fuel Oil	55.46	890.13
Gasoline	45.93	737.22
Iodine	307	4927
Methanol	49.3	791.3
Mercury	844	13546
Milk	63.6–65.4	1020 – 1050
Water	62.4	1000

Figure 7.33: Weight and Mass densities

Notes:

Example 223   **Computing work performed: pumping fluids**

A cylindrical storage tank with a radius of 10 ft and a height of 30 ft is filled with water, which weighs approximately 62.4 lb/ft^3. Compute the amount of work performed by pumping the water up to a point 5 feet above the top of the tank.

SOLUTION   We will refer often to Figure 7.34 which illustrates the salient aspects of this problem.

We start as we often do: we partition an interval into subintervals. We orient our tank vertically since this makes intuitive sense with the base of the tank at $y = 0$. Hence the top of the water is at $y = 30$, meaning we are interested in subdividing the y-interval $[0, 30]$ into n subintervals as

$$0 = y_1 < y_2 < \cdots < y_{n+1} = 30.$$

Consider the work W_i of pumping only the water residing in the ith subinterval, illustrated in Figure 7.34. The force required to move this water is equal to its weight which we calculate as volume $\times$ density. The volume of water in this subinterval is $V_i = 10^2 \pi \Delta y_i$; its density is 62.4 lb/ft^3. Thus the required force is $6240\pi \Delta y_i$ lb.

We approximate the distance the force is applied by using any y-value contained in the ith subinterval; for simplicity, we arbitrarily use y_i for now (it will not matter later on). The water will be pumped to a point 5 feet above the top of the tank, that is, to the height of $y = 35$ ft. Thus the distance the water at height y_i travels is $35 - y_i$ ft.

In all, the approximate work W_i peformed in moving the water in the ith subinterval to a point 5 feet above the tank is

$$W_i \approx 6240\pi \Delta y_i (35 - y_i).$$

To approximate the total work performed in pumping out all the water from the tank, we sum all the work W_i performed in pumping the water from each of the n subintervals of $[0, 30]$:

$$W \approx \sum_{i=1}^{n} W_i = \sum_{i=1}^{n} 6240\pi \Delta y_i (35 - y_i).$$

This is a Riemann sum. Taking the limit as the subinterval length goes to 0 gives

$$W = \int_0^{30} 6240\pi (35 - y) \, dy$$
$$= \left(6240\pi \left(35y - 1/2 y^2 \right) \right) \Big|_0^{30}$$
$$= 11,762,123 \text{ ft–lb}$$
$$\approx 1.176 \times 10^7 \text{ ft–lb}.$$

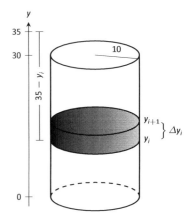

Figure 7.34: Illustrating a water tank in order to compute the work required to empty it in Example 223.

Chapter 7 Applications of Integration

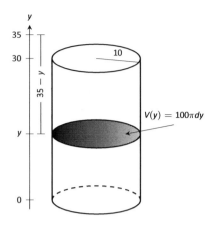

Figure 7.35: A simplified illustration for computing work.

We can "streamline" the above process a bit as we may now recognize what the important features of the problem are. Figure 7.35 shows the tank from Example 223 without the ith subinterval identified. Instead, we just draw one differential element. This helps establish the height a small amount of water must travel along with the force required to move it (where the force is volume × density).

We demonstrate the concepts again in the next examples.

Example 224   **Computing work performed: pumping fluids**
A conical water tank has its top at ground level and its base 10 feet below ground. The radius of the cone at ground level is 2 ft. It is filled with water weighing 62.4 lb/ft^3 and is to be emptied by pumping the water to a spigot 3 feet above ground level. Find the total amount of work performed in emptying the tank.

SOLUTION   The conical tank is sketched in Figure 7.36. We can orient the tank in a variety of ways; we could let $y = 0$ represent the base of the tank and $y = 10$ represent the top of the tank, but we choose to keep the convention of the wording given in the problem and let $y = 0$ represent ground level and hence $y = -10$ represents the bottom of the tank. The actual "height" of the water does not matter; rather, we are concerned with the distance the water travels.

The figure also sketches a differential element, a cross–sectional circle. The radius of this circle is variable, depending on y. When $y = -10$, the circle has radius 0; when $y = 0$, the circle has radius 2. These two points, $(-10, 0)$ and $(0, 2)$, allow us to find the equation of the line that gives the radius of the cross–sectional circle, which is $r(y) = 1/5y + 2$. Hence the volume of water at this height is $V(y) = \pi(1/5y + 2)^2 dy$, where dy represents a very small height of the differential element. The force required to move the water at height y is $F(y) = 62.4 \times V(y)$.

The distance the water at height y travels is given by $h(y) = 3 - y$. Thus the total work done in pumping the water from the tank is

$$W = \int_{-10}^{0} 62.4\pi(1/5y + 2)^2(3 - y)\, dy$$

$$= 62.4\pi \int_{-10}^{0} \left(-\frac{1}{25}y^3 - \frac{17}{25}y^2 - \frac{8}{5}y + 12\right) dy$$

$$= 62.2\pi \cdot \frac{220}{3} \approx 14{,}376 \text{ ft–lb.}$$

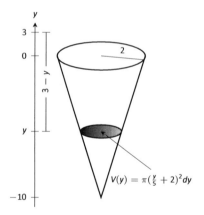

Figure 7.36: A graph of the conical water tank in Example 224.

Notes:

7.5 Work

Example 225 **Computing work performed: pumping fluids**

A rectangular swimming pool is 20 ft wide and has a 3 ft "shallow end" and a 6 ft "deep end." It is to have its water pumped out to a point 2 ft above the current top of the water. The cross–sectional dimensions of the water in the pool are given in Figure 7.37; note that the dimensions are for the water, not the pool itself. Compute the amount of work performed in draining the pool.

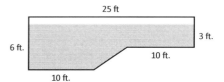

Figure 7.37: The cross–section of a swimming pool filled with water in Example 225.

SOLUTION For the purposes of this problem we choose to set $y = 0$ to represent the bottom of the pool, meaning the top of the water is at $y = 6$. Figure 7.38 shows the pool oriented with this y-axis, along with 2 differential elements as the pool must be split into two different regions.

The top region lies in the y-interval of $[3, 6]$, where the length of the differential element is 25 ft as shown. As the pool is 20 ft wide, this differential element represents a this slice of water with volume $V(y) = 20 \cdot 25 \cdot dy$. The water is to be pumped to a height of $y = 8$, so the height function is $h(y) = 8 - y$. The work done in pumping this top region of water is

$$W_t = 62.4 \int_3^6 500(8 - y)\, dy = 327,600 \text{ ft–lb}.$$

The bottom region lies in the y-interval of $[0, 3]$; we need to compute the length of the differential element in this interval.

One end of the differential element is at $x = 0$ and the other is along the line segment joining the points $(10, 0)$ and $(15, 3)$. The equation of this line is $y = 3/5(x - 10)$; as we will be integrating with respect to y, we rewrite this equation as $x = 5/3y + 10$. So the length of the differential element is a difference of x-values: $x = 0$ and $x = 5/3y + 10$, giving a length of $x = 5/3y + 10$.

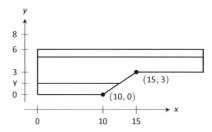

Figure 7.38: Orienting the pool and showing differential elements for Example 225.

Again, as the pool is 20 ft wide, this differential element represents a thin slice of water with volume $V(y) = 20 \cdot (5/3y + 10) \cdot dy$; the height function is the same as before at $h(y) = 8 - y$. The work performed in emptying this part of the pool is

$$W_b = 62.4 \int_0^3 20(5/3y + 10)(8 - y)\, dy = 299,520 \text{ ft–lb}.$$

The total work in empyting the pool is

$$W = W_b + W_t = 327,600 + 299,520 = 627,120 \text{ ft–lb}.$$

Notice how the emptying of the bottom of the pool performs almost as much work as emptying the top. The top portion travels a shorter distance but has more water. In the end, this extra water produces more work.

The next section introduces one final application of the definite integral, the calculation of fluid force on a plate.

Notes:

Exercises 7.5

Terms and Concepts

1. What are the typical units of work?

2. If a man has a mass of 80 kg on Earth, will his mass on the moon be bigger, smaller, or the same?

3. If a woman weighs 130 lb on Earth, will her weight on the moon be bigger, smaller, or the same?

Problems

4. A 100 ft rope, weighing 0.1 lb/ft, hangs over the edge of a tall building.

    (a) How much work is done pulling the entire rope to the top of the building?

    (b) How much rope is pulled in when half of the total work is done?

5. A 50 m rope, with a mass density of 0.2 kg/m, hangs over the edge of a tall building.

    (a) How much work is done pulling the entire rope to the top of the building?

    (b) How much work is done pulling in the first 20 m?

6. A rope of length ℓ ft hangs over the edge of tall cliff. (Assume the cliff is taller than the length of the rope.) The rope has a weight density of d lb/ft.

    (a) How much work is done pulling the entire rope to the top of the cliff?

    (b) What percentage of the total work is done pulling in the first half of the rope?

    (c) How much rope is pulled in when half of the total work is done?

7. A 20 m rope with mass density of 0.5 kg/m hangs over the edge of a 10 m building. How much work is done pulling the rope to the top?

8. A crane lifts a 2,000 lb load vertically 30 ft with a 1" cable weighing 1.68 lb/ft.

    (a) How much work is done lifting the cable alone?

    (b) How much work is done lifting the load alone?

    (c) Could one conclude that the work done lifting the cable is negligible compared to the work done lifting the load?

9. A 100 lb bag of sand is lifted uniformly 120 ft in one minute. Sand leaks from the bag at a rate of 1/4 lb/s. What is the total work done in lifting the bag?

10. A box weighing 2 lb lifts 10 lb of sand vertically 50 ft. A crack in the box allows the sand to leak out such that 9 lb of sand is in the box at the end of the trip. Assume the sand leaked out at a uniform rate. What is the total work done in lifting the box and sand?

11. A force of 1000 lb compresses a spring 3 in. How much work is performed in compressing the spring?

12. A force of 2 N stretches a spring 5 cm. How much work is performed in stretching the spring?

13. A force of 50 lb compresses a spring from a natural length of 18 in to 12 in. How much work is performed in compressing the spring?

14. A force of 20 lb stretches a spring from a natural length of 6 in to 8 in. How much work is performed in stretching the spring?

15. A force of 7 N stretches a spring from a natural length of 11 cm to 21 cm. How much work is performed in stretching the spring from a length of 16 cm to 21 cm?

16. A force of f N stretches a spring d m from its natural length. How much work is performed in stretching the spring?

17. A 20 lb weight is attached to a spring. The weight rests on the spring, compressing the spring from a natural length of 1 ft to 6 in.

    How much work is done in lifting the box 1.5 ft (i.e, the spring will be stretched 1 ft beyond its natural length)?

18. A 20 lb weight is attached to a spring. The weight rests on the spring, compressing the spring from a natural length of 1 ft to 6 in.

    How much work is done in lifting the box 6 in (i.e, bringing the spring back to its natural length)?

19. A 5 m tall cylindrical tank with radius of 2 m is filled with 3 m of gasoline, with a mass density of 737.22 kg/m^3. Compute the total work performed in pumping all the gasoline to the top of the tank.

20. A 6 ft cylindrical tank with a radius of 3 ft is filled with water, which has a weight density of 62.4 lb/ft^3. The water is to be pumped to a point 2 ft above the top of the tank.

    (a) How much work is performed in pumping all the water from the tank?

    (b) How much work is performed in pumping 3 ft of water from the tank?

    (c) At what point is 1/2 of the total work done?

21. A gasoline tanker is filled with gasoline with a weight density of 45.93 lb/ft^3. The dispensing valve at the base is jammed shut, forcing the operator to empty the tank via

pumping the gas to a point 1 ft above the top of the tank. Assume the tank is a perfect cylinder, 20 ft long with a diameter of 7.5 ft. How much work is performed in pumping all the gasoline from the tank?

22. A fuel oil storage tank is 10 ft deep with trapezoidal sides, 5 ft at the top and 2 ft at the bottom, and is 15 ft wide (see diagram below). Given that fuel oil weighs 55.46 lb/ft^3, find the work performed in pumping all the oil from the tank to a point 3 ft above the top of the tank.

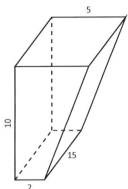

23. A conical water tank is 5 m deep with a top radius of 3 m. (This is similar to Example 224.) The tank is filled with pure water, with a mass density of 1000 kg/m^3.

    (a) Find the work performed in pumping all the water to the top of the tank.
    (b) Find the work performed in pumping the top 2.5 m of water to the top of the tank.
    (c) Find the work performed in pumping the top half of the water, by volume, to the top of the tank.

24. A water tank has the shape of a truncated cone, with dimensions given below, and is filled with water with a weight density of 62.4 lb/ft^3. Find the work performed in pumping all water to a point 1 ft above the top of the tank.

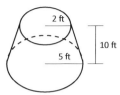

25. A water tank has the shape of an inverted pyramid, with dimensions given below, and is filled with water with a mass density of 1000 kg/m^3. Find the work performed in pumping all water to a point 5 m above the top of the tank.

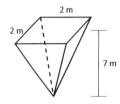

26. A water tank has the shape of an truncated, inverted pyramid, with dimensions given below, and is filled with water with a mass density of 1000 kg/m^3. Find the work performed in pumping all water to a point 1 m above the top of the tank.

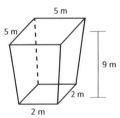

7.6 Fluid Forces

In the unfortunate situation of a car driving into a body of water, the conventional wisdom is that the water pressure on the doors will quickly be so great that they will be effectively unopenable. (Survival techniques suggest immediately opening the door, rolling down or breaking the window, or waiting until the water fills up the interior at which point the pressure is equalized and the door will open. See Mythbusters episode #72 to watch Adam Savage test these options.)

How can this be true? How much force does it take to open the door of a submerged car? In this section we will find the answer to this question by examining the forces exerted by fluids.

We start with **pressure**, which is related to **force** by the following equations:

$$\text{Pressure} = \frac{\text{Force}}{\text{Area}} \quad \Leftrightarrow \quad \text{Force} = \text{Pressure} \times \text{Area}.$$

In the context of fluids, we have the following definition.

> **Definition 26**   **Fluid Pressure**
>
> Let w be the weight–density of a fluid. The **pressure** p exerted on an object at depth d in the fluid is $p = w \cdot d$.

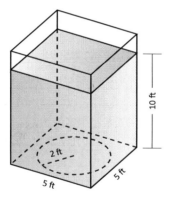

Figure 7.39: A cylindrical tank in Example 226.

We use this definition to find the **force** exerted on a horizontal sheet by considering the sheet's area.

Example 226   **Computing fluid force**

1. A cylindrical storage tank has a radius of 2 ft and holds 10 ft of a fluid with a weight–density of 50 lb/ft^3. (See Figure 7.39.) What is the force exerted on the base of the cylinder by the fluid?

2. A rectangular tank whose base is a 5 ft square has a circular hatch at the bottom with a radius of 2 ft. The tank holds 10 ft of a fluid with a weight–density of 50 lb/ft^3. (See Figure 7.40.) What is the force exerted on the hatch by the fluid?

Figure 7.40: A rectangular tank in Example 226.

SOLUTION

1. Using Definition 26, we calculate that the pressure exerted on the cylinder's base is $w \cdot d = 50$ lb/ft$^3 \times 10$ ft $= 500$ lb/ft^2. The area of the base is

Notes:

$\pi \cdot 2^2 = 4\pi$ ft^2. So the force exerted by the fluid is

$$F = 500 \times 4\pi = 6283 \text{ lb}.$$

Note that we effectively just computed the *weight* of the fluid in the tank.

2. The dimensions of the tank in this problem are irrelevant. All we are concerned with are the dimensions of the hatch and the depth of the fluid. Since the dimensions of the hatch are the same as the base of the tank in the previous part of this example, as is the depth, we see that the fluid force is the same. That is, $F = 6283$ lb.

   A key concept to understand here is that we are effectively measuring the weight of a 10 ft column of water above the hatch. The size of the tank holding the fluid does not matter.

The previous example demonstrates that computing the force exerted on a horizontally oriented plate is relatively easy to compute. What about a vertically oriented plate? For instance, suppose we have a circular porthole located on the side of a submarine. How do we compute the fluid force exerted on it?

Pascal's Principle states that the pressure exerted by a fluid at a depth is equal in all directions. Thus the pressure on any portion of a plate that is 1 ft below the surface of water is the same no matter how the plate is oriented. (Thus a hollow cube submerged at a great depth will not simply be "crushed" from above, but the sides will also crumple in. The fluid will exert force on *all* sides of the cube.)

So consider a vertically oriented plate as shown in Figure 7.41 submerged in a fluid with weight–density w. What is the total fluid force exerted on this plate? We find this force by first approximating the force on small horizontal strips.

Let the top of the plate be at depth b and let the bottom be at depth a. (For now we assume that surface of the fluid is at depth 0, so if the bottom of the plate is 3 ft under the surface, we have $a = -3$. We will come back to this later.) We partition the interval $[a, b]$ into n subintervals

$$a = y_1 < y_2 < \cdots < y_{n+1} = b,$$

with the ith subinterval having length Δy_i. The force F_i exerted on the plate in the ith subinterval is $F_i = $ Pressure $\times$ Area.

The pressure is depth $\times w$. We approximate the depth of this thin strip by choosing any value d_i in $[y_i, y_{i+1}]$; the depth is approximately $-d_i$. (Our convention has d_i being a negative number, so $-d_i$ is positive.) For convenience, we let d_i be an endpoint of the subinterval; we let $d_i = y_i$.

The area of the thin strip is approximately length $\times$ width. The width is Δy_i. The length is a function of some y-value c_i in the ith subinterval. We state the

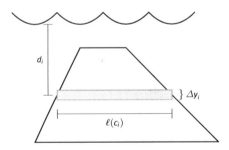

Figure 7.41: A thin, vertically oriented plate submerged in a fluid with weight–density w.

length is $\ell(c_i)$. Thus

$$F_i = \text{Pressure} \times \text{Area}$$
$$= -y_i \cdot w \times \ell(c_i) \cdot \Delta y_i.$$

To approximate the total force, we add up the approximate forces on each of the n thin strips:

$$F = \sum_{i=1}^{n} F_i \approx \sum_{i=1}^{n} -w \cdot y_i \cdot \ell(c_i) \cdot \Delta y_i.$$

This is, of course, another Riemann Sum. We can find the exact force by taking a limit as the subinterval lengths go to 0; we evaluate this limit with a definite integral.

Key Idea 30   **Fluid Force on a Vertically Oriented Plate**

Let a vertically oriented plate be submerged in a fluid with weight–density w where the top of the plate is at $y = b$ and the bottom is at $y = a$. Let $\ell(y)$ be the length of the plate at y.

1. If $y = 0$ corresponds to the surface of the fluid, then the force exerted on the plate by the fluid is

$$F = \int_a^b w \cdot (-y) \cdot \ell(y)\, dy.$$

2. In general, let $d(y)$ represent the distance between the surface of the fluid and the plate at y. Then the force exerted on the plate by the fluid is

$$F = \int_a^b w \cdot d(y) \cdot \ell(y)\, dy.$$

Figure 7.42: A thin plate in the shape of an isosceles triangle in Example 227.

Example 227   **Finding fluid force**
Consider a thin plate in the shape of an isosceles triangle as shown in Figure 7.42 submerged in water with a weight–density of 62.4 lb/ft^3. If the bottom of the plate is 10 ft below the surface of the water, what is the total fluid force exerted on this plate?

Solution   We approach this problem in two different ways to illustrate the different ways Key Idea 30 can be implemented. First we will let $y = 0$ represent the surface of the water, then we will consider an alternate convention.

Notes:

1. We let $y = 0$ represent the surface of the water; therefore the bottom of the plate is at $y = -10$. We center the triangle on the y-axis as shown in Figure 7.43. The depth of the plate at y is $-y$ as indicated by the Key Idea. We now consider the length of the plate at y.

   We need to find equations of the left and right edges of the plate. The right hand side is a line that connects the points $(0, -10)$ and $(2, -6)$: that line has equation $x = 1/2(y + 10)$. (Find the equation in the familiar $y = mx + b$ format and solve for x.) Likewise, the left hand side is described by the line $x = -1/2(y + 10)$. The total length is the distance between these two lines: $\ell(y) = 1/2(y + 10) - (-1/2(y + 10)) = y + 10$.

   The total fluid force is then:

   $$F = \int_{-10}^{-6} 62.4(-y)(y + 10)\, dy$$
   $$= 62.4 \cdot \frac{176}{3} \approx 3660.8 \text{ lb}.$$

2. Sometimes it seems easier to orient the thin plate nearer the origin. For instance, consider the convention that the bottom of the triangular plate is at $(0, 0)$, as shown in Figure 7.44. The equations of the left and right hand sides are easy to find. They are $y = 2x$ and $y = -2x$, respectively, which we rewrite as $x = 1/2y$ and $x = -1/2y$. Thus the length function is $\ell(y) = 1/2y - (-1/2y) = y$.

   As the surface of the water is 10 ft above the base of the plate, we have that the surface of the water is at $y = 10$. Thus the depth function is the distance between $y = 10$ and y; $d(y) = 10 - y$. We compute the total fluid force as:

   $$F = \int_0^4 62.4(10 - y)(y)\, dy$$
   $$\approx 3660.8 \text{ lb}.$$

The correct answer is, of course, independent of the placement of the plate in the coordinate plane as long as we are consistent.

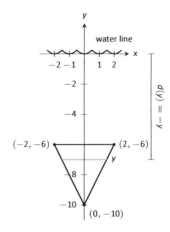

Figure 7.43: Sketching the triangular plate in Example 227 with the convention that the water level is at $y = 0$.

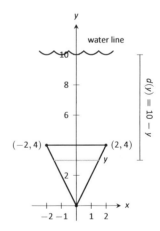

Figure 7.44: Sketching the triangular plate in Example 227 with the convention that the base of the triangle is at $(0, 0)$.

Example 228   **Finding fluid force**
Find the total fluid force on a car door submerged up to the bottom of its window in water, where the car door is a rectangle 40" long and 27" high (based on the dimensions of a 2005 Fiat Grande Punto.)

SOLUTION   The car door, as a rectangle, is drawn in Figure 7.45. Its length is $10/3$ ft and its height is 2.25 ft. We adopt the convention that the

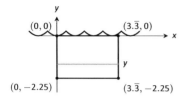

Figure 7.45: Sketching a submerged car door in Example 228.

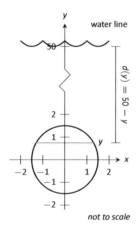

Figure 7.46: Measuring the fluid force on an underwater porthole in Example 229.

top of the door is at the surface of the water, both of which are at $y = 0$. Using the weight–density of water of 62.4 lb/ft³, we have the total force as

$$F = \int_{-2.25}^{0} 62.4(-y)10/3 \, dy$$
$$= \int_{-2.25}^{0} -208y \, dy$$
$$= -104y^2 \Big|_{-2.25}^{0}$$
$$= 526.5 \text{ lb.}$$

Most adults would find it very difficult to apply over 500 lb of force to a car door while seated inside, making the door effectively impossible to open. This is counter–intuitive as most assume that the door would be relatively easy to open. The truth is that it is not, hence the survival tips mentioned at the beginning of this section.

Example 229 **Finding fluid force**
An underwater observation tower is being built with circular viewing portholes enabling visitors to see underwater life. Each vertically oriented porthole is to have a 3 ft diameter whose center is to be located 50 ft underwater. Find the total fluid force exerted on each porthole. Also, compute the fluid force on a horizontally oriented porthole that is under 50 ft of water.

Solution We place the center of the porthole at the origin, meaning the surface of the water is at $y = 50$ and the depth function will be $d(y) = 50 - y$; see Figure 7.46

The equation of a circle with a radius of 1.5 is $x^2 + y^2 = 2.25$; solving for x we have $x = \pm\sqrt{2.25 - y^2}$, where the positive square root corresponds to the right side of the circle and the negative square root corresponds to the left side of the circle. Thus the length function at depth y is $\ell(y) = 2\sqrt{2.25 - y^2}$. Integrating on $[-1.5, 1.5]$ we have:

$$F = 62.4 \int_{-1.5}^{1.5} 2(50 - y)\sqrt{2.25 - y^2} \, dy$$
$$= 62.4 \int_{-1.5}^{1.5} \left(100\sqrt{2.25 - y^2} - 2y\sqrt{2.25 - y^2}\right) dy$$
$$= 6240 \int_{-1.5}^{1.5} \left(\sqrt{2.25 - y^2}\right) dy - 62.4 \int_{-1.5}^{1.5} \left(2y\sqrt{2.25 - y^2}\right) dy.$$

Notes:

The second integral above can be evaluated using Substitution. Let $u = 2.25 - y^2$ with $du = -2y\,dy$. The new bounds are: $u(-1.5) = 0$ and $u(1.5) = 0$; the new integral will integrate from $u = 0$ to $u = 0$, hence the integral is 0.

The first integral above finds the area of half a circle of radius 1.5, thus the first integral evaluates to $6240 \cdot \pi \cdot 1.5^2/2 = 22{,}054$. Thus the total fluid force on a vertically oriented porthole is 22,054 lb.

Finding the force on a horizontally oriented porthole is more straightforward:

$$F = \text{Pressure} \times \text{Area} = 62.4 \cdot 50 \times \pi \cdot 1.5^2 = 22{,}054 \text{ lb.}$$

That these two forces are equal is not coincidental; it turns out that the fluid force applied to a vertically oriented circle whose center is at depth d is the same as force applied to a horizontally oriented circle at depth d.

We end this chapter with a reminder of the true skills meant to be developed here. We are not truly concerned with an ability to find fluid forces or the volumes of solids of revolution. Work done by a variable force is important, though measuring the work done in pulling a rope up a cliff is probably not.

What we are actually concerned with is the ability to solve certain problems by first approximating the solution, then refining the approximation, then recognizing if/when this refining process results in a definite integral through a limit. Knowing the formulas found inside the special boxes within this chapter is beneficial as it helps solve problems found in the exercises, and other mathematical skills are strengthened by properly applying these formulas. However, more importantly, understand how each of these formulas was constructed. Each is the result of a summation of approximations; each summation was a Riemann sum, allowing us to take a limit and find the exact answer through a definite integral.

The next chapter addresses an entirely different topic: sequences and series. In short, a sequence is a list of numbers, where a series is the summation of a list of numbers. These seemingly–simple ideas lead to very powerful mathematics.

Notes:

Exercises 7.6

Terms and Concepts

1. State in your own words Pascal's Principle.

2. State in your own words how pressure is different from force.

Problems

In Exercises 3 – 12, find the fluid force exerted on the given plate, submerged in water with a weight density of 62.4 lb/ft^3.

3.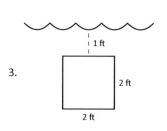

4.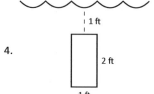

5.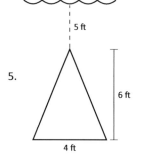

6.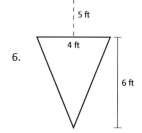

7.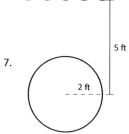

8.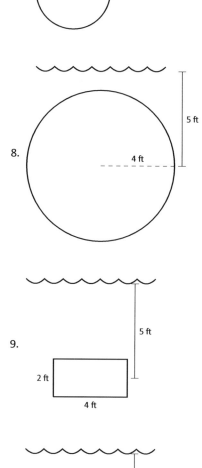

9.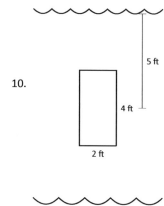

10.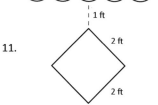

11.

12.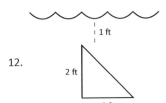

In Exercises 13 – 18, the side of a container is pictured. Find the fluid force exerted on this plate when the container is full of:

1. water, with a weight density of 62.4 lb/ft³, and

2. concrete, with a weight density of 150 lb/ft³.

13.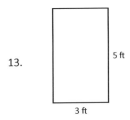

14.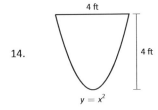

15.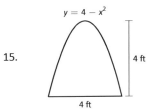

16.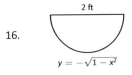

17.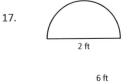

18.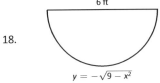

19. How deep must the center of a vertically oriented circular plate with a radius of 1 ft be submerged in water, with a weight density of 62.4 lb/ft³, for the fluid force on the plate to reach 1,000 lb?

20. How deep must the center of a vertically oriented square plate with a side length of 2 ft be submerged in water, with a weight density of 62.4 lb/ft³, for the fluid force on the plate to reach 1,000 lb?

8: Sequences and Series

This chapter introduces **sequences** and **series**, important mathematical constructions that are useful when solving a large variety of mathematical problems. The content of this chapter is considerably different from the content of the chapters before it. While the material we learn here definitely falls under the scope of "calculus," we will make very little use of derivatives or integrals. Limits are extremely important, though, especially limits that involve infinity.

One of the problems addressed by this chapter is this: suppose we know information about a function and its derivatives at a point, such as $f(1) = 3$, $f'(1) = 1, f''(1) = -2, f'''(1) = 7$, and so on. What can I say about $f(x)$ itself? Is there any reasonable approximation of the value of $f(2)$? The topic of Taylor Series addresses this problem, and allows us to make excellent approximations of functions when limited knowledge of the function is available.

Notation: We use $\mathbb{N}$ to describe the set of natural numbers, that is, the integers 1, 2, 3, ...

Factorial: The expression 3! refers to the number $3 \cdot 2 \cdot 1 = 6$.

In general, $n! = n \cdot (n-1) \cdot (n-2) \cdots 2 \cdot 1$, where n is a natural number.

We define $0! = 1$. While this does not immediately make sense, it makes many mathematical formulas work properly.

8.1 Sequences

We commonly refer to a set of events that occur one after the other as a *sequence* of events. In mathematics, we use the word *sequence* to refer to an ordered set of numbers, i.e., a set of numbers that "occur one after the other."

For instance, the numbers 2, 4, 6, 8, ..., form a sequence. The order is important; the first number is 2, the second is 4, etc. It seems natural to seek a formula that describes a given sequence, and often this can be done. For instance, the sequence above could be described by the function $a(n) = 2n$, for the values of $n = 1, 2, \ldots$ To find the 10th term in the sequence, we would compute $a(10)$. This leads us to the following, formal definition of a sequence.

Definition 27   **Sequence**

A **sequence** is a function $a(n)$ whose domain is $\mathbb{N}$. The **range** of a sequence is the set of all distinct values of $a(n)$.

The **terms** of a sequence are the values $a(1), a(2), \ldots$, which are usually denoted with subscripts as $a_1, a_2, \ldots$.

A sequence $a(n)$ is often denoted as $\{a_n\}$.

Chapter 8 Sequences and Series

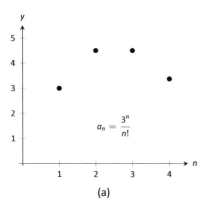

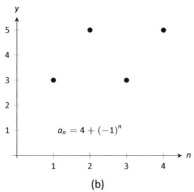

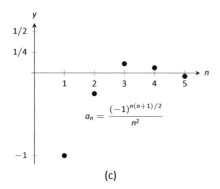

Figure 8.1: Plotting sequences in Example 230.

Example 230   **Listing terms of a sequence**
List the first four terms of the following sequences.

1. $\{a_n\} = \left\{\dfrac{3^n}{n!}\right\}$ 2. $\{a_n\} = \{4+(-1)^n\}$ 3. $\{a_n\} = \left\{\dfrac{(-1)^{n(n+1)/2}}{n^2}\right\}$

Solution

1. $a_1 = \dfrac{3^1}{1!} = 3;$ $a_2 = \dfrac{3^2}{2!} = \dfrac{9}{2};$ $a_3 = \dfrac{3^3}{3!} = \dfrac{9}{2};$ $a_4 = \dfrac{3^4}{4!} = \dfrac{27}{8}$

We can plot the terms of a sequence with a scatter plot. The "x"-axis is used for the values of n, and the values of the terms are plotted on the y-axis. To visualize this sequence, see Figure 8.1(a).

2. $a_1 = 4 + (-1)^1 = 3;$ $a_2 = 4 + (-1)^2 = 5;$

$a_3 = 4+(-1)^3 = 3;$ $a_4 = 4+(-1)^4 = 5.$ Note that the range of this sequence is finite, consisting of only the values 3 and 5. This sequence is plotted in Figure 8.1(b).

3. $a_1 = \dfrac{(-1)^{1(2)/2}}{1^2} = -1;$ $a_2 = \dfrac{(-1)^{2(3)/2}}{2^2} = -\dfrac{1}{4}$

$a_3 = \dfrac{(-1)^{3(4)/2}}{3^2} = \dfrac{1}{9}$ $a_4 = \dfrac{(-1)^{4(5)/2}}{4^2} = \dfrac{1}{16};$

$a_5 = \dfrac{(-1)^{5(6)/2}}{5^2} = -\dfrac{1}{25}.$

We gave one extra term to begin to show the pattern of signs is "$-, -, +, +, -, -, \ldots$", due to the fact that the exponent of -1 is a special quadratic. This sequence is plotted in Figure 8.1(c).

Example 231   **Determining a formula for a sequence**
Find the nth term of the following sequences, i.e., find a function that describes each of the given sequences.

1. $2, 5, 8, 11, 14, \ldots$

2. $2, -5, 10, -17, 26, -37, \ldots$

3. $1, 1, 2, 6, 24, 120, 720, \ldots$

4. $\dfrac{5}{2}, \dfrac{5}{2}, \dfrac{15}{8}, \dfrac{5}{4}, \dfrac{25}{32}, \ldots$

Notes:

SOLUTION   We should first note that there is never exactly one function that describes a finite set of numbers as a sequence. There are many sequences that start with 2, then 5, as our first example does. We are looking for a simple formula that describes the terms given, knowing there is possibly more than one answer.

1. Note how each term is 3 more than the previous one. This implies a linear function would be appropriate: $a(n) = a_n = 3n + b$ for some appropriate value of b. As we want $a_1 = 2$, we set $b = -1$. Thus $a_n = 3n - 1$.

2. First notice how the sign changes from term to term. This is most commonly accomplished by multiplying the terms by either $(-1)^n$ or $(-1)^{n+1}$. Using $(-1)^n$ multiplies the odd terms by (-1); using $(-1)^{n+1}$ multiplies the even terms by (-1). As this sequence has negative even terms, we will multiply by $(-1)^{n+1}$.

    After this, we might feel a bit stuck as to how to proceed. At this point, we are just looking for a pattern of some sort: what do the numbers 2, 5, 10, 17, etc., have in common? There are many correct answers, but the one that we'll use here is that each is one more than a perfect square. That is, $2 = 1^2 + 1$, $5 = 2^2 + 1$, $10 = 3^2 + 1$, etc. Thus our formula is $a_n = (-1)^{n+1}(n^2 + 1)$.

3. One who is familiar with the factorial function will readily recognize these numbers. They are 0!, 1!, 2!, 3!, etc. Since our sequences start with $n = 1$, we cannot write $a_n = n!$, for this misses the 0! term. Instead, we shift by 1, and write $a_n = (n-1)!$.

4. This one may appear difficult, especially as the first two terms are the same, but a little "sleuthing" will help. Notice how the terms in the numerator are always multiples of 5, and the terms in the denominator are always powers of 2. Does something as simple as $a_n = \frac{5n}{2^n}$ work?

    When $n = 1$, we see that we indeed get 5/2 as desired. When $n = 2$, we get $10/4 = 5/2$. Further checking shows that this formula indeed matches the other terms of the sequence.

A common mathematical endeavor is to create a new mathematical object (for instance, a sequence) and then apply previously known mathematics to the new object. We do so here. The fundamental concept of calculus is the limit, so we will investigate what it means to find the limit of a sequence.

Notes:

Chapter 8 Sequences and Series

> **Definition 28**   **Limit of a Sequence, Convergent, Divergent**
>
> Let $\{a_n\}$ be a sequence and let L be a real number. Given any $\varepsilon > 0$, if an m can be found such that $|a_n - L| < \varepsilon$ for all $n > m$, then we say the **limit of $\{a_n\}$, as n approaches infinity, is L**, denoted
>
> $$\lim_{n\to\infty} a_n = L.$$
>
> If $\lim_{n\to\infty} a_n$ exists, we say the sequence **converges**; otherwise, the sequence **diverges**.

This definition states, informally, that if the limit of a sequence is L, then if you go far enough out along the sequence, all subsequent terms will be *really close* to L. Of course, the terms "far enough" and "really close" are subjective terms, but hopefully the intent is clear.

This definition is reminiscent of the ε–δ proofs of Chapter 1. In that chapter we developed other tools to evaluate limits apart from the formal definition; we do so here as well.

> **Theorem 55**   **Limit of a Sequence**
>
> Let $\{a_n\}$ be a sequence and let $f(x)$ be a function whose domain contains the positive real numbers where $f(n) = a_n$ for all n in $\mathbb{N}$.
>
> If $\lim_{x\to\infty} f(x) = L$, then $\lim_{n\to\infty} a_n = L$.

Theorem 55 allows us, in certain cases, to apply the tools developed in Chapter 1 to limits of sequences. Note two things *not* stated by the theorem:

1. If $\lim_{x\to\infty} f(x)$ does not exist, we cannot conclude that $\lim_{n\to\infty} a_n$ does not exist. It may, or may not, exist. For instance, we can define a sequence $\{a_n\} = \{\cos(2\pi n)\}$. Let $f(x) = \cos(2\pi x)$. Since the cosine function oscillates over the real numbers, the limit $\lim_{x\to\infty} f(x)$ does not exist.

   However, for every positive integer n, $\cos(2\pi n) = 1$, so $\lim_{n\to\infty} a_n = 1$.

2. If we cannot find a function $f(x)$ whose domain contains the positive real numbers where $f(n) = a_n$ for all n in $\mathbb{N}$, we cannot conclude $\lim_{n\to\infty} a_n$ does not exist. It may, or may not, exist.

Notes:

8.1 Sequences

Example 232   **Determining convergence/divergence of a sequence**
Determine the convergence or divergence of the following sequences.

1. $\{a_n\} = \left\{ \dfrac{3n^2 - 2n + 1}{n^2 - 1000} \right\}$ 2. $\{a_n\} = \{\cos n\}$ 3. $\{a_n\} = \left\{ \dfrac{(-1)^n}{n} \right\}$

SOLUTION

1. Using Theorem 11, we can state that $\displaystyle\lim_{x \to \infty} \dfrac{3x^2 - 2x + 1}{x^2 - 1000} = 3$. (We could have also directly applied l'Hôpital's Rule.) Thus the sequence $\{a_n\}$ converges, and its limit is 3. A scatter plot of every 5 values of a_n is given in Figure 8.2 (a). The values of a_n vary widely near $n = 30$, ranging from about -73 to 125, but as n grows, the values approach 3.

2. The limit $\displaystyle\lim_{x \to \infty} \cos x$ does not exist, as $\cos x$ oscillates (and takes on every value in $[-1, 1]$ infinitely many times). Thus we cannot apply Theorem 55.

   The fact that the cosine function oscillates strongly hints that $\cos n$, when n is restricted to $\mathbb{N}$, will also oscillate. Figure 8.2 (b), where the sequence is plotted, shows that this is true. Because only discrete values of cosine are plotted, it does not bear strong resemblance to the familiar cosine wave.

   We conclude that $\displaystyle\lim_{n \to \infty} a_n$ does not exist.

3. We cannot actually apply Theorem 55 here, as the function $f(x) = (-1)^x/x$ is not well defined. (What does $(-1)^{\sqrt{2}}$ mean? In actuality, there is an answer, but it involves *complex analysis*, beyond the scope of this text.) So for now we say that we cannot determine the limit. (But we will be able to very soon.) By looking at the plot in Figure 8.2 (c), we would like to conclude that the sequence converges to 0. That is true, but at this point we are unable to decisively say so.

It seems that $\{(-1)^n/n\}$ converges to 0 but we lack the formal tool to prove it. The following theorem gives us that tool.

Theorem 56   **Absolute Value Theorem**

Let $\{a_n\}$ be a sequence. If $\displaystyle\lim_{n \to \infty} |a_n| = 0$, then $\displaystyle\lim_{n \to \infty} a_n = 0$

Example 233   **Determining the convergence/divergence of a sequence**
Determine the convergence or divergence of the following sequences.

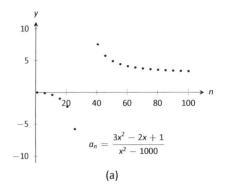

(a)

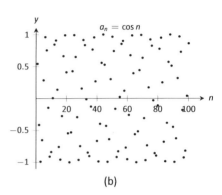

(b)

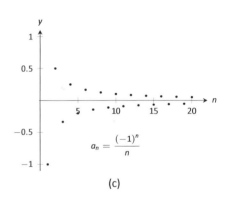

(c)

Figure 8.2: Scatter plots of the sequences in Example 232.

Notes:

1. $\{a_n\} = \left\{\dfrac{(-1)^n}{n}\right\}$ 2. $\{a_n\} = \left\{\dfrac{(-1)^n(n+1)}{n}\right\}$

Solution

1. This appeared in Example 232. We want to apply Theorem 56, so consider the limit of $\{|a_n|\}$:

$$\lim_{n\to\infty} |a_n| = \lim_{n\to\infty} \left|\dfrac{(-1)^n}{n}\right|$$
$$= \lim_{n\to\infty} \dfrac{1}{n}$$
$$= 0.$$

Since this limit is 0, we can apply Theorem 56 and state that $\lim\limits_{n\to\infty} a_n = 0$.

2. Because of the alternating nature of this sequence (i.e., every other term is multiplied by -1), we cannot simply look at the limit $\lim\limits_{x\to\infty} \dfrac{(-1)^x(x+1)}{x}$. We can try to apply the techniques of Theorem 56:

$$\lim_{n\to\infty} |a_n| = \lim_{n\to\infty} \left|\dfrac{(-1)^n(n+1)}{n}\right|$$
$$= \lim_{n\to\infty} \dfrac{n+1}{n}$$
$$= 1.$$

We have concluded that when we ignore the alternating sign, the sequence approaches 1. This means we cannot apply Theorem 56; it states the the limit must be 0 in order to conclude anything.

Since we know that the signs of the terms alternate *and* we know that the limit of $|a_n|$ is 1, we know that as n approaches infinity, the terms will alternate between values close to 1 and -1, meaning the sequence diverges. A plot of this sequence is given in Figure 8.3.

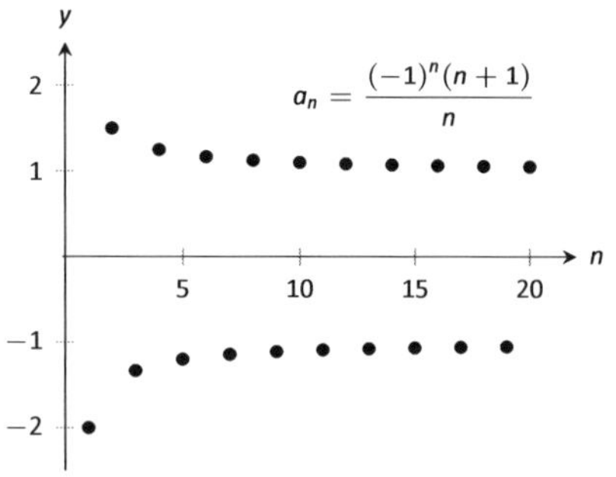

We continue our study of the limits of sequences by considering some of the properties of these limits.

Figure 8.3: A plot of a sequence in Example 233, part 2.

Notes:

8.1 Sequences

> **Theorem 57**   **Properties of the Limits of Sequences**
>
> Let $\{a_n\}$ and $\{b_n\}$ be sequences such that $\lim\limits_{n\to\infty} a_n = L$, $\lim\limits_{n\to\infty} b_n = K$, and let c be a real number.
>
> 1. $\lim\limits_{n\to\infty} (a_n \pm b_n) = L \pm K$
> 2. $\lim\limits_{n\to\infty} (a_n \cdot b_n) = L \cdot K$
> 3. $\lim\limits_{n\to\infty} (a_n/b_n) = L/K,\ K \neq 0$
> 4. $\lim\limits_{n\to\infty} c \cdot a_n = c \cdot L$

Example 234   **Applying properties of limits of sequences**

Let the following sequences, and their limits, be given:

- $\{a_n\} = \left\{\dfrac{n+1}{n^2}\right\}$, and $\lim\limits_{n\to\infty} a_n = 0$;
- $\{b_n\} = \left\{\left(1 + \dfrac{1}{n}\right)^n\right\}$, and $\lim\limits_{n\to\infty} b_n = e$; and
- $\{c_n\} = \{n \cdot \sin(5/n)\}$, and $\lim\limits_{n\to\infty} c_n = 5$.

Evaluate the following limits.

1. $\lim\limits_{n\to\infty} (a_n + b_n)$ 2. $\lim\limits_{n\to\infty} (b_n \cdot c_n)$ 3. $\lim\limits_{n\to\infty} (1000 \cdot a_n)$

SOLUTION   We will use Theorem 57 to answer each of these.

1. Since $\lim\limits_{n\to\infty} a_n = 0$ and $\lim\limits_{n\to\infty} b_n = e$, we conclude that $\lim\limits_{n\to\infty} (a_n + b_n) = 0 + e = e$. So even though we are adding something to each term of the sequence b_n, we are adding something so small that the final limit is the same as before.

2. Since $\lim\limits_{n\to\infty} b_n = e$ and $\lim\limits_{n\to\infty} c_n = 5$, we conclude that $\lim\limits_{n\to\infty} (b_n \cdot c_n) = e \cdot 5 = 5e$.

3. Since $\lim\limits_{n\to\infty} a_n = 0$, we have $\lim\limits_{n\to\infty} 1000 a_n = 1000 \cdot 0 = 0$. It does not matter that we multiply each term by 1000; the sequence still approaches 0. (It just takes longer to get close to 0.)

There is more to learn about sequences than just their limits. We will also study their range and the relationships terms have with the terms that follow. We start with some definitions describing properties of the range.

Notes:

Chapter 8 Sequences and Series

> **Definition 29**   **Bounded and Unbounded Sequences**
>
> A sequence $\{a_n\}$ is said to be **bounded** if there exists real numbers m and M such that $m < a_n < M$ for all n in $\mathbb{N}$.
>
> A sequence $\{a_n\}$ is said to be **unbounded** if it is not bounded.
>
> A sequence $\{a_n\}$ is said to be **bounded above** if there exists an M such that $a_n < M$ for all n in $\mathbb{N}$; it is **bounded below** if there exists an m such that $m < a_n$ for all n in $\mathbb{N}$.

It follows from this definition that an unbounded sequence may be bounded above or bounded below; a sequence that is both bounded above and below is simply a bounded sequence.

Example 235   **Determining boundedness of sequences**
Determine the boundedness of the following sequences.

1. $\{a_n\} = \left\{\dfrac{1}{n}\right\}$ 2. $\{a_n\} = \{2^n\}$

SOLUTION

1. The terms of this sequence are always positive but are decreasing, so we have $0 < a_n < 2$ for all n. Thus this sequence is bounded. Figure 8.4(a) illustrates this.

2. The terms of this sequence obviously grow without bound. However, it is also true that these terms are all positive, meaning $0 < a_n$. Thus we can say the sequence is unbounded, but also bounded below. Figure 8.4(b) illustrates this.

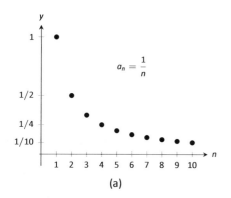

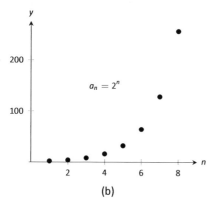

Figure 8.4: A plot of $\{a_n\} = \{1/n\}$ and $\{a_n\} = \{2^n\}$ from Example 235.

The previous example produces some interesting concepts. First, we can recognize that the sequence $\{1/n\}$ converges to 0. This says, informally, that "most" of the terms of the sequence are "really close" to 0. This implies that the sequence is bounded, using the following logic. First, "most" terms are near 0, so we could find some sort of bound on these terms (using Definition 28, the bound is ε). That leaves a "few" terms that are not near 0 (i.e., a *finite* number of terms). A finite list of numbers is always bounded.

This logic implies that if a sequence converges, it must be bounded. This is indeed true, as stated by the following theorem.

Notes:

> **Theorem 58**   **Convergent Sequences are Bounded**
>
> Let $\{a_n\}$ be a convergent sequence. Then $\{a_n\}$ is bounded.

Note: Keep in mind what Theorem 58 does *not* say. It does not say that bounded sequences must converge, nor does it say that if a sequence does not converge, it is not bounded.

In Example 234 we saw the sequence $\{b_n\} = \{(1+1/n)^n\}$, where it was stated that $\lim\limits_{n\to\infty} b_n = e$. (Note that this is simply restating part of Theorem 5.) Even though it may be difficult to intuitively grasp the behavior of this sequence, we know immediately that it is bounded.

Another interesting concept to come out of Example 235 again involves the sequence $\{1/n\}$. We stated, without proof, that the terms of the sequence were decreasing. That is, that $a_{n+1} < a_n$ for all n. (This is easy to show. Clearly $n < n+1$. Taking reciprocals flips the inequality: $1/n > 1/(n+1)$. This is the same as $a_n > a_{n+1}$.) Sequences that either steadily increase or decrease are important, so we give this property a name.

> **Definition 30**   **Monotonic Sequences**
>
> 1. A sequence $\{a_n\}$ is **monotonically increasing** if $a_n \leq a_{n+1}$ for all n, i.e.,
> $$a_1 \leq a_2 \leq a_3 \leq \cdots a_n \leq a_{n+1} \cdots$$
>
> 2. A sequence $\{a_n\}$ is **monotonically decreasing** if $a_n \geq a_{n+1}$ for all n, i.e.,
> $$a_1 \geq a_2 \geq a_3 \geq \cdots a_n \geq a_{n+1} \cdots$$
>
> 3. A sequence is **monotonic** if it is monotonically increasing or monotonically decreasing.

Note: It is sometimes useful to call a monotonically increasing sequence *strictly increasing* if $a_n < a_{n+1}$ for all n; i.e, we remove the possibility that subsequent terms are equal.
A similar statement holds for *strictly decreasing*.

Example 236   **Determining monotonicity**

Determine the monotonicity of the following sequences.

1. $\{a_n\} = \left\{\dfrac{n+1}{n}\right\}$
2. $\{a_n\} = \left\{\dfrac{n^2+1}{n+1}\right\}$
3. $\{a_n\} = \left\{\dfrac{n^2-9}{n^2-10n+26}\right\}$
4. $\{a_n\} = \left\{\dfrac{n^2}{n!}\right\}$

SOLUTION   In each of the following, we will examine $a_{n+1} - a_n$. If $a_{n+1} - a_n > 0$, we conclude that $a_n < a_{n+1}$ and hence the sequence is increasing. If

Notes:

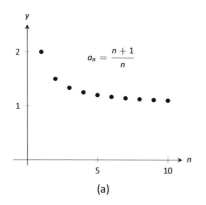

(a)

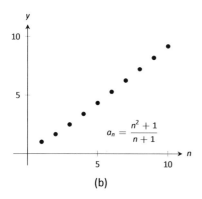

(b)

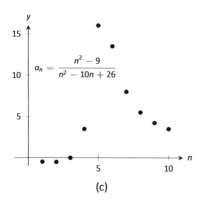

(c)

Figure 8.5: Plots of sequences in Example 236.

$a_{n+1} - a_n < 0$, we conclude that $a_n > a_{n+1}$ and the sequence is decreasing. Of course, a sequence need not be monotonic and perhaps neither of the above will apply.

We also give a scatter plot of each sequence. These are useful as they suggest a pattern of monotonicity, but analytic work should be done to confirm a graphical trend.

1. $$a_{n+1} - a_n = \frac{n+2}{n+1} - \frac{n+1}{n}$$
$$= \frac{(n+2)(n) - (n+1)^2}{(n+1)n}$$
$$= \frac{-1}{n(n+1)}$$
$$< 0 \quad \text{for all } n.$$

Since $a_{n+1} - a_n < 0$ for all n, we conclude that the sequence is decreasing.

2. $$a_{n+1} - a_n = \frac{(n+1)^2 + 1}{n+2} - \frac{n^2+1}{n+1}$$
$$= \frac{((n+1)^2 + 1)(n+1) - (n^2+1)(n+2)}{(n+1)(n+2)}$$
$$= \frac{n^2 + 4n + 1}{(n+1)(n+2)}$$
$$> 0 \quad \text{for all } n.$$

Since $a_{n+1} - a_n > 0$ for all n, we conclude the sequence is increasing.

3. We can clearly see in Figure 8.5 (c), where the sequence is plotted, that it is not monotonic. However, it does seem that after the first 4 terms it is decreasing. To understand why, perform the same analysis as done before:

$$a_{n+1} - a_n = \frac{(n+1)^2 - 9}{(n+1)^2 - 10(n+1) + 26} - \frac{n^2 - 9}{n^2 - 10n + 26}$$
$$= \frac{n^2 + 2n - 8}{n^2 - 8n + 17} - \frac{n^2 - 9}{n^2 - 10n + 26}$$
$$= \frac{(n^2 + 2n - 8)(n^2 - 10n + 26) - (n^2 - 9)(n^2 - 8n + 17)}{(n^2 - 8n + 17)(n^2 - 10n + 26)}$$
$$= \frac{-10n^2 + 60n - 55}{(n^2 - 8n + 17)(n^2 - 10n + 26)}.$$

We want to know when this is greater than, or less than, 0. The denominator is always positive, therefore we are only concerned with the numerator. Using the quadratic formula, we can determine that $-10n^2 + 60n -$

Notes:

$55 = 0$ when $n \approx 1.13, 4.87$. So for $n < 1.13$, the sequence is decreasing. Since we are only dealing with the natural numbers, this means that $a_1 > a_2$.

Between 1.13 and 4.87, i.e., for $n = 2, 3$ and 4, we have that $a_{n+1} > a_n$ and the sequence is increasing. (That is, when $n = 2, 3$ and 4, the numerator $-10n^2 + 60n + 55$ from the fraction above is > 0.)

When $n > 4.87$, i.e, for $n \geq 5$, we have that $-10n^2 + 60n + 55 < 0$, hence $a_{n+1} - a_n < 0$, so the sequence is decreasing.

In short, the sequence is simply not monotonic. However, it is useful to note that for $n \geq 5$, the sequence is monotonically decreasing.

4. Again, the plot in Figure 8.6 shows that the sequence is not monotonic, but it suggests that it is monotonically decreasing after the first term. We perform the usual analysis to confirm this.

$$a_{n+1} - a_n = \frac{(n+1)^2}{(n+1)!} - \frac{n^2}{n!}$$
$$= \frac{(n+1)^2 - n^2(n+1)}{(n+1)!}$$
$$= \frac{-n^3 + 2n + 1}{(n+1)!}$$

When $n = 1$, the above expression is > 0; for $n \geq 2$, the above expression is < 0. Thus this sequence is not monotonic, but it is monotonically decreasing after the first term.

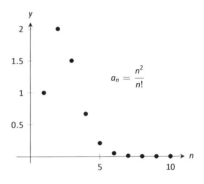

Figure 8.6: A plot of $\{a_n\} = \{n^2/n!\}$ in Example 236.

Knowing that a sequence is monotonic can be useful. In particular, if we know that a sequence is bounded and monotonic, we can conclude it converges! Consider, for example, a sequence that is monotonically decreasing and is bounded below. We know the sequence is always getting smaller, but that there is a bound to how small it can become. This is enough to prove that the sequence will converge, as stated in the following theorem.

Theorem 59   **Bounded Monotonic Sequences are Convergent**

1. Let $\{a_n\}$ be a bounded, monotonic sequence. Then $\{a_n\}$ converges; i.e., $\lim\limits_{n \to \infty} a_n$ exists.

2. Let $\{a_n\}$ be a monotonically increasing sequence that is bounded above. Then $\{a_n\}$ converges.

3. Let $\{a_n\}$ be a monotonically decreasing sequence that is bounded below. Then $\{a_n\}$ converges.

Notes:

Consider once again the sequence $\{a_n\} = \{1/n\}$. It is easy to show it is monotonically decreasing and that it is always positive (i.e., bounded below by 0). Therefore we can conclude by Theorem 59 that the sequence converges. We already knew this by other means, but in the following section this theorem will become very useful.

Sequences are a great source of mathematical inquiry. The On-Line Encyclopedia of Integer Sequences (http://oeis.org) contains thousands of sequences and their formulae. (As of this writing, there are 257,537 sequences in the database.) Perusing this database quickly demonstrates that a single sequence can represent several different "real life" phenomena.

Interesting as this is, our interest actually lies elsewhere. We are more interested in the *sum* of a sequence. That is, given a sequence $\{a_n\}$, we are very interested in $a_1 + a_2 + a_3 + \cdots$. Of course, one might immediately counter with "Doesn't this just add up to 'infinity'?" Many times, yes, but there are many important cases where the answer is no. This is the topic of *series*, which we begin to investigate in the next section.

Notes:

Exercises 8.1

Terms and Concepts

1. Use your own words to define a *sequence*.

2. The domain of a sequence is the _____ numbers.

3. Use your own words to describe the *range* of a sequence.

4. Describe what it means for a sequence to be *bounded*.

Problems

In Exercises 5 – 8, give the first five terms of the given sequence.

5. $\{a_n\} = \left\{\dfrac{4^n}{(n+1)!}\right\}$

6. $\{b_n\} = \left\{\left(-\dfrac{3}{2}\right)^n\right\}$

7. $\{c_n\} = \left\{-\dfrac{n^{n+1}}{n+2}\right\}$

8. $\{d_n\} = \left\{\dfrac{1}{\sqrt{5}}\left(\left(\dfrac{1+\sqrt{5}}{2}\right)^n - \left(\dfrac{1-\sqrt{5}}{2}\right)^n\right)\right\}$

In Exercises 9 – 12, determine the nth term of the given sequence.

9. 4, 7, 10, 13, 16, ...

10. $3, -\dfrac{3}{2}, \dfrac{3}{4}, -\dfrac{3}{8}, \ldots$

11. 10, 20, 40, 80, 160, ...

12. $1, 1, \dfrac{1}{2}, \dfrac{1}{6}, \dfrac{1}{24}, \dfrac{1}{120}, \ldots$

In Exercises 13 – 16, use the following information to determine the limit of the given sequences.

- $\{a_n\} = \left\{\dfrac{2^n - 20}{2^n}\right\}$; $\lim\limits_{n\to\infty} a_n = 1$
- $\{b_n\} = \left\{\left(1+\dfrac{2}{n}\right)^n\right\}$; $\lim\limits_{n\to\infty} b_n = e^2$
- $\{c_n\} = \{\sin(3/n)\}$; $\lim\limits_{n\to\infty} c_n = 0$

13. $\{a_n\} = \left\{\dfrac{2^n - 20}{7 \cdot 2^n}\right\}$

14. $\{a_n\} = \{3b_n - a_n\}$

15. $\{a_n\} = \left\{\sin(3/n)\left(1+\dfrac{2}{n}\right)^n\right\}$

16. $\{a_n\} = \left\{\left(1+\dfrac{2}{n}\right)^{2n}\right\}$

In Exercises 17 – 28, determine whether the sequence converges or diverges. If convergent, give the limit of the sequence.

17. $\{a_n\} = \left\{(-1)^n \dfrac{n}{n+1}\right\}$

18. $\{a_n\} = \left\{\dfrac{4n^2 - n + 5}{3n^2 + 1}\right\}$

19. $\{a_n\} = \left\{\dfrac{4^n}{5^n}\right\}$

20. $\{a_n\} = \left\{\dfrac{n-1}{n} - \dfrac{n}{n-1}\right\}, n \geq 2$

21. $\{a_n\} = \{\ln(n)\}$

22. $\{a_n\} = \left\{\dfrac{3n}{\sqrt{n^2+1}}\right\}$

23. $\{a_n\} = \left\{\left(1+\dfrac{1}{n}\right)^n\right\}$

24. $\{a_n\} = \left\{5 - \dfrac{1}{n}\right\}$

25. $\{a_n\} = \left\{\dfrac{(-1)^{n+1}}{n}\right\}$

26. $\{a_n\} = \left\{\dfrac{1.1^n}{n}\right\}$

27. $\{a_n\} = \left\{\dfrac{2n}{n+1}\right\}$

28. $\{a_n\} = \left\{(-1)^n \dfrac{n^2}{2^n - 1}\right\}$

In Exercises 29 – 34, determine whether the sequence is bounded, bounded above, bounded below, or none of the above.

29. $\{a_n\} = \{\sin n\}$

30. $\{a_n\} = \{\tan n\}$

31. $\{a_n\} = \left\{(-1)^n \dfrac{3n-1}{n}\right\}$

32. $\{a_n\} = \left\{\dfrac{3n^2 - 1}{n}\right\}$

33. $\{a_n\} = \{n \cos n\}$

34. $\{a_n\} = \{2^n - n!\}$

In Exercises 35 – 38, determine whether the sequence is monotonically increasing or decreasing. If it is not, determine if there is an *m* such that it is monotonic for all $n \geq m$.

35. $\{a_n\} = \left\{\dfrac{n}{n+2}\right\}$

36. $\{a_n\} = \left\{\dfrac{n^2 - 6n + 9}{n}\right\}$

37. $\{a_n\} = \left\{(-1)^n \dfrac{1}{n^3}\right\}$

38. $\{a_n\} = \left\{\dfrac{n^2}{2^n}\right\}$

39. Prove Theorem 56; that is, use the definition of the limit of a sequence to show that if $\lim\limits_{n\to\infty} |a_n| = 0$, then $\lim\limits_{n\to\infty} a_n = 0$.

40. Let $\{a_n\}$ and $\{b_n\}$ be sequences such that $\lim\limits_{n\to\infty} a_n = L$ and $\lim\limits_{n\to\infty} b_n = K$.

    (a) Show that if $a_n < b_n$ for all n, then $L \leq K$.

    (b) Give an example where $L = K$.

41. Prove the Squeeze Theorem for sequences: Let $\{a_n\}$ and $\{b_n\}$ be such that $\lim\limits_{n\to\infty} a_n = L$ and $\lim\limits_{n\to\infty} b_n = L$, and let $\{c_n\}$ be such that $a_n \leq c_n \leq b_n$ for all n. Then $\lim\limits_{n\to\infty} c_n = L$

8.2 Infinite Series

Given the sequence $\{a_n\} = \{1/2^n\} = 1/2,\ 1/4,\ 1/8,\ \ldots$, consider the following sums:

$$\begin{aligned} a_1 &= 1/2 &= 1/2 \\ a_1 + a_2 &= 1/2 + 1/4 &= 3/4 \\ a_1 + a_2 + a_3 &= 1/2 + 1/4 + 1/8 &= 7/8 \\ a_1 + a_2 + a_3 + a_4 &= 1/2 + 1/4 + 1/8 + 1/16 &= 15/16 \end{aligned}$$

In general, we can show that

$$a_1 + a_2 + a_3 + \cdots + a_n = \frac{2^n - 1}{2^n} = 1 - \frac{1}{2^n}.$$

Let S_n be the sum of the first n terms of the sequence $\{1/2^n\}$. From the above, we see that $S_1 = 1/2$, $S_2 = 3/4$, etc. Our formula at the end shows that $S_n = 1 - 1/2^n$.

Now consider the following limit: $\lim_{n\to\infty} S_n = \lim_{n\to\infty} \left(1 - 1/2^n\right) = 1$. This limit can be interpreted as saying something amazing: *the sum of all the terms of the sequence $\{1/2^n\}$ is 1.*

This example illustrates some interesting concepts that we explore in this section. We begin this exploration with some definitions.

Definition 31 **Infinite Series, n^{th} Partial Sums, Convergence, Divergence**

Let $\{a_n\}$ be a sequence.

1. The sum $\displaystyle\sum_{n=1}^{\infty} a_n$ is an **infinite series** (or, simply **series**).

2. Let $\displaystyle S_n = \sum_{i=1}^{n} a_i$; the sequence $\{S_n\}$ is the sequence of nth **partial sums** of $\{a_n\}$.

3. If the sequence $\{S_n\}$ converges to L, we say the series $\displaystyle\sum_{n=1}^{\infty} a_n$ **converges** to L, and we write $\displaystyle\sum_{n=1}^{\infty} a_n = L$.

4. If the sequence $\{S_n\}$ diverges, the series $\displaystyle\sum_{n=1}^{\infty} a_n$ **diverges**.

Notes:

Chapter 8 Sequences and Series

Using our new terminology, we can state that the series $\sum_{n=1}^{\infty} 1/2^n$ converges, and $\sum_{n=1}^{\infty} 1/2^n = 1$.

We will explore a variety of series in this section. We start with two series that diverge, showing how we might discern divergence.

Example 237   **Showing series diverge**

1. Let $\{a_n\} = \{n^2\}$. Show $\sum_{n=1}^{\infty} a_n$ diverges.

2. Let $\{b_n\} = \{(-1)^{n+1}\}$. Show $\sum_{n=1}^{\infty} b_n$ diverges.

SOLUTION

1. Consider S_n, the nth partial sum.
$$S_n = a_1 + a_2 + a_3 + \cdots + a_n$$
$$= 1^2 + 2^2 + 3^2 \cdots + n^2.$$

    By Theorem 37, this is
    $$= \frac{n(n+1)(2n+1)}{6}.$$

    Since $\lim_{n\to\infty} S_n = \infty$, we conclude that the series $\sum_{n=1}^{\infty} n^2$ diverges. It is instructive to write $\sum_{n=1}^{\infty} n^2 = \infty$ for this tells us *how* the series diverges: it grows without bound.

    A scatter plot of the sequences $\{a_n\}$ and $\{S_n\}$ is given in Figure 8.7(a). The terms of $\{a_n\}$ are growing, so the terms of the partial sums $\{S_n\}$ are growing even faster, illustrating that the series diverges.

2. The sequence $\{b_n\}$ starts with 1, −1, 1, −1, ... Consider some of the

Notes:

partial sums S_n of $\{b_n\}$:

$$S_1 = 1$$
$$S_2 = 0$$
$$S_3 = 1$$
$$S_4 = 0$$

This pattern repeats; we find that $S_n = \begin{cases} 1 & n \text{ is odd} \\ 0 & n \text{ is even} \end{cases}$. As $\{S_n\}$ oscillates, repeating 1, 0, 1, 0, ..., we conclude that $\lim\limits_{n\to\infty} S_n$ does not exist, hence $\sum\limits_{n=1}^{\infty} (-1)^{n+1}$ diverges.

A scatter plot of the sequence $\{b_n\}$ and the partial sums $\{S_n\}$ is given in Figure 8.7(b). When n is odd, $b_n = S_n$ so the marks for b_n are drawn oversized to show they coincide.

While it is important to recognize when a series diverges, we are generally more interested in the series that converge. In this section we will demonstrate a few general techniques for determining convergence; later sections will delve deeper into this topic.

Geometric Series

One important type of series is a *geometric series*.

Definition 32   **Geometric Series**

A **geometric series** is a series of the form

$$\sum_{n=0}^{\infty} r^n = 1 + r + r^2 + r^3 + \cdots + r^n + \cdots$$

Note that the index starts at $n = 0$, not $n = 1$.

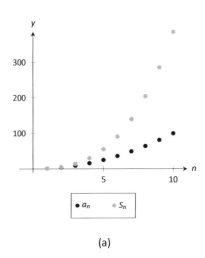

(a)

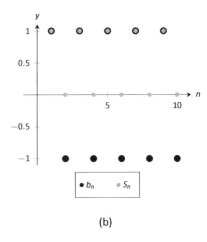

(b)

Figure 8.7: Scatter plots relating to Example 237.

We started this section with a geometric series, although we dropped the first term of 1. One reason geometric series are important is that they have nice convergence properties.

Notes:

Chapter 8 Sequences and Series

> **Theorem 60**   **Convergence of Geometric Series**
>
> Consider the geometric series $\sum_{n=0}^{\infty} r^n$.
>
> 1. The nth partial sum is: $S_n = \dfrac{1 - r^{n+1}}{1 - r}$.
>
> 2. The series converges if, and only if, $|r| < 1$. When $|r| < 1$,
>
> $$\sum_{n=0}^{\infty} r^n = \frac{1}{1-r}.$$

According to Theorem 60, the series

$$\sum_{n=0}^{\infty} \frac{1}{2^n} = \sum_{n=0}^{\infty} \left(\frac{1}{2}\right)^2 = 1 + \frac{1}{2} + \frac{1}{4} + \cdots$$

converges as $r = 1/2$, and $\sum_{n=0}^{\infty} \dfrac{1}{2^n} = \dfrac{1}{1 - 1/2} = 2$. This concurs with our introductory example; while there we got a sum of 1, we skipped the first term of 1.

Example 238   **Exploring geometric series**
Check the convergence of the following series. If the series converges, find its sum.

1. $\displaystyle\sum_{n=2}^{\infty} \left(\frac{3}{4}\right)^n$ 2. $\displaystyle\sum_{n=0}^{\infty} \left(\frac{-1}{2}\right)^n$ 3. $\displaystyle\sum_{n=0}^{\infty} 3^n$

SOLUTION

1. Since $r = 3/4 < 1$, this series converges. By Theorem 60, we have that

$$\sum_{n=0}^{\infty} \left(\frac{3}{4}\right)^n = \frac{1}{1 - 3/4} = 4.$$

However, note the subscript of the summation in the given series: we are to start with $n = 2$. Therefore we subtract off the first two terms, giving:

$$\sum_{n=2}^{\infty} \left(\frac{3}{4}\right)^n = 4 - 1 - \frac{3}{4} = \frac{9}{4}.$$

This is illustrated in Figure 8.8.

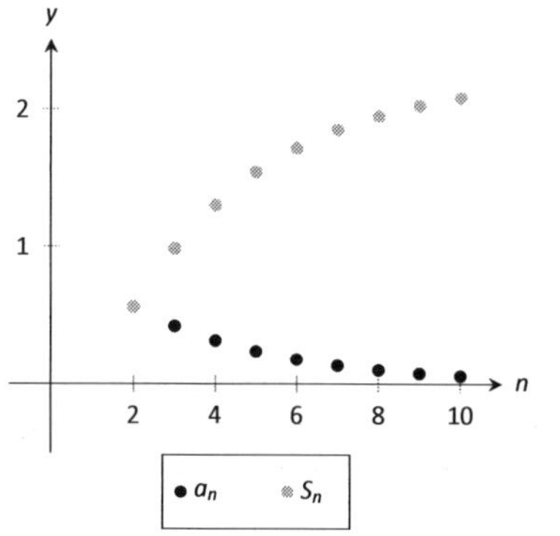

Figure 8.8: Scatter plots relating to the series in Example 238.

Notes:

2. Since $|r| = 1/2 < 1$, this series converges, and by Theorem 60,

$$\sum_{n=0}^{\infty} \left(\frac{-1}{2}\right)^n = \frac{1}{1-(-1/2)} = \frac{2}{3}.$$

The partial sums of this series are plotted in Figure 8.9(a). Note how the partial sums are not purely increasing as some of the terms of the sequence $\{(-1/2)^n\}$ are negative.

3. Since $r > 1$, the series diverges. (This makes "common sense"; we expect the sum

$$1 + 3 + 9 + 27 + 81 + 243 + \cdots$$

to diverge.) This is illustrated in Figure 8.9(b).

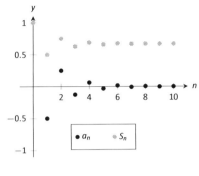

(a)

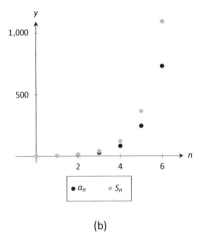

(b)

Figure 8.9: Scatter plots relating to the series in Example 238.

p–Series

Another important type of series is the *p-series*.

Definition 33   **p–Series, General p–Series**

1. A **p–series** is a series of the form

$$\sum_{n=1}^{\infty} \frac{1}{n^p}, \quad \text{where } p > 0.$$

2. A **general p–series** is a series of the form

$$\sum_{n=1}^{\infty} \frac{1}{(an+b)^p}, \quad \text{where } p > 0 \text{ and } a, b \text{ are real numbers.}$$

Like geometric series, one of the nice things about p–series is that they have easy to determine convergence properties.

Theorem 61   **Convergence of General p–Series**

A general p–series $\sum_{n=1}^{\infty} \frac{1}{(an+b)^p}$ will converge if, and only if, $p > 1$.

Note: Theorem 61 assumes that $an+b \neq 0$ for all n. If $an+b = 0$ for some n, then of course the series does not converge regardless of p as not all of the terms of the sequence are defined.

Notes:

Example 239   **Determining convergence of series**
Determine the convergence of the following series.

1. $\displaystyle\sum_{n=1}^{\infty} \frac{1}{n}$

2. $\displaystyle\sum_{n=1}^{\infty} \frac{1}{n^2}$

3. $\displaystyle\sum_{n=1}^{\infty} \frac{1}{\sqrt{n}}$

4. $\displaystyle\sum_{n=1}^{\infty} \frac{(-1)^n}{n}$

5. $\displaystyle\sum_{n=10}^{\infty} \frac{1}{(\frac{1}{2}n - 5)^3}$

6. $\displaystyle\sum_{n=1}^{\infty} \frac{1}{2^n}$

SOLUTION

1. This is a *p*–series with $p = 1$. By Theorem 61, this series diverges.

    This series is a famous series, called the *Harmonic Series*, so named because of its relationship to *harmonics* in the study of music and sound.

2. This is a *p*–series with $p = 2$. By Theorem 61, it converges. Note that the theorem does not give a formula by which we can determine *what* the series converges to; we just know it converges. A famous, unexpected result is that this series converges to $\pi^2/6$.

3. This is a *p*–series with $p = 1/2$; the theorem states that it diverges.

4. This is not a *p*–series; the definition does not allow for alternating signs. Therefore we cannot apply Theorem 61. (Another famous result states that this series, the *Alternating Harmonic Series*, converges to ln 2.)

5. This is a general *p*–series with $p = 3$, therefore it converges.

6. This is not a *p*–series, but a geometric series with $r = 1/2$. It converges.

Later sections will provide tests by which we can determine whether or not a given series converges. This, in general, is much easier than determining *what* a given series converges to. There are many cases, though, where the sum can be determined.

Example 240   **Telescoping series**
Evaluate the sum $\displaystyle\sum_{n=1}^{\infty} \left(\frac{1}{n} - \frac{1}{n+1}\right)$.

Notes:

8.2 Infinite Series

SOLUTION It will help to write down some of the first few partial sums of this series.

$$S_1 = \frac{1}{1} - \frac{1}{2} \hspace{4cm} = 1 - \frac{1}{2}$$

$$S_2 = \left(\frac{1}{1} - \frac{1}{2}\right) + \left(\frac{1}{2} - \frac{1}{3}\right) \hspace{2cm} = 1 - \frac{1}{3}$$

$$S_3 = \left(\frac{1}{1} - \frac{1}{2}\right) + \left(\frac{1}{2} - \frac{1}{3}\right) + \left(\frac{1}{3} - \frac{1}{4}\right) \hspace{1cm} = 1 - \frac{1}{4}$$

$$S_4 = \left(\frac{1}{1} - \frac{1}{2}\right) + \left(\frac{1}{2} - \frac{1}{3}\right) + \left(\frac{1}{3} - \frac{1}{4}\right) + \left(\frac{1}{4} - \frac{1}{5}\right) = 1 - \frac{1}{5}$$

Note how most of the terms in each partial sum are canceled out! In general, we see that $S_n = 1 - \frac{1}{n+1}$. The sequence $\{S_n\}$ converges, as $\lim\limits_{n\to\infty} S_n = \lim\limits_{n\to\infty}\left(1 - \frac{1}{n+1}\right) = 1$, and so we conclude that $\sum\limits_{n=1}^{\infty} \left(\frac{1}{n} - \frac{1}{n+1}\right) = 1$. Partial sums of the series are plotted in Figure 8.10.

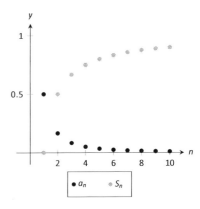

Figure 8.10: Scatter plots relating to the series of Example 240.

The series in Example 240 is an example of a **telescoping series**. Informally, a telescoping series is one in which the partial sums reduce to just a finite number of terms. The partial sum S_n did not contain n terms, but rather just two: 1 and $1/(n+1)$.

When possible, seek a way to write an explicit formula for the nth partial sum S_n. This makes evaluating the limit $\lim\limits_{n\to\infty} S_n$ much more approachable. We do so in the next example.

Example 241   **Evaluating series**
Evaluate each of the following infinite series.

1. $\sum\limits_{n=1}^{\infty} \frac{2}{n^2 + 2n}$ 2. $\sum\limits_{n=1}^{\infty} \ln\left(\frac{n+1}{n}\right)$

SOLUTION

1. We can decompose the fraction $2/(n^2 + 2n)$ as

$$\frac{2}{n^2 + 2n} = \frac{1}{n} - \frac{1}{n+2}.$$

(See Section 6.5, Partial Fraction Decomposition, to recall how this is done, if necessary.)

Notes:

Expressing the terms of $\{S_n\}$ is now more instructive:

$$S_1 = 1 - \frac{1}{3} \qquad\qquad\qquad\qquad\qquad\qquad\qquad\qquad = 1 - \frac{1}{3}$$

$$S_2 = \left(1 - \frac{1}{3}\right) + \left(\frac{1}{2} - \frac{1}{4}\right) \qquad\qquad\qquad\qquad = 1 + \frac{1}{2} - \frac{1}{3} - \frac{1}{4}$$

$$S_3 = \left(1 - \frac{1}{3}\right) + \left(\frac{1}{2} - \frac{1}{4}\right) + \left(\frac{1}{3} - \frac{1}{5}\right) \qquad\qquad = 1 + \frac{1}{2} - \frac{1}{4} - \frac{1}{5}$$

$$S_4 = \left(1 - \frac{1}{3}\right) + \left(\frac{1}{2} - \frac{1}{4}\right) + \left(\frac{1}{3} - \frac{1}{5}\right) + \left(\frac{1}{4} - \frac{1}{6}\right) \qquad = 1 + \frac{1}{2} - \frac{1}{5} - \frac{1}{6}$$

$$S_5 = \left(1 - \frac{1}{3}\right) + \left(\frac{1}{2} - \frac{1}{4}\right) + \left(\frac{1}{3} - \frac{1}{5}\right) + \left(\frac{1}{4} - \frac{1}{6}\right) + \left(\frac{1}{5} - \frac{1}{7}\right) = 1 + \frac{1}{2} - \frac{1}{6} - \frac{1}{7}$$

We again have a telescoping series. In each partial sum, most of the terms cancel and we obtain the formula $S_n = 1 + \dfrac{1}{2} - \dfrac{1}{n+1} - \dfrac{1}{n+2}$. Taking limits allows us to determine the convergence of the series:

$$\lim_{n\to\infty} S_n = \lim_{n\to\infty}\left(1 + \frac{1}{2} - \frac{1}{n+1} - \frac{1}{n+2}\right) = \frac{3}{2}, \quad \text{so} \quad \sum_{n=1}^{\infty} \frac{1}{n^2+2n} = \frac{3}{2}.$$

This is illustrated in Figure 8.11(a).

2. We begin by writing the first few partial sums of the series:

$$S_1 = \ln(2)$$

$$S_2 = \ln(2) + \ln\left(\frac{3}{2}\right)$$

$$S_3 = \ln(2) + \ln\left(\frac{3}{2}\right) + \ln\left(\frac{4}{3}\right)$$

$$S_4 = \ln(2) + \ln\left(\frac{3}{2}\right) + \ln\left(\frac{4}{3}\right) + \ln\left(\frac{5}{4}\right)$$

At first, this does not seem helpful, but recall the logarithmic identity: $\ln x + \ln y = \ln(xy)$. Applying this to S_4 gives:

$$S_4 = \ln(2) + \ln\left(\frac{3}{2}\right) + \ln\left(\frac{4}{3}\right) + \ln\left(\frac{5}{4}\right) = \ln\left(\frac{2}{1}\cdot\frac{3}{2}\cdot\frac{4}{3}\cdot\frac{5}{4}\right) = \ln(5).$$

We can conclude that $\{S_n\} = \{\ln(n+1)\}$. This sequence does not converge, as $\lim_{n\to\infty} S_n = \infty$. Therefore $\sum_{n=1}^{\infty} \ln\left(\dfrac{n+1}{n}\right) = \infty$; the series diverges. Note in Figure 8.11(b) how the sequence of partial sums grows

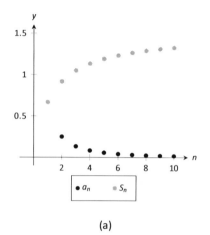

(a)

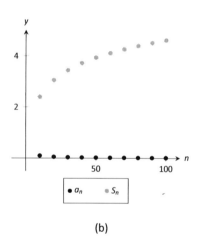

(b)

Figure 8.11: Scatter plots relating to the series in Example 241.

slowly; after 100 terms, it is not yet over 5. Graphically we may be fooled into thinking the series converges, but our analysis above shows that it does not.

We are learning about a new mathematical object, the series. As done before, we apply "old" mathematics to this new topic.

Theorem 62 Properties of Infinite Series

Let $\sum_{n=1}^{\infty} a_n = L$, $\sum_{n=1}^{\infty} b_n = K$, and let c be a constant.

1. Constant Multiple Rule: $\sum_{n=1}^{\infty} c \cdot a_n = c \cdot \sum_{n=1}^{\infty} a_n = c \cdot L$.

2. Sum/Difference Rule: $\sum_{n=1}^{\infty} (a_n \pm b_n) = \sum_{n=1}^{\infty} a_n \pm \sum_{n=1}^{\infty} b_n = L \pm K$.

Before using this theorem, we provide a few "famous" series.

Key Idea 31 Important Series

1. $\sum_{n=0}^{\infty} \dfrac{1}{n!} = e.$ (Note that the index starts with $n = 0$.)

2. $\sum_{n=1}^{\infty} \dfrac{1}{n^2} = \dfrac{\pi^2}{6}.$

3. $\sum_{n=1}^{\infty} \dfrac{(-1)^{n+1}}{n^2} = \dfrac{\pi^2}{12}.$

4. $\sum_{n=0}^{\infty} \dfrac{(-1)^n}{2n+1} = \dfrac{\pi}{4}.$

5. $\sum_{n=1}^{\infty} \dfrac{1}{n}$ diverges. (This is called the *Harmonic Series*.)

6. $\sum_{n=1}^{\infty} \dfrac{(-1)^{n+1}}{n} = \ln 2.$ (This is called the *Alternating Harmonic Series*.)

Notes:

Example 242   **Evaluating series**

Evaluate the given series.

1. $\displaystyle\sum_{n=1}^{\infty} \frac{(-1)^{n+1}(n^2-n)}{n^3}$ 2. $\displaystyle\sum_{n=1}^{\infty} \frac{1000}{n!}$ 3. $\dfrac{1}{16}+\dfrac{1}{25}+\dfrac{1}{36}+\dfrac{1}{49}+\cdots$

Solution

1. We start by using algebra to break the series apart:

$$\sum_{n=1}^{\infty} \frac{(-1)^{n+1}(n^2-n)}{n^3} = \sum_{n=1}^{\infty}\left(\frac{(-1)^{n+1}n^2}{n^3} - \frac{(-1)^{n+1}n}{n^3}\right)$$
$$= \sum_{n=1}^{\infty}\frac{(-1)^{n+1}}{n} - \sum_{n=1}^{\infty}\frac{(-1)^{n+1}}{n^2}$$
$$= \ln(2) - \frac{\pi^2}{12} \approx -0.1293.$$

This is illustrated in Figure 8.12(a).

2. This looks very similar to the series that involves e in Key Idea 31. Note, however, that the series given in this example starts with $n=1$ and not $n=0$. The first term of the series in the Key Idea is $1/0! = 1$, so we will subtract this from our result below:

$$\sum_{n=1}^{\infty}\frac{1000}{n!} = 1000 \cdot \sum_{n=1}^{\infty}\frac{1}{n!}$$
$$= 1000 \cdot (e-1) \approx 1718.28.$$

This is illustrated in Figure 8.12(b). The graph shows how this particular series converges very rapidly.

3. The denominators in each term are perfect squares; we are adding $\displaystyle\sum_{n=4}^{\infty}\frac{1}{n^2}$ (note we start with $n=4$, not $n=1$). This series will converge. Using the

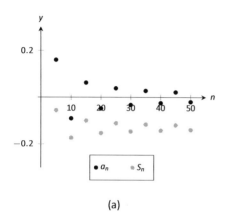

(a)

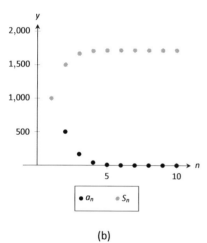

(b)

Figure 8.12: Scatter plots relating to the series in Example 242.

formula from Key Idea 31, we have the following:

$$\sum_{n=1}^{\infty} \frac{1}{n^2} = \sum_{n=1}^{3} \frac{1}{n^2} + \sum_{n=4}^{\infty} \frac{1}{n^2}$$

$$\sum_{n=1}^{\infty} \frac{1}{n^2} - \sum_{n=1}^{3} \frac{1}{n^2} = \sum_{n=4}^{\infty} \frac{1}{n^2}$$

$$\frac{\pi^2}{6} - \left(\frac{1}{1} + \frac{1}{4} + \frac{1}{9}\right) = \sum_{n=4}^{\infty} \frac{1}{n^2}$$

$$\frac{\pi^2}{6} - \frac{49}{36} = \sum_{n=4}^{\infty} \frac{1}{n^2}$$

$$0.2838 \approx \sum_{n=4}^{\infty} \frac{1}{n^2}$$

It may take a while before one is comfortable with this statement, whose truth lies at the heart of the study of infinite series: *it is possible that the sum of an infinite list of nonzero numbers is finite.* We have seen this repeatedly in this section, yet it still may "take some getting used to."

As one contemplates the behavior of series, a few facts become clear.

1. In order to add an infinite list of nonzero numbers and get a finite result, "most" of those numbers must be "very near" 0.

2. If a series diverges, it means that the sum of an infinite list of numbers is not finite (it may approach $\pm\infty$ or it may oscillate), and:

    (a) The series will still diverge if the first term is removed.

    (b) The series will still diverge if the first 10 terms are removed.

    (c) The series will still diverge if the first $1,000,000$ terms are removed.

    (d) The series will still diverge if any finite number of terms from anywhere in the series are removed.

These concepts are very important and lie at the heart of the next two theorems.

Notes:

> **Theorem 63** n^{th}–Term Test for Convergence/Divergence
>
> Consider the series $\sum_{n=1}^{\infty} a_n$.
>
> 1. If $\sum_{n=1}^{\infty} a_n$ converges, then $\lim_{n \to \infty} a_n = 0$.
>
> 2. If $\lim_{n \to \infty} a_n \neq 0$, then $\sum_{n=1}^{\infty} a_n$ diverges.

Note that the two statements in Theorem 63 are really the same. In order to converge, the limit of the terms of the sequence must approach 0; if they do not, the series will not converge.

Looking back, we can apply this theorem to the series in Example 237. In that example, the nth terms of both sequences do not converge to 0, therefore we can quickly conclude that each series diverges.

Important! This theorem *does not state* that if $\lim_{n \to \infty} a_n = 0$ then $\sum_{n=1}^{\infty} a_n$ converges. The standard example of this is the Harmonic Series, as given in Key Idea 31. The Harmonic Sequence, $\{1/n\}$, converges to 0; the Harmonic Series, $\sum_{n=1}^{\infty} 1/n$, diverges.

> **Theorem 64**   **Infinite Nature of Series**
>
> The convergence or divergence remains unchanged by the addition or subtraction of any finite number of terms. That is:
>
> 1. A divergent series will remain divergent with the addition or subtraction of any finite number of terms.
>
> 2. A convergent series will remain convergent with the addition or subtraction of any finite number of terms. (Of course, the *sum* will likely change.)

Consider once more the Harmonic Series $\sum_{n=1}^{\infty} \frac{1}{n}$ which diverges; that is, the

Notes:

sequence of partial sums $\{S_n\}$ grows (very, very slowly) without bound. One might think that by removing the "large" terms of the sequence that perhaps the series will converge. This is simply not the case. For instance, the sum of the first 10 million terms of the Harmonic Series is about 16.7. Removing the first 10 million terms from the Harmonic Series changes the nth partial sums, effectively subtracting 16.7 from the sum. However, a sequence that is growing without bound will still grow without bound when 16.7 is subtracted from it.

The equations below illustrate this. The first line shows the infinite sum of the Harmonic Series split into the sum of the first 10 million terms plus the sum of "everything else." The next equation shows us subtracting these first 10 million terms from both sides. The final equation employs a bit of "psuedo–math": subtracting 16.7 from "infinity" still leaves one with "infinity."

$$\sum_{n=1}^{\infty} \frac{1}{n} = \sum_{n=1}^{10,000,000} \frac{1}{n} + \sum_{n=10,000,001}^{\infty} \frac{1}{n}$$

$$\sum_{n=1}^{\infty} \frac{1}{n} - \sum_{n=1}^{10,000,000} \frac{1}{n} = \sum_{n=10,000,001}^{\infty} \frac{1}{n}$$

$$\infty \quad - \quad 16.7 \quad = \quad \infty.$$

This section introduced us to series and defined a few special types of series whose convergence properties are well known: we know when a *p*-series or a geometric series converges or diverges. Most series that we encounter are not one of these types, but we are still interested in knowing whether or not they converge. The next three sections introduce tests that help us determine whether or not a given series converges.

Notes:

Exercises 8.2

Terms and Concepts

1. Use your own words to describe how sequences and series are related.

2. Use your own words to define a *partial sum*.

3. Given a series $\sum_{n=1}^{\infty} a_n$, describe the two sequences related to the series that are important.

4. Use your own words to explain what a geometric series is.

5. T/F: If $\{a_n\}$ is convergent, then $\sum_{n=1}^{\infty} a_n$ is also convergent.

Problems

In Exercises 6 – 13, a series $\sum_{n=1}^{\infty} a_n$ is given.

(a) Give the first 5 partial sums of the series.

(b) Give a graph of the first 5 terms of a_n and S_n on the same axes.

6. $\sum_{n=1}^{\infty} \frac{(-1)^n}{n}$

7. $\sum_{n=1}^{\infty} \frac{1}{n^2}$

8. $\sum_{n=1}^{\infty} \cos(\pi n)$

9. $\sum_{n=1}^{\infty} n$

10. $\sum_{n=1}^{\infty} \frac{1}{n!}$

11. $\sum_{n=1}^{\infty} \frac{1}{3^n}$

12. $\sum_{n=1}^{\infty} \left(-\frac{9}{10}\right)^n$

13. $\sum_{n=1}^{\infty} \left(\frac{1}{10}\right)^n$

In Exercises 14 – 19, use Theorem 63 to show the given series diverges.

14. $\sum_{n=1}^{\infty} \frac{3n^2}{n(n+2)}$

15. $\sum_{n=1}^{\infty} \frac{2^n}{n^2}$

16. $\sum_{n=1}^{\infty} \frac{n!}{10^n}$

17. $\sum_{n=1}^{\infty} \frac{5^n - n^5}{5^n + n^5}$

18. $\sum_{n=1}^{\infty} \frac{2^n + 1}{2^{n+1}}$

19. $\sum_{n=1}^{\infty} \left(1 + \frac{1}{n}\right)^n$

In Exercises 20 – 29, state whether the given series converges or diverges.

20. $\sum_{n=1}^{\infty} \frac{1}{n^5}$

21. $\sum_{n=0}^{\infty} \frac{1}{5^n}$

22. $\sum_{n=0}^{\infty} \frac{6^n}{5^n}$

23. $\sum_{n=1}^{\infty} n^{-4}$

24. $\sum_{n=1}^{\infty} \sqrt{n}$

25. $\sum_{n=1}^{\infty} \frac{10}{n!}$

26. $\sum_{n=1}^{\infty} \left(\frac{1}{n!} + \frac{1}{n}\right)$

27. $\sum_{n=1}^{\infty} \frac{2}{(2n+8)^2}$

28. $\sum_{n=1}^{\infty} \frac{1}{2n}$

29. $\sum_{n=1}^{\infty} \dfrac{1}{2n-1}$

In Exercises 30 – 44, a series is given.

(a) Find a formula for S_n, the nth partial sum of the series.

(b) Determine whether the series converges or diverges. If it converges, state what it converges to.

30. $\sum_{n=0}^{\infty} \dfrac{1}{4^n}$

31. $1^3 + 2^3 + 3^3 + 4^3 + \cdots$

32. $\sum_{n=1}^{\infty} (-1)^n n$

33. $\sum_{n=0}^{\infty} \dfrac{5}{2^n}$

34. $\sum_{n=1}^{\infty} e^{-n}$

35. $1 - \dfrac{1}{3} + \dfrac{1}{9} - \dfrac{1}{27} + \dfrac{1}{81} + \cdots$

36. $\sum_{n=1}^{\infty} \dfrac{1}{n(n+1)}$

37. $\sum_{n=1}^{\infty} \dfrac{3}{n(n+2)}$

38. $\sum_{n=1}^{\infty} \dfrac{1}{(2n-1)(2n+1)}$

39. $\sum_{n=1}^{\infty} \ln\left(\dfrac{n}{n+1}\right)$

40. $\sum_{n=1}^{\infty} \dfrac{2n+1}{n^2(n+1)^2}$

41. $\dfrac{1}{1\cdot 4} + \dfrac{1}{2\cdot 5} + \dfrac{1}{3\cdot 6} + \dfrac{1}{4\cdot 7} + \cdots$

42. $2 + \left(\dfrac{1}{2} + \dfrac{1}{3}\right) + \left(\dfrac{1}{4} + \dfrac{1}{9}\right) + \left(\dfrac{1}{8} + \dfrac{1}{27}\right) + \cdots$

43. $\sum_{n=2}^{\infty} \dfrac{1}{n^2-1}$

44. $\sum_{n=0}^{\infty} (\sin 1)^n$

45. Break the Harmonic Series into the sum of the odd and even terms:
$$\sum_{n=1}^{\infty} \dfrac{1}{n} = \sum_{n=1}^{\infty} \dfrac{1}{2n-1} + \sum_{n=1}^{\infty} \dfrac{1}{2n}.$$

The goal is to show that each of the series on the right diverge.

(a) Show why $\sum_{n=1}^{\infty} \dfrac{1}{2n-1} > \sum_{n=1}^{\infty} \dfrac{1}{2n}$.

(Compare each nth partial sum.)

(b) Show why $\sum_{n=1}^{\infty} \dfrac{1}{2n-1} < 1 + \sum_{n=1}^{\infty} \dfrac{1}{2n}$

(c) Explain why (a) and (b) demonstrate that the series of odd terms is convergent, if, and only if, the series of even terms is also convergent. (That is, show both converge or both diverge.)

(d) Explain why knowing the Harmonic Series is divergent determines that the even and odd series are also divergent.

46. Show the series $\sum_{n=1}^{\infty} \dfrac{n}{(2n-1)(2n+1)}$ diverges.

8.3 Integral and Comparison Tests

Knowing whether or not a series converges is very important, especially when we discuss Power Series in Section 8.6. Theorems 60 and 61 give criteria for when Geometric and p-series converge, and Theorem 63 gives a quick test to determine if a series diverges. There are many important series whose convergence cannot be determined by these theorems, though, so we introduce a set of tests that allow us to handle a broad range of series. We start with the Integral Test.

Integral Test

We stated in Section 8.1 that a sequence $\{a_n\}$ is a function $a(n)$ whose domain is $\mathbb{N}$, the set of natural numbers. If we can extend $a(n)$ to $\mathbb{R}$, the real numbers, and it is both positive and decreasing on $[1, \infty)$, then the convergence of $\sum_{n=1}^{\infty} a_n$ is the same as $\int_1^{\infty} a(x)\, dx$.

Note: Theorem 65 does not state that the integral and the summation have the same value.

Theorem 65   **Integral Test**

Let a sequence $\{a_n\}$ be defined by $a_n = a(n)$, where $a(n)$ is continuous, positive and decreasing on $[1, \infty)$. Then $\sum_{n=1}^{\infty} a_n$ converges, if and only if, $\int_1^{\infty} a(x)\, dx$ converges.

We can demonstrate the truth of the Integral Test with two simple graphs. In Figure 8.13(a), the height of each rectangle is $a(n) = a_n$ for $n = 1, 2, \ldots$, and clearly the rectangles enclose more area than the area under $y = a(x)$. Therefore we can conclude that

$$\int_1^{\infty} a(x)\, dx < \sum_{n=1}^{\infty} a_n. \tag{8.1}$$

In Figure 8.13(b), we draw rectangles under $y = a(x)$ with the Right-Hand rule, starting with $n = 2$. This time, the area of the rectangles is less than the area under $y = a(x)$, so $\sum_{n=2}^{\infty} a_n < \int_1^{\infty} a(x)\, dx$. Note how this summation starts with $n = 2$; adding a_1 to both sides lets us rewrite the summation starting with

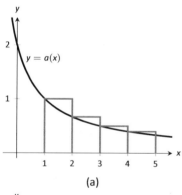

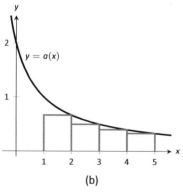

Figure 8.13: Illustrating the truth of the Integral Test.

8.3 Integral and Comparison Tests

$n = 1$:

$$\sum_{n=1}^{\infty} a_n < a_1 + \int_1^{\infty} a(x)\, dx. \qquad (8.2)$$

Combining Equations (8.1) and (8.2), we have

$$\sum_{n=1}^{\infty} a_n < a_1 + \int_1^{\infty} a(x)\, dx < a_1 + \sum_{n=1}^{\infty} a_n. \qquad (8.3)$$

From Equation (8.3) we can make the following two statements:

1. If $\sum_{n=1}^{\infty} a_n$ diverges, so does $\int_1^{\infty} a(x)\, dx$ (because $\sum_{n=1}^{\infty} a_n < a_1 + \int_1^{\infty} a(x)\, dx$)

2. If $\sum_{n=1}^{\infty} a_n$ converges, so does $\int_1^{\infty} a(x)\, dx$ (because $\int_1^{\infty} a(x)\, dx < \sum_{n=1}^{\infty} a_n$.)

Therefore the series and integral either both converge or both diverge. Theorem 64 allows us to extend this theorem to series where $a(n)$ is positive and decreasing on $[b, \infty)$ for some $b > 1$.

Example 243 **Using the Integral Test**

Determine the convergence of $\sum_{n=1}^{\infty} \dfrac{\ln n}{n^2}$. (The terms of the sequence $\{a_n\} = \{\ln n / n^2\}$ and the nth partial sums are given in Figure 8.14.)

SOLUTION Figure 8.14 implies that $a(n) = (\ln n)/n^2$ is positive and decreasing on $[2, \infty)$. We can determine this analytically, too. We know $a(n)$ is positive as both $\ln n$ and n^2 are positive on $[2, \infty)$. To determine that $a(n)$ is decreasing, consider $a'(n) = (1 - 2\ln n)/n^3$, which is negative for $n \geq 2$. Since $a'(n)$ is negative, $a(n)$ is decreasing.

Applying the Integral Test, we test the convergence of $\int_1^{\infty} \dfrac{\ln x}{x^2}\, dx$. Integrating this improper integral requires the use of Integration by Parts, with $u = \ln x$ and $dv = 1/x^2\, dx$.

$$\int_1^{\infty} \frac{\ln x}{x^2}\, dx = \lim_{b \to \infty} \int_1^b \frac{\ln x}{x^2}\, dx$$

$$= \lim_{b \to \infty} -\frac{1}{x} \ln x \Big|_1^b + \int_1^b \frac{1}{x^2}\, dx$$

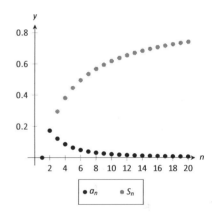

Figure 8.14: Plotting the sequence and series in Example 243.

Notes:

Chapter 8 Sequences and Series

$$= \lim_{b\to\infty} -\frac{1}{x}\ln x - \frac{1}{x}\Big|_1^b$$

$$= \lim_{b\to\infty} 1 - \frac{1}{b} - \frac{\ln b}{b}. \quad \text{Apply L'Hôpital's Rule:}$$

$$= 1.$$

Since $\int_1^\infty \frac{\ln x}{x^2}\, dx$ converges, so does $\sum_{n=1}^\infty \frac{\ln n}{n^2}$.

Theorem 61 was given without justification, stating that the general p-series $\sum_{n=1}^\infty \frac{1}{(an+b)^p}$ converges if, and only if, $p > 1$. In the following example, we prove this to be true by applying the Integral Test.

Example 244   **Using the Integral Test to establish Theorem 61.**
Use the Integral Test to prove that $\sum_{n=1}^\infty \frac{1}{(an+b)^p}$ converges if, and only if, $p > 1$.

SOLUTION   Consider the integral $\int_1^\infty \frac{1}{(ax+b)^p}\, dx$; assuming $p \neq 1$,

$$\int_1^\infty \frac{1}{(ax+b)^p}\, dx = \lim_{c\to\infty} \int_1^c \frac{1}{(ax+b)^p}\, dx$$

$$= \lim_{c\to\infty} \frac{1}{a(1-p)}(ax+b)^{1-p}\Big|_1^c$$

$$= \lim_{c\to\infty} \frac{1}{a(1-p)}\left((ac+b)^{1-p} - (a+b)^{1-p}\right).$$

This limit converges if, and only if, $p > 1$. It is easy to show that the integral also diverges in the case of $p = 1$. (This result is similar to the work preceding Key Idea 21.)

Therefore $\sum_{n=1}^\infty \frac{1}{(an+b)^p}$ converges if, and only if, $p > 1$.

We consider two more convergence tests in this section, both *comparison* tests. That is, we determine the convergence of one series by comparing it to another series with known convergence.

Notes:

Direct Comparison Test

Theorem 66   **Direct Comparison Test**

Let $\{a_n\}$ and $\{b_n\}$ be positive sequences where $a_n \leq b_n$ for all $n \geq N$, for some $N \geq 1$.

1. If $\sum_{n=1}^{\infty} b_n$ converges, then $\sum_{n=1}^{\infty} a_n$ converges.

2. If $\sum_{n=1}^{\infty} a_n$ diverges, then $\sum_{n=1}^{\infty} b_n$ diverges.

Note: A sequence $\{a_n\}$ is a **positive sequence** if $a_n > 0$ for all n.

Because of Theorem 64, any theorem that relies on a positive sequence still holds true when $a_n > 0$ for all but a finite number of values of n.

Example 245   **Applying the Direct Comparison Test**

Determine the convergence of $\sum_{n=1}^{\infty} \dfrac{1}{3^n + n^2}$.

SOLUTION   This series is neither a geometric or p-series, but seems related. We predict it will converge, so we look for a series with larger terms that converges. (Note too that the Integral Test seems difficult to apply here.)

Since $3^n < 3^n + n^2$, $\dfrac{1}{3^n} > \dfrac{1}{3^n + n^2}$ for all $n \geq 1$. The series $\sum_{n=1}^{\infty} \dfrac{1}{3^n}$ is a convergent geometric series; by Theorem 66, $\sum_{n=1}^{\infty} \dfrac{1}{3^n + n^2}$ converges.

Example 246   **Applying the Direct Comparison Test**

Determine the convergence of $\sum_{n=1}^{\infty} \dfrac{1}{n - \ln n}$.

SOLUTION   We know the Harmonic Series $\sum_{n=1}^{\infty} \dfrac{1}{n}$ diverges, and it seems that the given series is closely related to it, hence we predict it will diverge.

Since $n \geq n - \ln n$ for all $n \geq 1$, $\dfrac{1}{n} \leq \dfrac{1}{n - \ln n}$ for all $n \geq 1$.

The Harmonic Series diverges, so we conclude that $\sum_{n=1}^{\infty} \dfrac{1}{n - \ln n}$ diverges as well.

Notes:

The concept of direct comparison is powerful and often relatively easy to apply. Practice helps one develop the necessary intuition to quickly pick a proper series with which to compare. However, it is easy to construct a series for which it is difficult to apply the Direct Comparison Test.

Consider $\sum_{n=1}^{\infty} \frac{1}{n+\ln n}$. It is very similar to the divergent series given in Example 246. We suspect that it also diverges, as $\frac{1}{n} \approx \frac{1}{n+\ln n}$ for large n. However, the inequality that we naturally want to use "goes the wrong way": since $n \leq n+\ln n$ for all $n \geq 1$, $\frac{1}{n} \geq \frac{1}{n+\ln n}$ for all $n \geq 1$. The given series has terms *less than* the terms of a divergent series, and we cannot conclude anything from this.

Fortunately, we can apply another test to the given series to determine its convergence.

Limit Comparison Test

Theorem 67 **Limit Comparison Test**

Let $\{a_n\}$ and $\{b_n\}$ be positive sequences.

1. If $\lim_{n\to\infty} \frac{a_n}{b_n} = L$, where L is a positive real number, then $\sum_{n=1}^{\infty} a_n$ and $\sum_{n=1}^{\infty} b_n$ either both converge or both diverge.

2. If $\lim_{n\to\infty} \frac{a_n}{b_n} = 0$, then if $\sum_{n=1}^{\infty} b_n$ converges, then so does $\sum_{n=1}^{\infty} a_n$.

3. If $\lim_{n\to\infty} \frac{a_n}{b_n} = \infty$, then if $\sum_{n=1}^{\infty} b_n$ diverges, then so does $\sum_{n=1}^{\infty} a_n$.

Theorem 67 is most useful when the convergence of the series from $\{b_n\}$ is known and we are trying to determine the convergence of the series from $\{a_n\}$.

We use the Limit Comparison Test in the next example to examine the series $\sum_{n=1}^{\infty} \frac{1}{n+\ln n}$ which motivated this new test.

Notes:

8.3 Integral and Comparison Tests

Example 247   **Applying the Limit Comparison Test**

Determine the convergence of $\sum_{n=1}^{\infty} \frac{1}{n + \ln n}$ using the Limit Comparison Test.

SOLUTION   We compare the terms of $\sum_{n=1}^{\infty} \frac{1}{n + \ln n}$ to the terms of the Harmonic Sequence $\sum_{n=1}^{\infty} \frac{1}{n}$:

$$\lim_{n \to \infty} \frac{1/(n + \ln n)}{1/n} = \lim_{n \to \infty} \frac{n}{n + \ln n}$$
$$= 1 \quad \text{(after applying L'Hôpital's Rule)}.$$

Since the Harmonic Series diverges, we conclude that $\sum_{n=1}^{\infty} \frac{1}{n + \ln n}$ diverges as well.

Example 248   **Applying the Limit Comparison Test**

Determine the convergence of $\sum_{n=1}^{\infty} \frac{1}{3^n - n^2}$

SOLUTION   This series is similar to the one in Example 245, but now we are considering "$3^n - n^2$" instead of "$3^n + n^2$." This difference makes applying the Direct Comparison Test difficult.

Instead, we use the Limit Comparison Test and compare with the series $\sum_{n=1}^{\infty} \frac{1}{3^n}$:

$$\lim_{n \to \infty} \frac{1/(3^n - n^2)}{1/3^n} = \lim_{n \to \infty} \frac{3^n}{3^n - n^2}$$
$$= 1 \quad \text{(after applying L'Hôpital's Rule twice)}.$$

We know $\sum_{n=1}^{\infty} \frac{1}{3^n}$ is a convergent geometric series, hence $\sum_{n=1}^{\infty} \frac{1}{3^n - n^2}$ converges as well.

As mentioned before, practice helps one develop the intuition to quickly choose a series with which to compare. A general rule of thumb is to pick a series based on the dominant term in the expression of $\{a_n\}$. It is also helpful to note that factorials dominate exponentials, which dominate algebraic functions (e.g., polynomials), which dominate logarithms. In the previous example,

Notes:

the dominant term of $\frac{1}{3^n - n^2}$ was 3^n, so we compared the series to $\sum_{n=1}^{\infty} \frac{1}{3^n}$. It is hard to apply the Limit Comparison Test to series containing factorials, though, as we have not learned how to apply L'Hôpital's Rule to $n!$.

> **Example 249**   **Applying the Limit Comparison Test**
> Determine the convergence of $\sum_{n=1}^{\infty} \frac{\sqrt{x} + 3}{x^2 - x + 1}$.
>
> **SOLUTION**   We naïvely attempt to apply the rule of thumb given above and note that the dominant term in the expression of the series is $1/x^2$. Knowing that $\sum_{n=1}^{\infty} \frac{1}{n^2}$ converges, we attempt to apply the Limit Comparison Test:
>
> $$\lim_{n\to\infty} \frac{(\sqrt{x}+3)/(x^2-x+1)}{1/x^2} = \lim_{n\to\infty} \frac{x^2(\sqrt{x}+3)}{x^2-x+1}$$
> $$= \infty \quad \text{(Apply L'Hôpital's Rule)}.$$
>
> Theorem 67 part (3) only applies when $\sum_{n=1}^{\infty} b_n$ diverges; in our case, it converges. Ultimately, our test has not revealed anything about the convergence of our series.
>
> The problem is that we chose a poor series with which to compare. Since the numerator and denominator of the terms of the series are both algebraic functions, we should have compared our series to the dominant term of the numerator divided by the dominant term of the denominator.
>
> The dominant term of the numerator is $x^{1/2}$ and the dominant term of the denominator is x^2. Thus we should compare the terms of the given series to $x^{1/2}/x^2 = 1/x^{3/2}$:
>
> $$\lim_{n\to\infty} \frac{(\sqrt{x}+3)/(x^2-x+1)}{1/x^{3/2}} = \lim_{n\to\infty} \frac{x^{3/2}(\sqrt{x}+3)}{x^2-x+1}$$
> $$= 1 \quad \text{(Apply L'Hôpital's Rule)}.$$
>
> Since the p-series $\sum_{n=1}^{\infty} \frac{1}{x^{3/2}}$ converges, we conclude that $\sum_{n=1}^{\infty} \frac{\sqrt{x}+3}{x^2-x+1}$ converges as well.

We mentioned earlier that the Integral Test did not work well with series containing factorial terms. The next section introduces the Ratio Test, which does handle such series well. We also introduce the Root Test, which is good for series where each term is raised to a power.

Notes:

Exercises 8.3

Terms and Concepts

1. In order to apply the Integral Test to a sequence $\{a_n\}$, the function $a(n) = a_n$ must be _____, _____ and _____.

2. T/F: The Integral Test can be used to determine the sum of a convergent series.

3. What test(s) in this section do not work well with factorials?

4. Suppose $\sum_{n=0}^{\infty} a_n$ is convergent, and there are sequences $\{b_n\}$ and $\{c_n\}$ such that $b_n \leq a_n \leq c_n$ for all n. What can be said about the series $\sum_{n=0}^{\infty} b_n$ and $\sum_{n=0}^{\infty} c_n$?

Problems

In Exercises 5 – 12, use the Integral Test to determine the convergence of the given series.

5. $\sum_{n=1}^{\infty} \dfrac{1}{2^n}$

6. $\sum_{n=1}^{\infty} \dfrac{1}{n^4}$

7. $\sum_{n=1}^{\infty} \dfrac{n}{n^2 + 1}$

8. $\sum_{n=2}^{\infty} \dfrac{1}{n \ln n}$

9. $\sum_{n=1}^{\infty} \dfrac{1}{n^2 + 1}$

10. $\sum_{n=2}^{\infty} \dfrac{1}{n(\ln n)^2}$

11. $\sum_{n=1}^{\infty} \dfrac{n}{2^n}$

12. $\sum_{n=1}^{\infty} \dfrac{\ln n}{n^3}$

In Exercises 13 – 22, use the Direct Comparison Test to determine the convergence of the given series; state what series is used for comparison.

13. $\sum_{n=1}^{\infty} \dfrac{1}{n^2 + 3n - 5}$

14. $\sum_{n=1}^{\infty} \dfrac{1}{4^n + n^2 - n}$

15. $\sum_{n=1}^{\infty} \dfrac{\ln n}{n}$

16. $\sum_{n=1}^{\infty} \dfrac{1}{n! + n}$

17. $\sum_{n=2}^{\infty} \dfrac{1}{\sqrt{n^2 - 1}}$

18. $\sum_{n=5}^{\infty} \dfrac{1}{\sqrt{n - 2}}$

19. $\sum_{n=1}^{\infty} \dfrac{n^2 + n + 1}{n^3 - 5}$

20. $\sum_{n=1}^{\infty} \dfrac{2^n}{5^n + 10}$

21. $\sum_{n=2}^{\infty} \dfrac{n}{n^2 - 1}$

22. $\sum_{n=2}^{\infty} \dfrac{1}{n^2 \ln n}$

In Exercises 23 – 32, use the Limit Comparison Test to determine the convergence of the given series; state what series is used for comparison.

23. $\sum_{n=1}^{\infty} \dfrac{1}{n^2 - 3n + 5}$

24. $\sum_{n=1}^{\infty} \dfrac{1}{4^n - n^2}$

25. $\sum_{n=4}^{\infty} \dfrac{\ln n}{n - 3}$

26. $\sum_{n=1}^{\infty} \dfrac{1}{\sqrt{n^2 + n}}$

27. $\sum_{n=1}^{\infty} \dfrac{1}{n + \sqrt{n}}$

28. $\sum_{n=1}^{\infty} \dfrac{n - 10}{n^2 + 10n + 10}$

29. $\sum_{n=1}^{\infty} \sin(1/n)$

30. $\sum_{n=1}^{\infty} \dfrac{n+5}{n^3-5}$

31. $\sum_{n=1}^{\infty} \dfrac{\sqrt{n}+3}{n^2+17}$

32. $\sum_{n=1}^{\infty} \dfrac{1}{\sqrt{n+100}}$

In Exercises 33 – 40, determine the convergence of the given series. State the test used; more than one test may be appropriate.

33. $\sum_{n=1}^{\infty} \dfrac{n^2}{2^n}$

34. $\sum_{n=1}^{\infty} \dfrac{1}{(2n+5)^3}$

35. $\sum_{n=1}^{\infty} \dfrac{n!}{10^n}$

36. $\sum_{n=1}^{\infty} \dfrac{\ln n}{n!}$

37. $\sum_{n=1}^{\infty} \dfrac{1}{3^n+n}$

38. $\sum_{n=1}^{\infty} \dfrac{n-2}{10n+5}$

39. $\sum_{n=1}^{\infty} \dfrac{3^n}{n^3}$

40. $\sum_{n=1}^{\infty} \dfrac{\cos(1/n)}{\sqrt{n}}$

41. Given that $\sum_{n=1}^{\infty} a_n$ converges, state which of the following series converges, may converge, or does not converge.

(a) $\sum_{n=1}^{\infty} \dfrac{a_n}{n}$

(b) $\sum_{n=1}^{\infty} a_n a_{n+1}$

(c) $\sum_{n=1}^{\infty} (a_n)^2$

(d) $\sum_{n=1}^{\infty} n a_n$

(e) $\sum_{n=1}^{\infty} \dfrac{1}{a_n}$

8.4 Ratio and Root Tests

The nth–Term Test of Theorem 63 states that in order for a series $\sum_{n=1}^{\infty} a_n$ to converge, $\lim_{n\to\infty} a_n = 0$. That is, the terms of $\{a_n\}$ must get very small. Not only must the terms approach 0, they must approach 0 "fast enough": while $\lim_{n\to\infty} 1/n = 0$, the Harmonic Series $\sum_{n=1}^{\infty} \frac{1}{n}$ diverges as the terms of $\{1/n\}$ do not approach 0 "fast enough."

The comparison tests of the previous section determine convergence by comparing terms of a series to terms of another series whose convergence is known. This section introduces the Ratio and Root Tests, which determine convergence by analyzing the terms of a series to see if they approach 0 "fast enough."

Ratio Test

Theorem 68 Ratio Test

Let $\{a_n\}$ be a positive sequence where $\lim_{n\to\infty} \frac{a_{n+1}}{a_n} = L$.

1. If $L < 1$, then $\sum_{n=1}^{\infty} a_n$ converges.

2. If $L > 1$ or $L = \infty$, then $\sum_{n=1}^{\infty} a_n$ diverges.

3. If $L = 1$, the Ratio Test is inconclusive.

Note: Theorem 64 allows us to apply the Ratio Test to series where $\{a_n\}$ is positive for all but a finite number of terms.

The principle of the Ratio Test is this: if $\lim_{n\to\infty} \frac{a_{n+1}}{a_n} = L < 1$, then for large n, each term of $\{a_n\}$ is significantly smaller than its previous term which is enough to ensure convergence.

Example 250 Applying the Ratio Test

Use the Ratio Test to determine the convergence of the following series:

1. $\sum_{n=1}^{\infty} \frac{2^n}{n!}$
2. $\sum_{n=1}^{\infty} \frac{3^n}{n^3}$
3. $\sum_{n=1}^{\infty} \frac{1}{n^2+1}$.

Notes:

SOLUTION

1. $\sum_{n=1}^{\infty} \frac{2^n}{n!}$:

$$\lim_{n\to\infty} \frac{2^{n+1}/(n+1)!}{2^n/n!} = \lim_{n\to\infty} \frac{2^{n+1}n!}{2^n(n+1)!}$$
$$= \lim_{n\to\infty} \frac{2}{n+1}$$
$$= 0.$$

Since the limit is $0 < 1$, by the Ratio Test $\sum_{n=1}^{\infty} \frac{2^n}{n!}$ converges.

2. $\sum_{n=1}^{\infty} \frac{3^n}{n^3}$:

$$\lim_{n\to\infty} \frac{3^{n+1}/(n+1)^3}{3^n/n^3} = \lim_{n\to\infty} \frac{3^{n+1}n^3}{3^n(n+1)^3}$$
$$= \lim_{n\to\infty} \frac{3n^3}{(n+1)^3}$$
$$= 3.$$

Since the limit is $3 > 1$, by the Ratio Test $\sum_{n=1}^{\infty} \frac{3^n}{n^3}$ diverges.

3. $\sum_{n=1}^{\infty} \frac{1}{n^2+1}$:

$$\lim_{n\to\infty} \frac{1/((n+1)^2+1)}{1/(n^2+1)} = \lim_{n\to\infty} \frac{n^2+1}{(n+1)^2+1}$$
$$= 1.$$

Since the limit is 1, the Ratio Test is inconclusive. We can easily show this series converges using the Direct or Limit Comparison Tests, with each comparing to the series $\sum_{n=1}^{\infty} \frac{1}{n^2}$.

Notes:

The Ratio Test is not effective when the terms of a series *only* contain algebraic functions (e.g., polynomials). It is most effective when the terms contain some factorials or exponentials. The previous example also reinforces our developing intuition: factorials dominate exponentials, which dominate algebraic functions, which dominate logarithmic functions. In Part 1 of the example, the factorial in the denominator dominated the exponential in the numerator, causing the series to converge. In Part 2, the exponential in the numerator dominated the algebraic function in the denominator, causing the series to diverge.

While we have used factorials in previous sections, we have not explored them closely and one is likely to not yet have a strong intuitive sense for how they behave. The following example gives more practice with factorials.

Example 251   **Applying the Ratio Test**
Determine the convergence of $\sum_{n=1}^{\infty} \frac{n!n!}{(2n)!}$.

SOLUTION   Before we begin, be sure to note the difference between $(2n)!$ and $2n!$. When $n = 4$, the former is $8! = 8 \cdot 7 \cdot \ldots \cdot 2 \cdot 1 = 40,320$, whereas the latter is $2(4 \cdot 3 \cdot 2 \cdot 1) = 48$.

Applying the Ratio Test:

$$\lim_{n \to \infty} \frac{(n+1)!(n+1)!/(2(n+1))!}{n!n!/(2n)!} = \lim_{n \to \infty} \frac{(n+1)!(n+1)!(2n)!}{n!n!(2n+2)!}$$

Noting that $(2n+2)! = (2n+2) \cdot (2n+1) \cdot (2n)!$, we have

$$= \lim_{n \to \infty} \frac{(n+1)(n+1)}{(2n+2)(2n+1)}$$
$$= 1/4.$$

Since the limit is $1/4 < 1$, by the Ratio Test we conclude $\sum_{n=1}^{\infty} \frac{n!n!}{(2n)!}$ converges.

Root Test

The final test we introduce is the Root Test, which works particularly well on series where each term is raised to a power, and does not work well with terms containing factorials.

Notes:

Note: Theorem 64 allows us to apply the Root Test to series where $\{a_n\}$ is positive for all but a finite number of terms.

> **Theorem 69   Root Test**
>
> Let $\{a_n\}$ be a positive sequence, and let $\lim_{n\to\infty} (a_n)^{1/n} = L$.
>
> 1. If $L < 1$, then $\sum_{n=1}^{\infty} a_n$ converges.
>
> 2. If $L > 1$ or $L = \infty$, then $\sum_{n=1}^{\infty} a_n$ diverges.
>
> 3. If $L = 1$, the Root Test is inconclusive.

Example 252   Applying the Root Test

Determine the convergence of the following series using the Root Test:

1. $\sum_{n=1}^{\infty} \left(\dfrac{3n+1}{5n-2}\right)^n$ 2. $\sum_{n=1}^{\infty} \dfrac{n^4}{(\ln n)^n}$ 3. $\sum_{n=1}^{\infty} \dfrac{2^n}{n^2}$.

Solution

1. $\lim_{n\to\infty} \left(\left(\dfrac{3n+1}{5n-2}\right)^n\right)^{1/n} = \lim_{n\to\infty} \dfrac{3n+1}{5n-2} = \dfrac{3}{5}$.

   Since the limit is less than 1, we conclude the series converges. Note: it is difficult to apply the Ratio Test to this series.

2. $\lim_{n\to\infty} \left(\dfrac{n^4}{(\ln n)^n}\right)^{1/n} = \lim_{n\to\infty} \dfrac{\left(n^{1/n}\right)^4}{\ln n}$.

   As n grows, the numerator approaches 1 (apply L'Hôpital's Rule) and the denominator grows to infinity. Thus

   $$\lim_{n\to\infty} \dfrac{\left(n^{1/n}\right)^4}{\ln n} = 0.$$

   Since the limit is less than 1, we conclude the series converges.

3. $\lim_{n\to\infty} \left(\dfrac{2^n}{n^2}\right)^{1/n} = \lim_{n\to\infty} \dfrac{2}{\left(n^{1/n}\right)^2} = 2$.

   Since this is greater than 1, we conclude the series diverges.

Each of the tests we have encountered so far has required that we analyze series from *positive* sequences. The next section relaxes this restriction by considering *alternating series*, where the underlying sequence has terms that alternate between being positive and negative.

Notes:

Exercises 8.4

Terms and Concepts

1. The Ratio Test is not effective when the terms of a sequence only contain _____ functions.

2. The Ratio Test is most effective when the terms of a sequence contains _____ and/or _____ functions.

3. What three convergence tests do not work well with terms containing factorials?

4. The Root Test works particularly well on series where each term is _____ to a _____.

Problems

In Exercises 5 – 14, determine the convergence of the given series using the Ratio Test. If the Ratio Test is inconclusive, state so and determine convergence with another test.

5. $\sum_{n=0}^{\infty} \dfrac{2n}{n!}$

6. $\sum_{n=0}^{\infty} \dfrac{5^n - 3n}{4^n}$

7. $\sum_{n=0}^{\infty} \dfrac{n! 10^n}{(2n)!}$

8. $\sum_{n=1}^{\infty} \dfrac{5^n + n^4}{7^n + n^2}$

9. $\sum_{n=1}^{\infty} \dfrac{1}{n}$

10. $\sum_{n=1}^{\infty} \dfrac{1}{3n^3 + 7}$

11. $\sum_{n=1}^{\infty} \dfrac{10 \cdot 5^n}{7^n - 3}$

12. $\sum_{n=1}^{\infty} n \cdot \left(\dfrac{3}{5}\right)^n$

13. $\sum_{n=1}^{\infty} \dfrac{2 \cdot 4 \cdot 6 \cdot 8 \cdots 2n}{3 \cdot 6 \cdot 9 \cdot 12 \cdots 3n}$

14. $\sum_{n=1}^{\infty} \dfrac{n!}{5 \cdot 10 \cdot 15 \cdots (5n)}$

In Exercises 15 – 24, determine the convergence of the given series using the Root Test. If the Root Test is inconclusive, state so and determine convergence with another test.

15. $\sum_{n=1}^{\infty} \left(\dfrac{2n+5}{3n+11}\right)^n$

16. $\sum_{n=1}^{\infty} \left(\dfrac{.9n^2 - n - 3}{n^2 + n + 3}\right)^n$

17. $\sum_{n=1}^{\infty} \dfrac{2^n n^2}{3^n}$

18. $\sum_{n=1}^{\infty} \dfrac{1}{n^n}$

19. $\sum_{n=1}^{\infty} \dfrac{3^n}{n^2 2^{n+1}}$

20. $\sum_{n=1}^{\infty} \dfrac{4^{n+7}}{7^n}$

21. $\sum_{n=1}^{\infty} \left(\dfrac{n^2 - n}{n^2 + n}\right)^n$

22. $\sum_{n=1}^{\infty} \left(\dfrac{1}{n} - \dfrac{1}{n^2}\right)^n$

23. $\sum_{n=1}^{\infty} \dfrac{1}{(\ln n)^n}$

24. $\sum_{n=1}^{\infty} \dfrac{n^2}{(\ln n)^n}$

In Exercises 25 – 34, determine the convergence of the given series. State the test used; more than one test may be appropriate.

25. $\sum_{n=1}^{\infty} \dfrac{n^2 + 4n - 2}{n^3 + 4n^2 - 3n + 7}$

26. $\sum_{n=1}^{\infty} \dfrac{n^4 4^n}{n!}$

27. $\sum_{n=1}^{\infty} \dfrac{n^2}{3^n + n}$

28. $\sum_{n=1}^{\infty} \dfrac{3^n}{n^n}$

29. $\sum_{n=1}^{\infty} \dfrac{n}{\sqrt{n^2 + 4n + 1}}$

30. $\sum_{n=1}^{\infty} \dfrac{n!n!n!}{(3n)!}$

31. $\sum_{n=1}^{\infty} \dfrac{1}{\ln n}$

32. $\sum_{n=1}^{\infty} \left(\dfrac{n+2}{n+1}\right)^n$

33. $\sum_{n=1}^{\infty} \dfrac{n^3}{(\ln n)^n}$

34. $\sum_{n=1}^{\infty} \left(\dfrac{1}{n} - \dfrac{1}{n+2}\right)$

8.5 Alternating Series and Absolute Convergence

All of the series convergence tests we have used require that the underlying sequence $\{a_n\}$ be a positive sequence. (We can relax this with Theorem 64 and state that there must be an $N > 0$ such that $a_n > 0$ for all $n > N$; that is, $\{a_n\}$ is positive for all but a finite number of values of n.)

In this section we explore series whose summation includes negative terms. We start with a very specific form of series, where the terms of the summation alternate between being positive and negative.

Definition 34   **Alternating Series**

Let $\{a_n\}$ be a positive sequence. An **alternating series** is a series of either the form

$$\sum_{n=1}^{\infty}(-1)^n a_n \quad \text{or} \quad \sum_{n=1}^{\infty}(-1)^{n+1} a_n.$$

Recall the terms of Harmonic Series come from the Harmonic Sequence $\{a_n\} = \{1/n\}$. An important alternating series is the **Alternating Harmonic Series**:

$$\sum_{n=1}^{\infty}(-1)^{n+1}\frac{1}{n} = 1 - \frac{1}{2} + \frac{1}{3} - \frac{1}{4} + \frac{1}{5} - \frac{1}{6} + \cdots$$

Geometric Series can also be alternating series when $r < 0$. For instance, if $r = -1/2$, the geometric series is

$$\sum_{n=0}^{\infty}\left(\frac{-1}{2}\right)^n = 1 - \frac{1}{2} + \frac{1}{4} - \frac{1}{8} + \frac{1}{16} - \frac{1}{32} + \cdots$$

Theorem 60 states that geometric series converge when $|r| < 1$ and gives the sum: $\sum_{n=0}^{\infty} r^n = \frac{1}{1-r}$. When $r = -1/2$ as above, we find

$$\sum_{n=0}^{\infty}\left(\frac{-1}{2}\right)^n = \frac{1}{1-(-1/2)} = \frac{1}{3/2} = \frac{2}{3}.$$

A powerful convergence theorem exists for other alternating series that meet a few conditions.

Notes:

> **Theorem 70**   **Alternating Series Test**
>
> Let $\{a_n\}$ be a positive, decreasing sequence where $\lim_{n\to\infty} a_n = 0$. Then
>
> $$\sum_{n=1}^{\infty} (-1)^n a_n \quad \text{and} \quad \sum_{n=1}^{\infty} (-1)^{n+1} a_n$$
>
> converge.

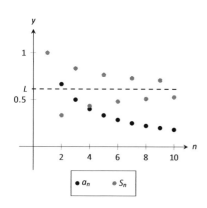

Figure 8.15: Illustrating convergence with the Alternating Series Test.

The basic idea behind Theorem 70 is illustrated in Figure 8.15. A positive, decreasing sequence $\{a_n\}$ is shown along with the partial sums

$$S_n = \sum_{i=1}^{n} (-1)^{i+1} a_i = a_1 - a_2 + a_3 - a_4 + \cdots + (-1)^{n+1} a_n.$$

Because $\{a_n\}$ is decreasing, the amount by which S_n bounces up/down decreases. Moreover, the odd terms of S_n form a decreasing, bounded sequence, while the even terms of S_n form an increasing, bounded sequence. Since bounded, monotonic sequences converge (see Theorem 59) and the terms of $\{a_n\}$ approach 0, one can show the odd and even terms of S_n converge to the same common limit L, the sum of the series.

Example 253   **Applying the Alternating Series Test**
Determine if the Alternating Series Test applies to each of the following series.

1. $\displaystyle\sum_{n=1}^{\infty} (-1)^{n+1} \frac{1}{n}$ 2. $\displaystyle\sum_{n=1}^{\infty} (-1)^{n} \frac{\ln n}{n}$ 3. $\displaystyle\sum_{n=1}^{\infty} (-1)^{n+1} \frac{|\sin n|}{n^2}$

Solution

1. This is the Alternating Harmonic Series as seen previously. The underlying sequence is $\{a_n\} = \{1/n\}$, which is positive, decreasing, and approaches 0 as $n \to \infty$. Therefore we can apply the Alternating Series Test and conclude this series converges.

    While the test does not state what the series converges to, we will see later that $\displaystyle\sum_{n=1}^{\infty} (-1)^{n+1} \frac{1}{n} = \ln 2$.

2. The underlying sequence is $\{a_n\} = \{\ln n/n\}$. This is positive and approaches 0 as $n \to \infty$ (use L'Hôpital's Rule). However, the sequence is not decreasing for all n. It is straightforward to compute $a_1 = 0$, $a_2 \approx 0.347$,

Notes:

$a_3 \approx 0.366$, and $a_4 \approx 0.347$: the sequence is increasing for at least the first 3 terms.

We do not immediately conclude that we cannot apply the Alternating Series Test. Rather, consider the long–term behavior of $\{a_n\}$. Treating $a_n = a(n)$ as a continuous function of n defined on $[1, \infty)$, we can take its derivative:
$$a'(n) = \frac{1 - \ln n}{n^2}.$$

The derivative is negative for all $n \geq 3$ (actually, for all $n > e$), meaning $a(n) = a_n$ is decreasing on $[3, \infty)$. We can apply the Alternating Series Test to the series when we start with $n = 3$ and conclude that $\sum_{n=3}^{\infty}(-1)^n \frac{\ln n}{n}$ converges; adding the terms with $n = 1$ and $n = 2$ do not change the convergence (i.e., we apply Theorem 64).

The important lesson here is that as before, if a series fails to meet the criteria of the Alternating Series Test on only a finite number of terms, we can still apply the test.

3. The underlying sequence is $\{a_n\} = |\sin n|/n$. This sequence is positive and approaches 0 as $n \to \infty$. However, it is not a decreasing sequence; the value of $|\sin n|$ oscillates between 0 and 1 as $n \to \infty$. We cannot remove a finite number of terms to make $\{a_n\}$ decreasing, therefore we cannot apply the Alternating Series Test.

Keep in mind that this does not mean we conclude the series diverges; in fact, it does converge. We are just unable to conclude this based on Theorem 70.

Key Idea 31 gives the sum of some important series. Two of these are

$$\sum_{n=1}^{\infty} \frac{1}{n^2} = \frac{\pi^2}{6} \approx 1.64493 \quad \text{and} \quad \sum_{n=1}^{\infty} \frac{(-1)^{n+1}}{n^2} = \frac{\pi^2}{12} \approx 0.82247.$$

These two series converge to their sums at different rates. To be accurate to two places after the decimal, we need 202 terms of the first series though only 13 of the second. To get 3 places of accuracy, we need 1069 terms of the first series though only 33 of the second. Why is it that the second series converges so much faster than the first?

While there are many factors involved when studying rates of convergence, the alternating structure of an alternating series gives us a powerful tool when approximating the sum of a convergent series.

Notes:

> **Theorem 71**   **The Alternating Series Approximation Theorem**
>
> Let $\{a_n\}$ be a sequence that satisfies the hypotheses of the Alternating Series Test, and let S_n and L be the nth partial sums and sum, respectively, of either $\sum_{n=1}^{\infty}(-1)^n a_n$ or $\sum_{n=1}^{\infty}(-1)^{n+1} a_n$. Then
>
> 1. $|S_n - L| < a_{n+1}$, and
>
> 2. L is between S_n and S_{n+1}.

Part 1 of Theorem 71 states that the nth partial sum of a convergent alternating series will be within a_{n+1} of its total sum. Consider the alternating series we looked at before the statement of the theorem, $\sum_{n=1}^{\infty} \frac{(-1)^{n+1}}{n^2}$. Since $a_{14} = 1/14^2 \approx 0.0051$, we know that S_{13} is within 0.0051 of the total sum.

Moreover, Part 2 of the theorem states that since $S_{13} \approx 0.8252$ and $S_{14} \approx 0.8201$, we know the sum L lies between 0.8201 and 0.8252. One use of this is the knowledge that S_{14} is accurate to two places after the decimal.

Some alternating series converge slowly. In Example 253 we determined the series $\sum_{n=1}^{\infty} (-1)^{n+1} \frac{\ln n}{n}$ converged. With $n = 1001$, we find $\ln n/n \approx 0.0069$, meaning that $S_{1000} \approx 0.1633$ is accurate to one, maybe two, places after the decimal. Since $S_{1001} \approx 0.1564$, we know the sum L is $0.1564 \le L \le 0.1633$.

Example 254   **Approximating the sum of convergent alternating series**
Approximate the sum of the following series, accurate to within 0.001.

1. $\sum_{n=1}^{\infty} (-1)^{n+1} \frac{1}{n^3}$ 2. $\sum_{n=1}^{\infty} (-1)^{n+1} \frac{\ln n}{n}$.

Solution

1. Using Theorem 71, we want to find n where $1/n^3 < 0.001$:

$$\frac{1}{n^3} \le 0.001 = \frac{1}{1000}$$
$$n^3 \ge 1000$$
$$n \ge \sqrt[3]{1000}$$
$$n \ge 10.$$

Notes:

8.5 Alternating Series and Absolute Convergence

Let L be the sum of this series. By Part 1 of the theorem, $|S_9 - L| < a_{10} = 1/1000$. We can compute $S_9 = 0.902116$, which our theorem states is within 0.001 of the total sum.

We can use Part 2 of the theorem to obtain an even more accurate result. As we know the 10th term of the series is $-1/1000$, we can easily compute $S_{10} = 0.901116$. Part 2 of the theorem states that L is between S_9 and S_{10}, so $0.901116 < L < 0.902116$.

2. We want to find n where $\ln(n)/n < 0.001$. We start by solving $\ln(n)/n = 0.001$ for n. This cannot be solved algebraically, so we will use Newton's Method to approximate a solution.

   Let $f(x) = \ln(x)/x - 0.001$; we want to know where $f(x) = 0$. We make a guess that x must be "large," so our initial guess will be $x_1 = 1000$. Recall how Newton's Method works: given an approximate solution x_n, our next approximation x_{n+1} is given by

   $$x_{n+1} = x_n - \frac{f(x_n)}{f'(x_n)}.$$

   We find $f'(x) = (1 - \ln(x))/x^2$. This gives

   $$x_2 = 1000 - \frac{\ln(1000)/1000 - 0.001}{(1 - \ln(1000))/1000^2}$$
   $$= 2000.$$

   Using a computer, we find that Newton's Method seems to converge to a solution $x = 9118.01$ after 8 iterations. Taking the next integer higher, we have $n = 9119$, where $\ln(9119)/9119 = 0.000999903 < 0.001$.

   Again using a computer, we find $S_{9118} = -0.160369$. Part 1 of the theorem states that this is within 0.001 of the actual sum L. Already knowing the 9,119th term, we can compute $S_{9119} = -0.159369$, meaning $-0.159369 < L < -0.160369$.

Notice how the first series converged quite quickly, where we needed only 10 terms to reach the desired accuracy, whereas the second series took over 9,000 terms.

One of the famous results of mathematics is that the Harmonic Series, $\sum_{n=1}^{\infty} \frac{1}{n}$ diverges, yet the Alternating Harmonic Series, $\sum_{n=1}^{\infty} (-1)^{n+1} \frac{1}{n}$, converges. The

Notes:

Chapter 8 Sequences and Series

notion that alternating the signs of the terms in a series can make a series converge leads us to the following definitions.

Definition 35   **Absolute and Conditional Convergence**

1. A series $\sum_{n=1}^{\infty} a_n$ **converges absolutely** if $\sum_{n=1}^{\infty} |a_n|$ converges.

2. A series $\sum_{n=1}^{\infty} a_n$ **converges conditionally** if $\sum_{n=1}^{\infty} a_n$ converges but $\sum_{n=1}^{\infty} |a_n|$ diverges.

Note: In Definition 35, $\sum_{n=1}^{\infty} a_n$ is not necessarily an alternating series; it just may have some negative terms.

Thus we say the Alternating Harmonic Series converges conditionally.

Example 255   **Determining absolute and conditional convergence.**
Determine if the following series converge absolutely, conditionally, or diverge.

1. $\sum_{n=1}^{\infty} (-1)^n \dfrac{n+3}{n^2+2n+5}$ 2. $\sum_{n=1}^{\infty} (-1)^n \dfrac{n^2+2n+5}{2^n}$ 3. $\sum_{n=3}^{\infty} (-1)^n \dfrac{3n-3}{5n-10}$

SOLUTION

1. We can show the series

$$\sum_{n=1}^{\infty} \left|(-1)^n \dfrac{n+3}{n^2+2n+5}\right| = \sum_{n=1}^{\infty} \dfrac{n+3}{n^2+2n+5}$$

diverges using the Limit Comparison Test, comparing with $1/n$.

The series $\sum_{n=1}^{\infty} (-1)^n \dfrac{n+3}{n^2+2n+5}$ converges using the Alternating Series Test; we conclude it converges conditionally.

2. We can show the series

$$\sum_{n=1}^{\infty} \left|(-1)^n \dfrac{n^2+2n+5}{2^n}\right| = \sum_{n=1}^{\infty} \dfrac{n^2+2n+5}{2^n}$$

converges using the Ratio Test.

Notes:

Therefore we conclude $\sum_{n=1}^{\infty}(-1)^n\dfrac{n^2+2n+5}{2^n}$ converges absolutely.

3. The series
$$\sum_{n=3}^{\infty}\left|(-1)^n\dfrac{3n-3}{5n-10}\right|=\sum_{n=3}^{\infty}\dfrac{3n-3}{5n-10}$$
diverges using the nth Term Test, so it does not converge absolutely.

The series $\sum_{n=3}^{\infty}(-1)^n\dfrac{3n-3}{5n-10}$ fails the conditions of the Alternating Series Test as $(3n-3)/(5n-10)$ does not approach 0 as $n\to\infty$. We can state further that this series diverges; as $n\to\infty$, the series effectively adds and subtracts $3/5$ over and over. This causes the sequence of partial sums to oscillate and not converge.

Therefore the series $\sum_{n=1}^{\infty}(-1)^n\dfrac{3n-3}{5n-10}$ diverges.

Knowing that a series converges absolutely allows us to make two important statements, given in the following theorem. The first is that absolute convergence is "stronger" than regular convergence. That is, just because $\sum_{n=1}^{\infty}a_n$ converges, we cannot conclude that $\sum_{n=1}^{\infty}|a_n|$ will converge, but knowing a series converges absolutely tells us that $\sum_{n=1}^{\infty}a_n$ will converge.

One reason this is important is that our convergence tests all require that the underlying sequence of terms be positive. By taking the absolute value of the terms of a series where not all terms are positive, we are often able to apply an appropriate test and determine absolute convergence. This, in turn, determines that the series we are given also converges.

The second statement relates to **rearrangements** of series. When dealing with a finite set of numbers, the sum of the numbers does not depend on the order which they are added. (So $1+2+3=3+1+2$.) One may be surprised to find out that when dealing with an infinite set of numbers, the same statement does not always hold true: some infinite lists of numbers may be rearranged in different orders to achieve different sums. The theorem states that the terms of an absolutely convergent series can be rearranged in any way without affecting the sum.

Notes:

> **Theorem 72**   **Absolute Convergence Theorem**
>
> Let $\sum_{n=1}^{\infty} a_n$ be a series that converges absolutely.
>
> 1. $\sum_{n=1}^{\infty} a_n$ converges.
>
> 2. Let $\{b_n\}$ be any rearrangement of the sequence $\{a_n\}$. Then
> $$\sum_{n=1}^{\infty} b_n = \sum_{n=1}^{\infty} a_n.$$

In Example 255, we determined the series in part 2 converges absolutely. Theorem 72 tells us the series converges (which we could also determine using the Alternating Series Test).

The theorem states that rearranging the terms of an absolutely convergent series does not affect its sum. This implies that perhaps the sum of a conditionally convergent series can change based on the arrangement of terms. Indeed, it can. The Riemann Rearrangement Theorem (named after Bernhard Riemann) states that any conditionally convergent series can have its terms rearranged so that the sum is any desired value, including ∞!

As an example, consider the Alternating Harmonic Series once more. We have stated that

$$\sum_{n=1}^{\infty} (-1)^{n+1} \frac{1}{n} = 1 - \frac{1}{2} + \frac{1}{3} - \frac{1}{4} + \frac{1}{5} - \frac{1}{6} + \frac{1}{7} \cdots = \ln 2,$$

(see Key Idea 31 or Example 253).

Consider the rearrangement where every positive term is followed by two negative terms:

$$1 - \frac{1}{2} - \frac{1}{4} + \frac{1}{3} - \frac{1}{6} - \frac{1}{8} + \frac{1}{5} - \frac{1}{10} - \frac{1}{12} \cdots$$

(Convince yourself that these are exactly the same numbers as appear in the Alternating Harmonic Series, just in a different order.) Now group some terms

Notes:

and simplify:

$$\left(1-\frac{1}{2}\right) - \frac{1}{4} + \left(\frac{1}{3}-\frac{1}{6}\right) - \frac{1}{8} + \left(\frac{1}{5}-\frac{1}{10}\right) - \frac{1}{12} + \cdots =$$
$$\frac{1}{2} - \frac{1}{4} + \frac{1}{6} - \frac{1}{8} + \frac{1}{10} - \frac{1}{12} + \cdots =$$
$$\frac{1}{2}\left(1 - \frac{1}{2} + \frac{1}{3} - \frac{1}{4} + \frac{1}{5} - \frac{1}{6} + \cdots\right) = \frac{1}{2}\ln 2.$$

By rearranging the terms of the series, we have arrived at a different sum! (One could *try* to argue that the Alternating Harmonic Series does not actually converge to ln 2, because rearranging the terms of the series *shouldn't* change the sum. However, the Alternating Series Test proves this series converges to L, for some number L, and if the rearrangement does not change the sum, then $L = L/2$, implying $L = 0$. But the Alternating Series Approximation Theorem quickly shows that $L > 0$. The only conclusion is that the rearrangement *did* change the sum.) This is an incredible result.

We end here our study of tests to determine convergence. The back cover of this text contains a table summarizing the tests that one may find useful.

While series are worthy of study in and of themselves, our ultimate goal within calculus is the study of Power Series, which we will consider in the next section. We will use power series to create functions where the output is the result of an infinite summation.

Notes:

Exercises 8.5

Terms and Concepts

1. Why is $\sum_{n=1}^{\infty} \sin n$ not an alternating series?

2. A series $\sum_{n=1}^{\infty} (-1)^n a_n$ converges when $\{a_n\}$ is _____, _____ and $\lim_{n \to \infty} a_n = $ _____.

3. Give an example of a series where $\sum_{n=0}^{\infty} a_n$ converges but $\sum_{n=0}^{\infty} |a_n|$ does not.

4. The sum of a _____ convergent series can be changed by rearranging the order of its terms.

Problems

In Exercises 5 – 20, an alternating series $\sum_{n=i}^{\infty} a_n$ is given.

(a) Determine if the series converges or diverges.

(b) Determine if $\sum_{n=0}^{\infty} |a_n|$ converges or diverges.

(c) If $\sum_{n=0}^{\infty} a_n$ converges, determine if the convergence is conditional or absolute.

5. $\sum_{n=1}^{\infty} \frac{(-1)^{n+1}}{n^2}$

6. $\sum_{n=1}^{\infty} \frac{(-1)^{n+1}}{\sqrt{n!}}$

7. $\sum_{n=0}^{\infty} (-1)^n \frac{n+5}{3n-5}$

8. $\sum_{n=1}^{\infty} (-1)^n \frac{2^n}{n^2}$

9. $\sum_{n=0}^{\infty} (-1)^{n+1} \frac{3n+5}{n^2 - 3n + 1}$

10. $\sum_{n=1}^{\infty} \frac{(-1)^n}{\ln n + 1}$

11. $\sum_{n=2}^{\infty} (-1)^n \frac{n}{\ln n}$

12. $\sum_{n=1}^{\infty} \frac{(-1)^{n+1}}{1 + 3 + 5 + \cdots + (2n-1)}$

13. $\sum_{n=1}^{\infty} \cos(\pi n)$

14. $\sum_{n=1}^{\infty} \frac{\sin((n+1/2)\pi)}{n \ln n}$

15. $\sum_{n=0}^{\infty} \left(-\frac{2}{3}\right)^n$

16. $\sum_{n=0}^{\infty} (-e)^{-n}$

17. $\sum_{n=0}^{\infty} \frac{(-1)^n n^2}{n!}$

18. $\sum_{n=0}^{\infty} (-1)^n 2^{-n^2}$

19. $\sum_{n=1}^{\infty} \frac{(-1)^n}{\sqrt{n}}$

20. $\sum_{n=1}^{\infty} \frac{(-1000)^n}{n!}$

Let S_n be the nth partial sum of a series. In Exercises 21 – 24, a convergent alternating series is given and a value of n. Compute S_n and S_{n+1} and use these values to find bounds on the sum of the series.

21. $\sum_{n=1}^{\infty} \frac{(-1)^n}{\ln(n+1)}$, $n = 5$

22. $\sum_{n=1}^{\infty} \frac{(-1)^{n+1}}{n^4}$, $n = 4$

23. $\sum_{n=0}^{\infty} \frac{(-1)^n}{n!}$, $n = 6$

24. $\sum_{n=0}^{\infty} \left(-\frac{1}{2}\right)^n$, $n = 9$

In Exercises 25 – 28, a convergent alternating series is given along with its sum and a value of ε. Use Theorem 71 to find n such that the nth partial sum of the series is within ε of the sum of the series.

25. $\sum_{n=1}^{\infty} \frac{(-1)^{n+1}}{n^4} = \frac{7\pi^4}{720}$, $\varepsilon = 0.001$

26. $\sum_{n=0}^{\infty} \frac{(-1)^n}{n!} = \frac{1}{e}, \quad \varepsilon = 0.0001$

27. $\sum_{n=0}^{\infty} \frac{(-1)^n}{2n+1} = \frac{\pi}{4}, \quad \varepsilon = 0.001$

28. $\sum_{n=0}^{\infty} \frac{(-1)^n}{(2n)!} = \cos 1, \quad \varepsilon = 10^{-8}$

8.6 Power Series

So far, our study of series has examined the question of "Is the sum of these infinite terms finite?," i.e., "Does the series converge?" We now approach series from a different perspective: as a function. Given a value of x, we evaluate $f(x)$ by finding the sum of a particular series that depends on x (assuming the series converges). We start this new approach to series with a definition.

> **Definition 36**   **Power Series**
>
> Let $\{a_n\}$ be a sequence, let x be a variable, and let c be a real number.
>
> 1. The **power series in** x is the series
>
> $$\sum_{n=0}^{\infty} a_n x^n = a_0 + a_1 x + a_2 x^2 + a_3 x^3 + \ldots$$
>
> 2. The **power series in** x **centered at** c is the series
>
> $$\sum_{n=0}^{\infty} a_n (x-c)^n = a_0 + a_1(x-c) + a_2(x-c)^2 + a_3(x-c)^3 + \ldots$$

Example 256   **Examples of power series**
Write out the first five terms of the following power series:

1. $\displaystyle\sum_{n=0}^{\infty} x^n$ 2. $\displaystyle\sum_{n=1}^{\infty} (-1)^{n+1} \frac{(x+1)^n}{n}$ 3. $\displaystyle\sum_{n=0}^{\infty} (-1)^{n+1} \frac{(x-\pi)^{2n}}{(2n)!}$.

SOLUTION

1. One of the conventions we adopt is that $x^0 = 1$ regardless of the value of x. Therefore

$$\sum_{n=0}^{\infty} x^n = 1 + x + x^2 + x^3 + x^4 + \ldots$$

This is a geometric series in x.

2. This series is centered at $c = -1$. Note how this series starts with $n = 1$. We could rewrite this series starting at $n = 0$ with the understanding that

Notes:

$a_0 = 0$, and hence the first term is 0.

$$\sum_{n=1}^{\infty}(-1)^{n+1}\frac{(x+1)^n}{n} = (x+1) - \frac{(x+1)^2}{2} + \frac{(x+1)^3}{3} - \frac{(x+1)^4}{4} + \frac{(x+1)^5}{5} \cdots$$

3. This series is centered at $c = \pi$. Recall that $0! = 1$.

$$\sum_{n=0}^{\infty}(-1)^{n+1}\frac{(x-\pi)^{2n}}{(2n)!} = -1 + \frac{(x-\pi)^2}{2} - \frac{(x-\pi)^4}{24} + \frac{(x-\pi)^6}{6!} - \frac{(x-\pi)^8}{8!} \cdots$$

We introduced power series as a type of function, where a value of x is given and the sum of a series is returned. Of course, not every series converges. For instance, in part 1 of Example 256, we recognized the series $\sum_{n=0}^{\infty} x^n$ as a geometric series in x. Theorem 60 states that this series converges only when $|x| < 1$.

This raises the question: "For what values of x will a given power series converge?," which leads us to a theorem and definition.

Theorem 73   **Convergence of Power Series**

Let a power series $\sum_{n=0}^{\infty} a_n(x-c)^n$ be given. Then one of the following is true:

1. The series converges only at $x = c$.

2. There is an $R > 0$ such that the series converges for all x in $(c-R, c+R)$ and diverges for all $x < c-R$ and $x > c+R$.

3. The series converges for all x.

The value of R is important when understanding a power series, hence it is given a name in the following definition. Also, note that part 2 of Theorem 73 makes a statement about the interval $(c-R, c+R)$, but the not the endpoints of that interval. A series may/may not converge at these endpoints.

Notes:

Chapter 8 Sequences and Series

> **Definition 37**   **Radius and Interval of Convergence**
>
> 1. The number R given in Theorem 73 is the **radius of convergence** of a given series. When a series converges for only $x = c$, we say the radius of convergence is 0, i.e., $R = 0$. When a series converges for all x, we say the series has an infinite radius of convergence, i.e., $R = \infty$.
>
> 2. The **interval of convergence** is the set of all values of x for which the series converges.

To find the values of x for which a given series converges, we will use the convergence tests we studied previously (especially the Ratio Test). However, the tests all required that the terms of a series be positive. The following theorem gives us a work–around to this problem.

> **Theorem 74**   **The Radius of Convergence of a Series and Absolute Convergence**
>
> The series $\sum_{n=0}^{\infty} a_n(x-c)^n$ and $\sum_{n=0}^{\infty} \left| a_n(x-c)^n \right|$ have the same radius of convergence R.

Theorem 74 allows us to find the radius of convergence R of a series by applying the Ratio Test (or any applicable test) to the absolute value of the terms of the series. We practice this in the following example.

Example 257   Determining the radius and interval of convergence.
Find the radius and interval of convergence for each of the following series:

1. $\sum_{n=0}^{\infty} \dfrac{x^n}{n!}$
2. $\sum_{n=1}^{\infty} (-1)^{n+1} \dfrac{x^n}{n}$
3. $\sum_{n=0}^{\infty} 2^n(x-3)^n$
4. $\sum_{n=0}^{\infty} n! x^n$

Solution

Notes:

1. We apply the Ratio Test to the series $\sum_{n=0}^{\infty} \left|\dfrac{x^n}{n!}\right|$:

$$\lim_{n\to\infty} \dfrac{|x^{n+1}/(n+1)!|}{|x^n/n!|} = \lim_{n\to\infty} \left|\dfrac{x^{n+1}}{x^n} \cdot \dfrac{n!}{(n+1)!}\right|$$

$$= \lim_{n\to\infty} \left|\dfrac{x}{n+1}\right|$$

$$= 0 \text{ for all } x.$$

The Ratio Test shows us that regardless of the choice of x, the series converges. Therefore the radius of convergence is $R = \infty$, and the interval of convergence is $(-\infty, \infty)$.

2. We apply the Ratio Test to the series $\sum_{n=1}^{\infty} \left|(-1)^{n+1}\dfrac{x^n}{n}\right| = \sum_{n=1}^{\infty} \left|\dfrac{x^n}{n}\right|$:

$$\lim_{n\to\infty} \dfrac{|x^{n+1}/(n+1)|}{|x^n/n|} = \lim_{n\to\infty} \left|\dfrac{x^{n+1}}{x^n} \cdot \dfrac{n}{n+1}\right|$$

$$= \lim_{n\to\infty} |x|\dfrac{n}{n+1}$$

$$= |x|.$$

The Ratio Test states a series converges if the limit of $|a_{n+1}/a_n| = L < 1$. We found the limit above to be $|x|$; therefore, the power series converges when $|x| < 1$, or when x is in $(-1, 1)$. Thus the radius of convergence is $R = 1$.

To determine the interval of convergence, we need to check the endpoints of $(-1, 1)$. When $x = -1$, we have the opposite of the Harmonic Series:

$$\sum_{n=1}^{\infty} (-1)^{n+1}\dfrac{(-1)^n}{n} = \sum_{n=1}^{\infty} \dfrac{-1}{n}$$

$$= -\infty.$$

The series diverges when $x = -1$.

When $x = 1$, we have the series $\sum_{n=1}^{\infty} (-1)^{n+1}\dfrac{(1)^n}{n}$, which is the Alternating Harmonic Series, which converges. Therefore the interval of convergence is $(-1, 1]$.

3. We apply the Ratio Test to the series $\sum_{n=0}^{\infty} \left|2^n(x-3)^n\right|$:

$$\lim_{n\to\infty} \frac{\left|2^{n+1}(x-3)^{n+1}\right|}{\left|2^n(x-3)^n\right|} = \lim_{n\to\infty} \left|\frac{2^{n+1}}{2^n} \cdot \frac{(x-3)^{n+1}}{(x-3)^n}\right|$$

$$= \lim_{n\to\infty} \left|2(x-3)\right|.$$

According to the Ratio Test, the series converges when $|2(x-3)| < 1 \implies |x-3| < 1/2$. The series is centered at 3, and x must be within $1/2$ of 3 in order for the series to converge. Therefore the radius of convergence is $R = 1/2$, and we know that the series converges absolutely for all x in $(3 - 1/2, 3 + 1/2) = (2.5, 3.5)$.

We check for convergence at the endpoints to find the interval of convergence. When $x = 2.5$, we have:

$$\sum_{n=0}^{\infty} 2^n(2.5-3)^n = \sum_{n=0}^{\infty} 2^n(-1/2)^n$$

$$= \sum_{n=0}^{\infty} (-1)^n,$$

which diverges. A similar process shows that the series also diverges at $x = 3.5$. Therefore the interval of convergence is $(2.5, 3.5)$.

4. We apply the Ratio Test to $\sum_{n=0}^{\infty} \left|n!x^n\right|$:

$$\lim_{n\to\infty} \frac{\left|(n+1)!x^{n+1}\right|}{\left|n!x^n\right|} = \lim_{n\to\infty} \left|(n+1)x\right|$$

$$= \infty \text{ for all } x, \text{ except } x = 0.$$

The Ratio Test shows that the series diverges for all x except $x = 0$. Therefore the radius of convergence is $R = 0$.

We can use a power series to define a function:

$$f(x) = \sum_{n=0}^{\infty} a_n x^n$$

where the domain of f is a subset of the interval of convergence of the power series. One can apply calculus techniques to such functions; in particular, we can find derivatives and antiderivatives.

Notes:

> **Theorem 75**   **Derivatives and Indefinite Integrals of Power Series Functions**
>
> Let $f(x) = \sum_{n=0}^{\infty} a_n(x-c)^n$ be a function defined by a power series, with radius of convergence R.
>
> 1. $f(x)$ is continuous and differentiable on $(c-R, c+R)$.
>
> 2. $f'(x) = \sum_{n=1}^{\infty} a_n \cdot n \cdot (x-c)^{n-1}$, with radius of convergence R.
>
> 3. $\int f(x)\,dx = C + \sum_{n=0}^{\infty} a_n \frac{(x-c)^{n+1}}{n+1}$, with radius of convergence R.

A few notes about Theorem 75:

1. The theorem states that differentiation and integration do not change the radius of convergence. It does not state anything about the *interval* of convergence. They are not always the same.

2. Notice how the summation for $f'(x)$ starts with $n=1$. This is because the constant term a_0 of $f(x)$ goes to 0.

3. Differentiation and integration are simply calculated term–by–term using the Power Rules.

Example 258   **Derivatives and indefinite integrals of power series**

Let $f(x) = \sum_{n=0}^{\infty} x^n$. Find $f'(x)$ and $F(x) = \int f(x)\,dx$, along with their respective intervals of convergence.

SOLUTION   We find the derivative and indefinite integral of $f(x)$, following Theorem 75.

1. $f'(x) = \sum_{n=1}^{\infty} nx^{n-1} = 1 + 2x + 3x^2 + 4x^3 + \cdots$.

    In Example 256, we recognized that $\sum_{n=0}^{\infty} x^n$ is a geometric series in x. We know that such a geometric series converges when $|x| < 1$; that is, the interval of convergence is $(-1, 1)$.

To determine the interval of convergence of $f'(x)$, we consider the endpoints of $(-1, 1)$:

$$f'(-1) = 1 - 2 + 3 - 4 + \cdots, \quad \text{which diverges.}$$
$$f'(1) = 1 + 2 + 3 + 4 + \cdots, \quad \text{which diverges.}$$

Therefore, the interval of convergence of $f'(x)$ is $(-1, 1)$.

2. $F(x) = \int f(x)\,dx = C + \sum_{n=0}^{\infty} \frac{x^{n+1}}{n+1} = C + x + \frac{x^2}{2} + \frac{x^3}{3} + \cdots$

To find the interval of convergence of $F(x)$, we again consider the endpoints of $(-1, 1)$:

$$F(-1) = C - 1 + 1/2 - 1/3 + 1/4 + \cdots$$

The value of C is irrelevant; notice that the rest of the series is an Alternating Series that whose terms converge to 0. By the Alternating Series Test, this series converges. (In fact, we can recognize that the terms of the series after C are the opposite of the Alternating Harmonic Series. We can thus say that $F(-1) = C - \ln 2$.)

$$F(1) = C + 1 + 1/2 + 1/3 + 1/4 + \cdots$$

Notice that this summation is $C\ +$ the Harmonic Series, which diverges. Since F converges for $x = -1$ and diverges for $x = 1$, the interval of convergence of $F(x)$ is $[-1, 1)$.

The previous example showed how to take the derivative and indefinite integral of a power series without motivation for why we care about such operations. We may care for the sheer mathematical enjoyment "that we can", which is motivation enough for many. However, we would be remiss to not recognize that we can learn a great deal from taking derivatives and indefinite integrals.

Recall that $f(x) = \sum_{n=0}^{\infty} x^n$ in Example 258 is a geometric series. According to Theorem 60, this series converges to $1/(1-x)$ when $|x| < 1$. Thus we can say

$$f(x) = \sum_{n=0}^{\infty} x^n = \frac{1}{1-x}, \quad \text{on} \quad (-1, 1).$$

Integrating the power series, (as done in Example 258,) we find

$$F(x) = C_1 + \sum_{n=0}^{\infty} \frac{x^{n+1}}{n+1}, \tag{8.4}$$

Notes:

while integrating the function $f(x) = 1/(1-x)$ gives

$$F(x) = -\ln|1-x| + C_2. \tag{8.5}$$

Equating Equations (8.4) and (8.5), we have

$$F(x) = C_1 + \sum_{n=0}^{\infty} \frac{x^{n+1}}{n+1} = -\ln|1-x| + C_2.$$

Letting $x = 0$, we have $F(0) = C_1 = C_2$. This implies that we can drop the constants and conclude

$$\sum_{n=0}^{\infty} \frac{x^{n+1}}{n+1} = -\ln|1-x|.$$

We established in Example 258 that the series on the left converges at $x = -1$; substituting $x = -1$ on both sides of the above equality gives

$$-1 + \frac{1}{2} - \frac{1}{3} + \frac{1}{4} - \frac{1}{5} + \cdots = -\ln 2.$$

On the left we have the opposite of the Alternating Harmonic Series; on the right, we have $-\ln 2$. We conclude that

$$1 - \frac{1}{2} + \frac{1}{3} - \frac{1}{4} + \cdots = \ln 2.$$

Important: We stated in Key Idea 31 (in Section 8.2) that the Alternating Harmonic Series converges to $\ln 2$, and referred to this fact again in Example 253 of Section 8.5. However, we never gave an argument for why this was the case. The work above finally shows how we conclude that the Alternating Harmonic Series converges to $\ln 2$.

We use this type of analysis in the next example.

Example 259   **Analyzing power series functions**

Let $f(x) = \sum_{n=0}^{\infty} \frac{x^n}{n!}$. Find $f'(x)$ and $\int f(x)\, dx$, and use these to analyze the behavior of $f(x)$.

Solution   We start by making two notes: first, in Example 257, we found the interval of convergence of this power series is $(-\infty, \infty)$. Second, we will find it useful later to have a few terms of the series written out:

$$\sum_{n=0}^{\infty} \frac{x^n}{n!} = 1 + x + \frac{x^2}{2} + \frac{x^3}{6} + \frac{x^4}{24} + \cdots \tag{8.6}$$

Notes:

We now find the derivative:

$$f'(x) = \sum_{n=1}^{\infty} n \frac{x^{n-1}}{n!}$$

$$= \sum_{n=1}^{\infty} \frac{x^{n-1}}{(n-1)!} = 1 + x + \frac{x^2}{2!} + \cdots.$$

Since the series starts at $n = 1$ and each term refers to $(n-1)$, we can re-index the series starting with $n = 0$:

$$= \sum_{n=0}^{\infty} \frac{x^n}{n!}$$

$$= f(x).$$

We found the derivative of $f(x)$ is $f(x)$. The only functions for which this is true are of the form $y = ce^x$ for some constant c. As $f(0) = 1$ (see Equation (8.6)), c must be 1. Therefore we conclude that

$$f(x) = \sum_{n=0}^{\infty} \frac{x^n}{n!} = e^x$$

for all x.

We can also find $\int f(x)\,dx$:

$$\int f(x)\,dx = C + \sum_{n=0}^{\infty} \frac{x^{n+1}}{n!(n+1)}$$

$$= C + \sum_{n=0}^{\infty} \frac{x^{n+1}}{(n+1)!}$$

We write out a few terms of this last series:

$$C + \sum_{n=0}^{\infty} \frac{x^{n+1}}{(n+1)!} = C + x + \frac{x^2}{2} + \frac{x^3}{6} + \frac{x^4}{24} + \cdots$$

The integral of $f(x)$ differs from $f(x)$ only by a constant, again indicating that $f(x) = e^x$.

Example 259 and the work following Example 258 established relationships between a power series function and "regular" functions that we have dealt with in the past. In general, given a power series function, it is difficult (if not

Notes:

impossible) to express the function in terms of elementary functions. We chose examples where things worked out nicely.

In this section's last example, we show how to solve a simple differential equation with a power series.

Example 260   **Solving a differential equation with a power series.**
Give the first 4 terms of the power series solution to $y' = 2y$, where $y(0) = 1$.

SOLUTION   The differential equation $y' = 2y$ describes a function $y = f(x)$ where the derivative of y is twice y and $y(0) = 1$. This is a rather simple differential equation; with a bit of thought one should realize that if $y = Ce^{2x}$, then $y' = 2Ce^{2x}$, and hence $y' = 2y$. By letting $C = 1$ we satisfy the initial condition of $y(0) = 1$.

Let's ignore the fact that we already know the solution and find a power series function that satisfies the equation. The solution we seek will have the form

$$f(x) = \sum_{n=0}^{\infty} a_n x^n = a_0 + a_1 x + a_2 x^2 + a_3 x^3 + \cdots$$

for unknown coefficients a_n. We can find $f'(x)$ using Theorem 75:

$$f'(x) = \sum_{n=1}^{\infty} a_n \cdot n \cdot x^{n-1} = a_1 + 2a_2 x + 3a_3 x^2 + 4a_4 x^3 \cdots .$$

Since $f'(x) = 2f(x)$, we have

$$a_1 + 2a_2 x + 3a_3 x^2 + 4a_4 x^3 \cdots = 2(a_0 + a_1 x + a_2 x^2 + a_3 x^3 + \cdots)$$
$$= 2a_0 + 2a_1 x + 2a_2 x^2 + 2a_3 x^3 + \cdots$$

The coefficients of like powers of x must be equal, so we find that

$$a_1 = 2a_0, \quad 2a_2 = 2a_1, \quad 3a_3 = 2a_2, \quad 4a_4 = 2a_3, \quad \text{etc.}$$

The initial condition $y(0) = f(0) = 1$ indicates that $a_0 = 1$; with this, we can find the values of the other coefficients:

$a_0 = 1$ and $a_1 = 2a_0 \Rightarrow a_1 = 2$;
$a_1 = 2$ and $2a_2 = 2a_1 \Rightarrow a_2 = 4/2 = 2$;
$a_2 = 2$ and $3a_3 = 2a_2 \Rightarrow a_3 = 8/(2 \cdot 3) = 4/3$;
$a_3 = 4/3$ and $4a_4 = 2a_3 \Rightarrow a_4 = 16/(2 \cdot 3 \cdot 4) = 2/3$.

Thus the first 5 terms of the power series solution to the differential equation $y' = 2y$ is

$$f(x) = 1 + 2x + 2x^2 + \frac{4}{3}x^3 + \frac{2}{3}x^4 + \cdots$$

Notes:

In Section 8.8, as we study Taylor Series, we will learn how to recognize this series as describing $y = e^{2x}$.

Our last example illustrates that it can be difficult to recognize an elementary function by its power series expansion. It is far easier to start with a known function, expressed in terms of elementary functions, and represent it as a power series function. One may wonder why we would bother doing so, as the latter function probably seems more complicated. In the next two sections, we show both *how* to do this and *why* such a process can be beneficial.

Notes:

Exercises 8.6

Terms and Concepts

1. We adopt the convention that $x^0 =$ _____, regardless of the value of x.

2. What is the difference between the radius of convergence and the interval of convergence?

3. If the radius of convergence of $\sum_{n=0}^{\infty} a_n x^n$ is 5, what is the radius of convergence of $\sum_{n=1}^{\infty} n \cdot a_n x^{n-1}$?

4. If the radius of convergence of $\sum_{n=0}^{\infty} a_n x^n$ is 5, what is the radius of convergence of $\sum_{n=0}^{\infty} (-1)^n a_n x^n$?

Problems

In Exercises 5 – 8, write out the sum of the first 5 terms of the given power series.

5. $\sum_{n=0}^{\infty} 2^n x^n$

6. $\sum_{n=1}^{\infty} \frac{1}{n^2} x^n$

7. $\sum_{n=0}^{\infty} \frac{1}{n!} x^n$

8. $\sum_{n=0}^{\infty} \frac{(-1)^n}{(2n)!} x^{2n}$

In Exercises 9 – 24, a power series is given.

(a) Find the radius of convergence.

(b) Find the interval of convergence.

9. $\sum_{n=0}^{\infty} \frac{(-1)^{n+1}}{n!} x^n$

10. $\sum_{n=0}^{\infty} n x^n$

11. $\sum_{n=1}^{\infty} \frac{(-1)^n (x-3)^n}{n}$

12. $\sum_{n=0}^{\infty} \frac{(x+4)^n}{n!}$

13. $\sum_{n=0}^{\infty} \frac{x^n}{2^n}$

14. $\sum_{n=0}^{\infty} \frac{(-1)^n (x-5)^n}{10^n}$

15. $\sum_{n=0}^{\infty} 5^n (x-1)^n$

16. $\sum_{n=0}^{\infty} (-2)^n x^n$

17. $\sum_{n=0}^{\infty} \sqrt{n} x^n$

18. $\sum_{n=0}^{\infty} \frac{n}{3^n} x^n$

19. $\sum_{n=0}^{\infty} \frac{3^n}{n!} (x-5)^n$

20. $\sum_{n=0}^{\infty} (-1)^n n! (x-10)^n$

21. $\sum_{n=1}^{\infty} \frac{x^n}{n^2}$

22. $\sum_{n=1}^{\infty} \frac{(x+2)^n}{n^3}$

23. $\sum_{n=0}^{\infty} n! \left(\frac{x}{10}\right)^n$

24. $\sum_{n=0}^{\infty} n^2 \left(\frac{x+4}{4}\right)^n$

In Exercises 25 – 30, a function $f(x) = \sum_{n=0}^{\infty} a_n x^n$ is given.

(a) Give a power series for $f'(x)$ and its interval of convergence.

(b) Give a power series for $\int f(x)\, dx$ and its interval of convergence.

25. $\sum_{n=0}^{\infty} n x^n$

26. $\sum_{n=1}^{\infty} \frac{x^n}{n}$

27. $\sum_{n=0}^{\infty} \left(\frac{x}{2}\right)^n$

28. $\sum_{n=0}^{\infty}(-3x)^n$

29. $\sum_{n=0}^{\infty} \frac{(-1)^n x^{2n}}{(2n)!}$

30. $\sum_{n=0}^{\infty} \frac{(-1)^n x^n}{n!}$

In Exercises 31 – 36, give the first 5 terms of the series that is a solution to the given differential equation.

31. $y' = 3y, \quad y(0) = 1$

32. $y' = 5y, \quad y(0) = 5$

33. $y' = y^2, \quad y(0) = 1$

34. $y' = y + 1, \quad y(0) = 1$

35. $y'' = -y, \quad y(0) = 0, y'(0) = 1$

36. $y'' = 2y, \quad y(0) = 1, y'(0) = 1$

8.7 Taylor Polynomials

Consider a function $y = f(x)$ and a point $\big(c, f(c)\big)$. The derivative, $f'(c)$, gives the instantaneous rate of change of f at $x = c$. Of all lines that pass through the point $\big(c, f(c)\big)$, the line that best approximates f at this point is the tangent line; that is, the line whose slope (rate of change) is $f'(c)$.

In Figure 8.16, we see a function $y = f(x)$ graphed. The table below the graph shows that $f(0) = 2$ and $f'(0) = 1$; therefore, the tangent line to f at $x = 0$ is $p_1(x) = 1(x-0)+2 = x+2$. The tangent line is also given in the figure. Note that "near" $x = 0$, $p_1(x) \approx f(x)$; that is, the tangent line approximates f well.

One shortcoming of this approximation is that the tangent line only matches the slope of f; it does not, for instance, match the concavity of f. We can find a polynomial, $p_2(x)$, that does match the concavity without much difficulty, though. The table in Figure 8.16 gives the following information:

$$f(0) = 2 \qquad f'(0) = 1 \qquad f''(0) = 2.$$

Therefore, we want our polynomial $p_2(x)$ to have these same properties. That is, we need

$$p_2(0) = 2 \qquad p_2'(0) = 1 \qquad p_2''(0) = 2.$$

This is simply an initial–value problem. We can solve this using the techniques first described in Section 5.1. To keep $p_2(x)$ as simple as possible, we'll assume that not only $p_2''(0) = 2$, but that $p_2''(x) = 2$. That is, the second derivative of p_2 is constant.

If $p_2''(x) = 2$, then $p_2'(x) = 2x + C$ for some constant C. Since we have determined that $p_2'(0) = 1$, we find that $C = 1$ and so $p_2'(x) = 2x + 1$. Finally, we can compute $p_2(x) = x^2 + x + C$. Using our initial values, we know $p_2(0) = 2$ so $C = 2$. We conclude that $p_2(x) = x^2 + x + 2$. This function is plotted with f in Figure 8.17.

We can repeat this approximation process by creating polynomials of higher degree that match more of the derivatives of f at $x = 0$. In general, a polynomial of degree n can be created to match the first n derivatives of f. Figure 8.17 also shows $p_4(x) = -x^4/2 - x^3/6 + x^2 + x + 2$, whose first four derivatives at 0 match those of f. (Using the table in Figure 8.16, start with $p_4^{(4)}(x) = -12$ and solve the related initial–value problem.)

As we use more and more derivatives, our polynomial approximation to f gets better and better. In this example, the interval on which the approximation is "good" gets bigger and bigger. Figure 8.18 shows $p_{13}(x)$; we can visually affirm that this polynomial approximates f very well on $[-2, 3]$. (The polynomial $p_{13}(x)$ is not particularly "nice". It is

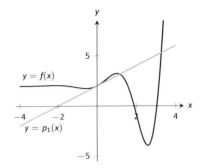

$f(0) = 2 \qquad f'''(0) = -1$
$f'(0) = 1 \qquad f^{(4)}(0) = -12$
$f''(0) = 2 \qquad f^{(5)}(0) = -19$

Figure 8.16: Plotting $y = f(x)$ and a table of derivatives of f evaluated at 0.

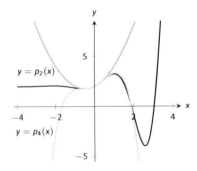

Figure 8.17: Plotting f, p_2 and p_4.

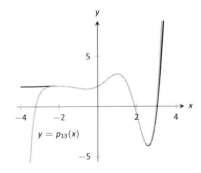

Figure 8.18: Plotting f and p_{13}.

The polynomials we have created are examples of *Taylor polynomials*, named after the British mathematician Brook Taylor who made important discoveries about such functions. While we created the above Taylor polynomials by solving initial–value problems, it can be shown that Taylor polynomials follow a general pattern that make their formation much more direct. This is described in the following definition.

Definition 38   **Taylor Polynomial, Maclaurin Polynomial**

Let f be a function whose first n derivatives exist at $x = c$.

1. The **Taylor polynomial of degree n of f at $x = c$** is

$$p_n(x) = f(c) + f'(c)(x-c) + \frac{f''(c)}{2!}(x-c)^2 + \frac{f'''(c)}{3!}(x-c)^3 + \cdots + \frac{f^{(n)}(c)}{n!}(x-c)^n.$$

2. A special case of the Taylor polynomial is the Maclaurin polynomial, where $c = 0$. That is, the **Maclaurin polynomial of degree n of f** is

$$p_n(x) = f(0) + f'(0)x + \frac{f''(0)}{2!}x^2 + \frac{f'''(0)}{3!}x^3 + \cdots + \frac{f^{(n)}(0)}{n!}x^n.$$

We will practice creating Taylor and Maclaurin polynomials in the following examples.

$$\begin{aligned}
f(x) &= e^x &&\Rightarrow& f(0) &= 1 \\
f'(x) &= e^x &&\Rightarrow& f'(0) &= 1 \\
f''(x) &= e^x &&\Rightarrow& f''(0) &= 1 \\
&\vdots &&& &\vdots \\
f^{(n)}(x) &= e^x &&\Rightarrow& f^{(n)}(0) &= 1
\end{aligned}$$

Figure 8.19: The derivatives of $f(x) = e^x$ evaluated at $x = 0$.

Example 261   **Finding and using Maclaurin polynomials**

1. Find the nth Maclaurin polynomial for $f(x) = e^x$.

2. Use $p_5(x)$ to approximate the value of e.

SOLUTION

1. We start with creating a table of the derivatives of e^x evaluated at $x = 0$. In this particular case, this is relatively simple, as shown in Figure 8.19. By the definition of the Maclaurin series, we have

---

Notes:

8.7 Taylor Polynomials

$$p_n(x) = f(0) + f'(0)x + \frac{f''(0)}{2!}x^2 + \frac{f'''(0)}{3!}x^3 + \cdots + \frac{f^n(0)}{n!}x^n$$

$$= 1 + x + \frac{1}{2}x^2 + \frac{1}{6}x^3 + \frac{1}{24}x^4 + \cdots + \frac{1}{n!}x^n.$$

2. Using our answer from part 1, we have

$$p_5 = 1 + x + \frac{1}{2}x^2 + \frac{1}{6}x^3 + \frac{1}{24}x^4 + \frac{1}{120}x^5.$$

To approximate the value of e, note that $e = e^1 = f(1) \approx p_5(1)$. It is very straightforward to evaluate $p_5(1)$:

$$p_5(1) = 1 + 1 + \frac{1}{2} + \frac{1}{6} + \frac{1}{24} + \frac{1}{120} = \frac{163}{60} \approx 2.71667.$$

A plot of $f(x) = e^x$ and $p_5(x)$ is given in Figure 8.20.

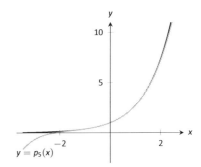

Figure 8.20: A plot of $f(x) = e^x$ and its 5th degree Maclaurin polynomial $p_5(x)$.

Example 262 Finding and using Taylor polynomials

1. Find the nth Taylor polynomial of $y = \ln x$ at $x = 1$.
2. Use $p_6(x)$ to approximate the value of $\ln 1.5$.
3. Use $p_6(x)$ to approximate the value of $\ln 2$.

SOLUTION

1. We begin by creating a table of derivatives of $\ln x$ evaluated at $x = 1$. While this is not as straightforward as it was in the previous example, a pattern does emerge, as shown in Figure 8.21.

Using Definition 38, we have

$$p_n(x) = f(c) + f'(c)(x-c) + \frac{f''(c)}{2!}(x-c)^2 + \frac{f'''(c)}{3!}(x-c)^3 + \cdots + \frac{f^n(c)}{n!}(x-c)^n$$

$$= 0 + (x-1) - \frac{1}{2}(x-1)^2 + \frac{1}{3}(x-1)^3 - \frac{1}{4}(x-1)^4 + \cdots + \frac{(-1)^{n+1}}{n}(x-1)^n.$$

Note how the coefficients of the $(x-1)$ terms turn out to be "nice."

$f(x) = \ln x \Rightarrow f(1) = 0$
$f'(x) = 1/x \Rightarrow f'(1) = 1$
$f''(x) = -1/x^2 \Rightarrow f''(1) = -1$
$f'''(x) = 2/x^3 \Rightarrow f'''(1) = 2$
$f^{(4)}(x) = -6/x^4 \Rightarrow f^{(4)}(1) = -6$
$\vdots \qquad\qquad \vdots$
$f^{(n)}(x) = \dfrac{(-1)^{n+1}(n-1)!}{x^n} \Rightarrow f^{(n)}(1) = (-1)^{n+1}(n-1)!$

Figure 8.21: Derivatives of $\ln x$ evaluated at $x = 1$.

2. We can compute $p_6(x)$ using our work above:

$$p_6(x) = (x-1) - \frac{1}{2}(x-1)^2 + \frac{1}{3}(x-1)^3 - \frac{1}{4}(x-1)^4 + \frac{1}{5}(x-1)^5 - \frac{1}{6}(x-1)^6.$$

Notes:

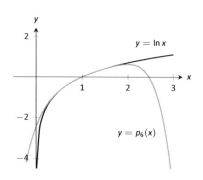

Figure 8.22: A plot of $y = \ln x$ and its 6th degree Taylor polynomial at $x = 1$.

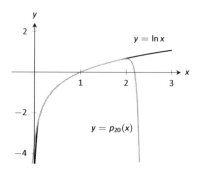

Figure 8.23: A plot of $y = \ln x$ and its 20th degree Taylor polynomial at $x = 1$.

Since $p_6(x)$ approximates $\ln x$ well near $x = 1$, we approximate $\ln 1.5 \approx p_6(1.5)$:

$$p_6(1.5) = (1.5 - 1) - \frac{1}{2}(1.5 - 1)^2 + \frac{1}{3}(1.5 - 1)^3 - \frac{1}{4}(1.5 - 1)^4 + \cdots$$
$$\cdots + \frac{1}{5}(1.5 - 1)^5 - \frac{1}{6}(1.5 - 1)^6$$
$$= \frac{259}{640}$$
$$\approx 0.404688.$$

This is a good approximation as a calculator shows that $\ln 1.5 \approx 0.4055$. Figure 8.22 plots $y = \ln x$ with $y = p_6(x)$. We can see that $\ln 1.5 \approx p_6(1.5)$.

3. We approximate $\ln 2$ with $p_6(2)$:

$$p_6(2) = (2 - 1) - \frac{1}{2}(2 - 1)^2 + \frac{1}{3}(2 - 1)^3 - \frac{1}{4}(2 - 1)^4 + \cdots$$
$$\cdots + \frac{1}{5}(2 - 1)^5 - \frac{1}{6}(2 - 1)^6$$
$$= 1 - \frac{1}{2} + \frac{1}{3} - \frac{1}{4} + \frac{1}{5} - \frac{1}{6}$$
$$= \frac{37}{60}$$
$$\approx 0.616667.$$

This approximation is not terribly impressive: a hand held calculator shows that $\ln 2 \approx 0.693147$. The graph in Figure 8.22 shows that $p_6(x)$ provides less accurate approximations of $\ln x$ as x gets close to 0 or 2.

Surprisingly enough, even the 20th degree Taylor polynomial fails to approximate $\ln x$ for $x > 2$, as shown in Figure 8.23. We'll soon discuss why this is.

Taylor polynomials are used to approximate functions $f(x)$ in mainly two situations:

1. When $f(x)$ is known, but perhaps "hard" to compute directly. For instance, we can define $y = \cos x$ as either the ratio of sides of a right triangle ("adjacent over hypotenuse") or with the unit circle. However, neither of these provides a convenient way of computing $\cos 2$. A Taylor polynomial of sufficiently high degree can provide a reasonable method of computing such values using only operations usually hard–wired into a computer ($+$, $-$, $\times$ and $\div$).

Notes:

2. When $f(x)$ is not known, but information about its derivatives is known. This occurs more often than one might think, especially in the study of differential equations.

In both situations, a critical piece of information to have is "How good is my approximation?" If we use a Taylor polynomial to compute cos 2, how do we know how accurate the approximation is?

We had the same problem when studying Numerical Integration. Theorem 43 provided bounds on the error when using, say, Simpson's Rule to approximate a definite integral. These bounds allowed us to determine that, for instance, using 10 subintervals provided an approximation within $\pm.01$ of the exact value. The following theorem gives similar bounds for Taylor (and hence Maclaurin) polynomials.

Note: Even though Taylor polynomials *could* be used in calculators and computers to calculate values of trigonometric functions, in practice they generally aren't. Other more efficient and accurate methods have been developed, such as the CORDIC algorithm.

Theorem 76   Taylor's Theorem

1. Let f be a function whose $n+1^{th}$ derivative exists on an interval I and let c be in I. Then, for each x in I, there exists z_x between x and c such that

$$f(x) = f(c) + f'(c)(x-c) + \frac{f''(c)}{2!}(x-c)^2 + \cdots + \frac{f^{(n)}(c)}{n!}(x-c)^n + R_n(x),$$

where $R_n(x) = \frac{f^{(n+1)}(z_x)}{(n+1)!}(x-c)^{(n+1)}$.

2. $\left|R_n(x)\right| \leq \frac{\max \left|f^{(n+1)}(z)\right|}{(n+1)!}\left|(x-c)^{(n+1)}\right|$

The first part of Taylor's Theorem states that $f(x) = p_n(x) + R_n(x)$, where $p_n(x)$ is the nth order Taylor polynomial and $R_n(x)$ is the remainder, or error, in the Taylor approximation. The second part gives bounds on how big that error can be. If the $(n+1)^{th}$ derivative is large, the error may be large; if x is far from c, the error may also be large. However, the $(n+1)!$ term in the denominator tends to ensure that the error gets smaller as n increases.

The following example computes error estimates for the approximations of ln 1.5 and ln 2 made in Example 262.

Example 263   Finding error bounds of a Taylor polynomial
Use Theorem 76 to find error bounds when approximating ln 1.5 and ln 2 with $p_6(x)$, the Taylor polynomial of degree 6 of $f(x) = \ln x$ at $x = 1$, as calculated in Example 262.

Notes:

SOLUTION

1. We start with the approximation of ln 1.5 with $p_6(1.5)$. The theorem references an open interval I that contains both x and c. The smaller the interval we use the better; it will give us a more accurate (and smaller!) approximation of the error. We let $I = (0.9, 1.6)$, as this interval contains both $c = 1$ and $x = 1.5$.

   The theorem references $\max\left|f^{(n+1)}(z)\right|$. In our situation, this is asking "How big can the 7th derivative of $y = \ln x$ be on the interval $(0.9, 1.6)$?" The seventh derivative is $y = -6!/x^7$. The largest value it attains on I is about 1506. Thus we can bound the error as:

   $$\begin{aligned}\left|R_6(1.5)\right| &\leq \frac{\max\left|f^{(7)}(z)\right|}{7!}\left|(1.5-1)^7\right| \\ &\leq \frac{1506}{5040} \cdot \frac{1}{2^7} \\ &\approx 0.0023.\end{aligned}$$

   We computed $p_6(1.5) = 0.404688$; using a calculator, we find ln 1.5 $\approx$ 0.405465, so the actual error is about 0.000778, which is less than our bound of 0.0023. This affirms Taylor's Theorem; the theorem states that our approximation would be within about 2 thousandths of the actual value, whereas the approximation was actually closer.

2. We again find an interval I that contains both $c = 1$ and $x = 2$; we choose $I = (0.9, 2.1)$. The maximum value of the seventh derivative of f on this interval is again about 1506 (as the largest values come near $x = 0.9$). Thus

   $$\begin{aligned}\left|R_6(2)\right| &\leq \frac{\max\left|f^{(7)}(z)\right|}{7!}\left|(2-1)^7\right| \\ &\leq \frac{1506}{5040} \cdot 1^7 \\ &\approx 0.30.\end{aligned}$$

   This bound is not as nearly as good as before. Using the degree 6 Taylor polynomial at $x = 1$ will bring us within 0.3 of the correct answer. As $p_6(2) \approx 0.61667$, our error estimate guarantees that the actual value of ln 2 is somewhere between 0.31667 and 0.91667. These bounds are not particularly useful.

   In reality, our approximation was only off by about 0.07. However, we are approximating ostensibly because we do not know the real answer. In order to be assured that we have a good approximation, we would have to resort to using a polynomial of higher degree.

Notes:

8.7 Taylor Polynomials

We practice again. This time, we use Taylor's theorem to find n that guarantees our approximation is within a certain amount.

Example 264   **Finding sufficiently accurate Taylor polynomials**
Find n such that the nth Taylor polynomial of $f(x) = \cos x$ at $x = 0$ approximates $\cos 2$ to within 0.001 of the actual answer. What is $p_n(2)$?

SOLUTION   Following Taylor's theorem, we need bounds on the size of the derivatives of $f(x) = \cos x$. In the case of this trigonometric function, this is easy. All derivatives of cosine are $\pm \sin x$ or $\pm \cos x$. In all cases, these functions are never greater than 1 in absolute value. We want the error to be less than 0.001. To find the appropriate n, consider the following inequalities:

$$\frac{\max \left|f^{(n+1)}(z)\right|}{(n+1)!}\left|(2-0)^{(n+1)}\right| \leq 0.001$$

$$\frac{1}{(n+1)!} \cdot 2^{(n+1)} \leq 0.001$$

We find an n that satisfies this last inequality with trial–and–error. When $n = 8$, we have $\dfrac{2^{8+1}}{(8+1)!} \approx 0.0014$; when $n = 9$, we have $\dfrac{2^{9+1}}{(9+1)!} \approx 0.000282 <$ 0.001. Thus we want to approximate $\cos 2$ with $p_9(2)$.

We now set out to compute $p_9(x)$. We again need a table of the derivatives of $f(x) = \cos x$ evaluated at $x = 0$. A table of these values is given in Figure 8.24. Notice how the derivatives, evaluated at $x = 0$, follow a certain pattern. All the odd powers of x in the Taylor polynomial will disappear as their coefficient is 0. While our error bounds state that we need $p_9(x)$, our work shows that this will be the same as $p_8(x)$.

Since we are forming our polynomial at $x = 0$, we are creating a Maclaurin polynomial, and:

$$p_8(x) = f(0) + f'(0)x + \frac{f''(0)}{2!}x^2 + \frac{f'''(0)}{3!}x^3 + \cdots + \frac{f^{(8)}}{8!}x^8$$

$$= 1 - \frac{1}{2!}x^2 + \frac{1}{4!}x^4 - \frac{1}{6!}x^6 + \frac{1}{8!}x^8$$

We finally approximate $\cos 2$:

$$\cos 2 \approx p_8(2) = -\frac{131}{315} \approx -0.41587.$$

Our error bound guarantee that this approximation is within 0.001 of the correct answer. Technology shows us that our approximation is actually within about 0.0003 of the correct answer.

Figure 8.25 shows a graph of $y = p_8(x)$ and $y = \cos x$. Note how well the two functions agree on about $(-\pi, \pi)$.

$$\begin{array}{ll}
f(x) = \cos x & \Rightarrow \quad f(0) = 1 \\
f'(x) = -\sin x & \Rightarrow \quad f'(0) = 0 \\
f''(x) = -\cos x & \Rightarrow \quad f''(0) = -1 \\
f'''(x) = \sin x & \Rightarrow \quad f'''(0) = 0 \\
f^{(4)}(x) = \cos x & \Rightarrow \quad f^{(4)}(0) = 1 \\
f^{(5)}(x) = -\sin x & \Rightarrow \quad f^{(5)}(0) = 0 \\
f^{(6)}(x) = -\cos x & \Rightarrow \quad f^{(6)}(0) = -1 \\
f^{(7)}(x) = \sin x & \Rightarrow \quad f^{(7)}(0) = 0 \\
f^{(8)}(x) = \cos x & \Rightarrow \quad f^{(8)}(0) = 1 \\
f^{(9)}(x) = -\sin x & \Rightarrow \quad f^{(9)}(0) = 0
\end{array}$$

Figure 8.24: A table of the derivatives of $f(x) = \cos x$ evaluated at $x = 0$.

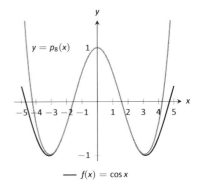

Figure 8.25: A graph of $f(x) = \cos x$ and its degree 8 Maclaurin polynomial.

Notes:

Chapter 8 Sequences and Series

$f(x) = \sqrt{x}$ $\Rightarrow$ $f(4) = 2$

$f'(x) = \dfrac{1}{2\sqrt{x}}$ $\Rightarrow$ $f'(4) = \dfrac{1}{4}$

$f''(x) = \dfrac{-1}{4x^{3/2}}$ $\Rightarrow$ $f''(4) = \dfrac{-1}{32}$

$f'''(x) = \dfrac{3}{8x^{5/2}}$ $\Rightarrow$ $f'''(4) = \dfrac{3}{256}$

$f^{(4)}(x) = \dfrac{-15}{16x^{7/2}}$ $\Rightarrow$ $f^{(4)}(4) = \dfrac{-15}{2048}$

Figure 8.26: A table of the derivatives of $f(x) = \sqrt{x}$ evaluated at $x = 4$.

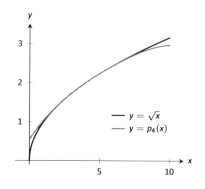

Figure 8.27: A graph of $f(x) = \sqrt{x}$ and its degree 4 Taylor polynomial at $x = 4$.

Example 265   **Finding and using Taylor polynomials**

1. Find the degree 4 Taylor polynomial, $p_4(x)$, for $f(x) = \sqrt{x}$ at $x = 4$.
2. Use $p_4(x)$ to approximate $\sqrt{3}$.
3. Find bounds on the error when approximating $\sqrt{3}$ with $p_4(3)$.

SOLUTION

1. We begin by evaluating the derivatives of f at $x = 4$. This is done in Figure 8.26. These values allow us to form the Taylor polynomial $p_4(x)$:

$$p_4(x) = 2 + \frac{1}{4}(x-4) + \frac{-1/32}{2!}(x-4)^2 + \frac{3/256}{3!}(x-4)^3 + \frac{-15/2048}{4!}(x-4)^4.$$

2. As $p_4(x) \approx \sqrt{x}$ near $x = 4$, we approximate $\sqrt{3}$ with $p_4(3) = 1.73212$.

3. To find a bound on the error, we need an open interval that contains $x = 3$ and $x = 4$. We set $I = (2.9, 4.1)$. The largest value the fifth derivative of $f(x) = \sqrt{x}$ takes on this interval is near $x = 2.9$, at about 0.0273. Thus

$$\left| R_4(3) \right| \leq \frac{0.0273}{5!} \left| (3-4)^5 \right| \approx 0.00023.$$

This shows our approximation is accurate to at least the first 2 places after the decimal. (It turns out that our approximation is actually accurate to 4 places after the decimal.) A graph of $f(x) = \sqrt{x}$ and $p_4(x)$ is given in Figure 8.27. Note how the two functions are nearly indistinguishable on $(2, 7)$.

Our final example gives a brief introduction to using Taylor polynomials to solve differential equations.

Example 266   **Approximating an unknown function**

A function $y = f(x)$ is unknown save for the following two facts.

1. $y(0) = f(0) = 1$, and
2. $y' = y^2$

(This second fact says that amazingly, the derivative of the function is actually the function squared!)

Find the degree 3 Maclaurin polynomial $p_3(x)$ of $y = f(x)$.

Notes:

8.7 Taylor Polynomials

SOLUTION One might initially think that not enough information is given to find $p_3(x)$. However, note how the second fact above actually lets us know what $y'(0)$ is:

$$y' = y^2 \Rightarrow y'(0) = y^2(0).$$

Since $y(0) = 1$, we conclude that $y'(0) = 1$.

Now we find information about y''. Starting with $y' = y^2$, take derivatives of both sides, *with respect to x*. That means we must use implicit differentiation.

$$y' = y^2$$
$$\frac{d}{dx}(y') = \frac{d}{dx}(y^2)$$
$$y'' = 2y \cdot y'.$$

Now evaluate both sides at $x = 0$:

$$y''(0) = 2y(0) \cdot y'(0)$$
$$y''(0) = 2$$

We repeat this once more to find $y'''(0)$. We again use implicit differentiation; this time the Product Rule is also required.

$$\frac{d}{dx}(y'') = \frac{d}{dx}(2yy')$$
$$y''' = 2y' \cdot y' + 2y \cdot y''.$$

Now evaluate both sides at $x = 0$:

$$y'''(0) = 2y'(0)^2 + 2y(0)y''(0)$$
$$y'''(0) = 2 + 4 = 6$$

In summary, we have:

$$y(0) = 1 \quad y'(0) = 1 \quad y''(0) = 2 \quad y'''(0) = 6.$$

We can now form $p_3(x)$:

$$p_3(x) = 1 + x + \frac{2}{2!}x^2 + \frac{6}{3!}x^3$$
$$= 1 + x + x^2 + x^3.$$

It turns out that the differential equation we started with, $y' = y^2$, where $y(0) = 1$, can be solved without too much difficulty: $y = \frac{1}{1-x}$. Figure 8.28 shows this function plotted with $p_3(x)$. Note how similar they are near $x = 0$.

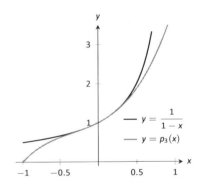

Figure 8.28: A graph of $y = -1/(x - 1)$ and $y = p_3(x)$ from Example 266.

It is beyond the scope of this text to pursue error analysis when using Taylor polynomials to approximate solutions to differential equations. This topic is often broached in introductory Differential Equations courses and usually covered in depth in Numerical Analysis courses. Such an analysis is very important; one needs to know how good their approximation is. We explored this example simply to demonstrate the usefulness of Taylor polynomials.

Most of this chapter has been devoted to the study of infinite series. This section has taken a step back from this study, focusing instead on finite summation of terms. In the next section, we explore **Taylor Series**, where we represent a function with an infinite series.

Notes:

Exercises 8.7

Terms and Concepts

1. What is the difference between a Taylor polynomial and a Maclaurin polynomial?

2. T/F: In general, $p_n(x)$ approximates $f(x)$ better and better as n gets larger.

3. For some function $f(x)$, the Maclaurin polynomial of degree 4 is $p_4(x) = 6 + 3x - 4x^2 + 5x^3 - 7x^4$. What is $p_2(x)$?

4. For some function $f(x)$, the Maclaurin polynomial of degree 4 is $p_4(x) = 6 + 3x - 4x^2 + 5x^3 - 7x^4$. What is $f'''(0)$?

Problems

In Exercises 5 – 12, find the Maclaurin polynomial of degree n for the given function.

5. $f(x) = e^{-x}$, $n = 3$

6. $f(x) = \sin x$, $n = 8$

7. $f(x) = x \cdot e^x$, $n = 5$

8. $f(x) = \tan x$, $n = 6$

9. $f(x) = e^{2x}$, $n = 4$

10. $f(x) = \dfrac{1}{1-x}$, $n = 4$

11. $f(x) = \dfrac{1}{1+x}$, $n = 4$

12. $f(x) = \dfrac{1}{1+x}$, $n = 7$

In Exercises 13 – 20, find the Taylor polynomial of degree n, at $x = c$, for the given function.

13. $f(x) = \sqrt{x}$, $n = 4$, $c = 1$

14. $f(x) = \ln(x+1)$, $n = 4$, $c = 1$

15. $f(x) = \cos x$, $n = 6$, $c = \pi/4$

16. $f(x) = \sin x$, $n = 5$, $c = \pi/6$

17. $f(x) = \dfrac{1}{x}$, $n = 5$, $c = 2$

18. $f(x) = \dfrac{1}{x^2}$, $n = 8$, $c = 1$

19. $f(x) = \dfrac{1}{x^2+1}$, $n = 3$, $c = -1$

20. $f(x) = x^2 \cos x$, $n = 2$, $c = \pi$

In Exercises 21 – 24, approximate the function value with the indicated Taylor polynomial and give approximate bounds on the error.

21. Approximate $\sin 0.1$ with the Maclaurin polynomial of degree 3.

22. Approximate $\cos 1$ with the Maclaurin polynomial of degree 4.

23. Approximate $\sqrt{10}$ with the Taylor polynomial of degree 2 centered at $x = 9$.

24. Approximate $\ln 1.5$ with the Taylor polynomial of degree 3 centered at $x = 1$.

Exercises 25 – 28 ask for an n to be found such that $p_n(x)$ approximates $f(x)$ within a certain bound of accuracy.

25. Find n such that the Maclaurin polynomial of degree n of $f(x) = e^x$ approximates e within 0.0001 of the actual value.

26. Find n such that the Taylor polynomial of degree n of $f(x) = \sqrt{x}$, centered at $x = 4$, approximates $\sqrt{3}$ within 0.0001 of the actual value.

27. Find n such that the Maclaurin polynomial of degree n of $f(x) = \cos x$ approximates $\cos \pi/3$ within 0.0001 of the actual value.

28. Find n such that the Maclaurin polynomial of degree n of $f(x) = \sin x$ approximates $\cos \pi$ within 0.0001 of the actual value.

In Exercises 29 – 33, find the nth term of the indicated Taylor polynomial.

29. Find a formula for the nth term of the Maclaurin polynomial for $f(x) = e^x$.

30. Find a formula for the nth term of the Maclaurin polynomial for $f(x) = \cos x$.

31. Find a formula for the nth term of the Maclaurin polynomial for $f(x) = \dfrac{1}{1-x}$.

32. Find a formula for the nth term of the Maclaurin polynomial for $f(x) = \dfrac{1}{1+x}$.

33. Find a formula for the nth term of the Taylor polynomial for $f(x) = \ln x$.

In Exercises 34 – 36, approximate the solution to the given differential equation with a degree 4 Maclaurin polynomial.

34. $y' = y$, $y(0) = 1$

35. $y' = 5y$, $y(0) = 3$

36. $y' = \dfrac{2}{y}$, $y(0) = 1$

8.8 Taylor Series

In Section 8.6, we showed how certain functions can be represented by a power series function. In 8.7, we showed how we can approximate functions with polynomials, given that enough derivative information is available. In this section we combine these concepts: if a function $f(x)$ is infinitely differentiable, we show how to represent it with a power series function.

Definition 39   **Taylor and Maclaurin Series**

Let $f(x)$ have derivatives of all orders at $x = c$.

1. The **Taylor Series of $f(x)$, centered at c** is

$$\sum_{n=0}^{\infty} \frac{f^{(n)}(c)}{n!}(x-c)^n.$$

2. Setting $c = 0$ gives the **Maclaurin Series of $f(x)$**:

$$\sum_{n=0}^{\infty} \frac{f^{(n)}(0)}{n!}x^n.$$

The difference between a Taylor polynomial and a Taylor series is the former is a polynomial, containing only a finite number of terms, whereas the latter is a series, a summation of an infinite set of terms. When creating the Taylor polynomial of degree n for a function $f(x)$ at $x = c$, we needed to evaluate f, and the first n derivatives of f, at $x = c$. When creating the Taylor series of f, it helps to find a pattern that describes the nth derivative of f at $x = c$. We demonstrate this in the next two examples.

Example 267   **The Maclaurin series of $f(x) = \cos x$**
Find the Maclaurin series of $f(x) = \cos x$.

SOLUTION   In Example 264 we found the 8th degree Maclaurin polynomial of $\cos x$. In doing so, we created the table shown in Figure 8.29. Notice how $f^{(n)}(0) = 0$ when n is odd, $f^{(n)}(0) = 1$ when n is divisible by 4, and $f^{(n)}(0) = -1$ when n is even but not divisible by 4. Thus the Maclaurin series of $\cos x$ is

$$1 - \frac{x^2}{2} + \frac{x^4}{4!} - \frac{x^6}{6!} + \frac{x^8}{8!} - \cdots$$

$f(x) = \cos x \quad \Rightarrow \quad f(0) = 1$
$f'(x) = -\sin x \quad \Rightarrow \quad f'(0) = 0$
$f''(x) = -\cos x \quad \Rightarrow \quad f''(0) = -1$
$f'''(x) = \sin x \quad \Rightarrow \quad f'''(0) = 0$
$f^{(4)}(x) = \cos x \quad \Rightarrow \quad f^{(4)}(0) = 1$
$f^{(5)}(x) = -\sin x \quad \Rightarrow \quad f^{(5)}(0) = 0$
$f^{(6)}(x) = -\cos x \quad \Rightarrow \quad f^{(6)}(0) = -1$
$f^{(7)}(x) = \sin x \quad \Rightarrow \quad f^{(7)}(0) = 0$
$f^{(8)}(x) = \cos x \quad \Rightarrow \quad f^{(8)}(0) = 1$
$f^{(9)}(x) = -\sin x \quad \Rightarrow \quad f^{(9)}(0) = 0$

Figure 8.29: A table of the derivatives of $f(x) = \cos x$ evaluated at $x = 0$.

Notes:

We can go further and write this as a summation. Since we only need the terms where the power of x is even, we write the power series in terms of x^{2n}:

$$\sum_{n=0}^{\infty}(-1)^n\frac{x^{2n}}{(2n)!}.$$

Example 268   **The Taylor series of $f(x) = \ln x$ at $x = 1$**
Find the Taylor series of $f(x) = \ln x$ centered at $x = 1$.

Solution   Figure 8.30 shows the nth derivative of $\ln x$ evaluated at $x = 1$ for $n = 0, \ldots, 5$, along with an expression for the nth term:

$$f^{(n)}(1) = (-1)^{n+1}(n-1)! \quad \text{for } n \geq 1.$$

Remember that this is what distinguishes Taylor series from Taylor polynomials; we are very interested in finding a pattern for the nth term, not just finding a finite set of coefficients for a polynomial. Since $f(1) = \ln 1 = 0$, we skip the first term and start the summation with $n = 1$, giving the Taylor series for $\ln x$, centered at $x = 1$, as

$$\sum_{n=1}^{\infty}(-1)^{n+1}(n-1)!\frac{1}{n!}(x-1)^n = \sum_{n=1}^{\infty}(-1)^{n+1}\frac{(x-1)^n}{n}.$$

It is important to note that Definition 39 defines a Taylor series given a function $f(x)$; however, we *cannot* yet state that $f(x)$ *is equal* to its Taylor series. We will find that "most of the time" they are equal, but we need to consider the conditions that allow us to conclude this.

Theorem 76 states that the error between a function $f(x)$ and its nth–degree Taylor polynomial $p_n(x)$ is $R_n(x)$, where

$$|R_n(x)| \leq \frac{\max\left|f^{(n+1)}(z)\right|}{(n+1)!}\left|(x-c)^{(n+1)}\right|.$$

If $R_n(x)$ goes to 0 for each x in an interval I as n approaches infinity, we conclude that the function is equal to its Taylor series expansion.

$f(x) = \ln x \quad \Rightarrow \quad f(1) = 0$
$f'(x) = 1/x \quad \Rightarrow \quad f'(1) = 1$
$f''(x) = -1/x^2 \quad \Rightarrow \quad f''(1) = -1$
$f'''(x) = 2/x^3 \quad \Rightarrow \quad f'''(1) = 2$
$f^{(4)}(x) = -6/x^4 \quad \Rightarrow \quad f^{(4)}(1) = -6$
$f^{(5)}(x) = 24/x^5 \quad \Rightarrow \quad f^{(5)}(1) = 24$
$\vdots \quad\quad\quad\quad\quad\quad \vdots$
$f^{(n)}(x) = \quad\quad \Rightarrow \quad f^{(n)}(1) =$
$\frac{(-1)^{n+1}(n-1)!}{x^n} \quad\quad (-1)^{n+1}(n-1)!$

Figure 8.30: Derivatives of $\ln x$ evaluated at $x = 1$.

Notes:

8.8 Taylor Series

> **Theorem 77**   **Function and Taylor Series Equality**
>
> Let $f(x)$ have derivatives of all orders at $x = c$, let $R_n(x)$ be as stated in Theorem 76, and let I be an interval on which the Taylor series of $f(x)$ converges. If $\lim_{n \to \infty} R_n(x) = 0$ for all x in I, then
>
> $$f(x) = \sum_{n=0}^{\infty} \frac{f^{(n)}(c)}{n!}(x-c)^n \quad \text{on } I.$$

We demonstrate the use of this theorem in an example.

Example 269   **Establishing equality of a function and its Taylor series**

Show that $f(x) = \cos x$ is equal to its Maclaurin series, as found in Example 267, for all x.

SOLUTION   Given a value x, the magnitude of the error term $R_n(x)$ is bounded by

$$\left|R_n(x)\right| \leq \frac{\max\left|f^{(n+1)}(z)\right|}{(n+1)!}\left|x^{n+1}\right|.$$

Since all derivatives of $\cos x$ are $\pm \sin x$ or $\pm \cos x$, whose magnitudes are bounded by 1, we can state

$$\left|R_n(x)\right| \leq \frac{1}{(n+1)!}\left|x^{n+1}\right|$$

which implies

$$-\frac{\left|x^{n+1}\right|}{(n+1)!} \leq R_n(x) \leq \frac{\left|x^{n+1}\right|}{(n+1)!}. \quad (8.7)$$

For any x, $\lim_{n \to \infty} \frac{x^{n+1}}{(n+1)!} = 0$. Applying the Squeeze Theorem to Equation (8.7), we conclude that $\lim_{n \to \infty} R_n(x) = 0$ for all x, and hence

$$\cos x = \sum_{n=0}^{\infty} (-1)^n \frac{x^{2n}}{(2n)!} \quad \text{for all } x.$$

It is natural to assume that a function is equal to its Taylor series on the series' interval of convergence, but this is not the case. In order to properly establish equality, one must use Theorem 77. This is a bit disappointing, as we developed beautiful techniques for determining the interval of convergence of a power series, and proving that $R_n(x) \to 0$ can be cumbersome as it deals with high order derivatives of the function.

Notes:

There is good news. A function $f(x)$ that is equal to its Taylor series, centered at any point the domain of $f(x)$, is said to be an **analytic function**, and most, if not all, functions that we encounter within this course are analytic functions. Generally speaking, any function that one creates with elementary functions (polynomials, exponentials, trigonometric functions, etc.) that is not piecewise defined is probably analytic. For most functions, we assume the function is equal to its Taylor series on the series' interval of convergence and only use Theorem 77 when we suspect something may not work as expected.

We develop the Taylor series for one more important function, then give a table of the Taylor series for a number of common functions.

Example 270   **The Binomial Series**
Find the Maclaurin series of $f(x) = (1+x)^k$, $k \neq 0$.

Solution   When k is a positive integer, the Maclaurin series is finite. For instance, when $k = 4$, we have

$$f(x) = (1+x)^4 = 1 + 4x + 6x^2 + 4x^3 + x^4.$$

The coefficients of x when k is a positive integer are known as the *binomial coefficients*, giving the series we are developing its name.

When $k = 1/2$, we have $f(x) = \sqrt{1+x}$. Knowing a series representation of this function would give a useful way of approximating $\sqrt{1.3}$, for instance.

To develop the Maclaurin series for $f(x) = (1+x)^k$ for any value of $k \neq 0$, we consider the derivatives of f evaluated at $x = 0$:

$$f(x) = (1+x)^k \qquad\qquad f(0) = 1$$
$$f'(x) = k(1+x)^{k-1} \qquad\qquad f'(0) = k$$
$$f''(x) = k(k-1)(1+x)^{k-2} \qquad\qquad f''(0) = k(k-1)$$
$$f'''(x) = k(k-1)(k-2)(1+x)^{k-3} \qquad\qquad f'''(0) = k(k-1)(k-2)$$
$$\vdots \qquad\qquad\qquad\qquad \vdots$$
$$f^{(n)}(x) = k(k-1)\cdots(k-(n-1))(1+x)^{k-n} \qquad f^{(n)}(0) = k(k-1)\cdots(k-(n-1))$$

Thus the Maclaurin series for $f(x) = (1+x)^k$ is

$$1 + k + \frac{k(k-1)}{2!} + \frac{k(k-1)(k-2)}{3!} + \ldots + \frac{k(k-1)\cdots(k-(n-1))}{n!} + \ldots$$

It is important to determine the interval of convergence of this series. With

$$a_n = \frac{k(k-1)\cdots(k-(n-1))}{n!} x^n,$$

Notes:

we apply the Ratio Test:

$$\lim_{n \to \infty} \frac{|a_{n+1}|}{|a_n|} = \lim_{n \to \infty} \left| \frac{k(k-1)\cdots(k-n)}{(n+1)!} x^{n+1} \right| \bigg/ \left| \frac{k(k-1)\cdots(k-(n-1))}{n!} x^n \right|$$

$$= \lim_{n \to \infty} \left| \frac{k-n}{n} x \right|$$

$$= |x|.$$

The series converges absolutely when the limit of the Ratio Test is less than 1; therefore, we have absolute convergence when $|x| < 1$.

While outside the scope of this text, the interval of convergence depends on the value of k. When $k > 0$, the interval of convergence is $[-1, 1]$. When $-1 < k < 0$, the interval of convergence is $[-1, 1)$. If $k \leq -1$, the interval of convergence is $(-1, 1)$.

We learned that Taylor polynomials offer a way of approximating a "difficult to compute" function with a polynomial. Taylor series offer a way of exactly representing a function with a series. One probably can see the use of a good approximation; is there any use of representing a function exactly as a series?

While we should not overlook the mathematical beauty of Taylor series (which is reason enough to study them), there are practical uses as well. They provide a valuable tool for solving a variety of problems, including problems relating to integration and differential equations.

In Key Idea 32 (on the following page) we give a table of the Taylor series of a number of common functions. We then give a theorem about the "algebra of power series," that is, how we can combine power series to create power series of new functions. This allows us to find the Taylor series of functions like $f(x) = e^x \cos x$ by knowing the Taylor series of e^x and $\cos x$.

Before we investigate combining functions, consider the Taylor series for the arctangent function (see Key Idea 32). Knowing that $\tan^{-1}(1) = \pi/4$, we can use this series to approximate the value of π:

$$\frac{\pi}{4} = \tan^{-1}(1) = 1 - \frac{1}{3} + \frac{1}{5} - \frac{1}{7} + \frac{1}{9} - \cdots$$

$$\pi = 4\left(1 - \frac{1}{3} + \frac{1}{5} - \frac{1}{7} + \frac{1}{9} - \cdots\right)$$

Unfortunately, this particular expansion of π converges very slowly. The first 100 terms approximate π as 3.13159, which is not particularly good.

Notes:

Chapter 8 Sequences and Series

Key Idea 32   Important Taylor Series Expansions

Function and Series	First Few Terms	Interval of Convergence
$e^x = \sum_{n=0}^{\infty} \dfrac{x^n}{n!}$	$1 + x + \dfrac{x^2}{2!} + \dfrac{x^3}{3!} + \cdots$	$(-\infty, \infty)$
$\sin x = \sum_{n=0}^{\infty} (-1)^n \dfrac{x^{2n+1}}{(2n+1)!}$	$x - \dfrac{x^3}{3!} + \dfrac{x^5}{5!} - \dfrac{x^7}{7!} + \cdots$	$(-\infty, \infty)$
$\cos x = \sum_{n=0}^{\infty} (-1)^n \dfrac{x^{2n}}{(2n)!}$	$1 - \dfrac{x^2}{2!} + \dfrac{x^4}{4!} - \dfrac{x^6}{6!} + \cdots$	$(-\infty, \infty)$
$\ln x = \sum_{n=1}^{\infty} (-1)^{n+1} \dfrac{(x-1)^n}{n}$	$(x-1) - \dfrac{(x-1)^2}{2} + \dfrac{(x-1)^3}{3} - \cdots$	$(0, 2]$
$\dfrac{1}{1-x} = \sum_{n=0}^{\infty} x^n$	$1 + x + x^2 + x^3 + \cdots$	$(-1, 1)$
$(1+x)^k = \sum_{n=0}^{\infty} \dfrac{k(k-1)\cdots(k-(n-1))}{n!} x^n$	$1 + kx + \dfrac{k(k-1)}{2!} x^2 + \cdots$	$(-1, 1)^a$
$\tan^{-1} x = \sum_{n=0}^{\infty} (-1)^n \dfrac{x^{2n+1}}{2n+1}$	$x - \dfrac{x^3}{3} + \dfrac{x^5}{5} - \dfrac{x^7}{7} + \cdots$	$[-1, 1]$

^{a}Convergence at $x = \pm 1$ depends on the value of k.

Theorem 78   Algebra of Power Series

Let $f(x) = \sum_{n=0}^{\infty} a_n x^n$ and $g(x) = \sum_{n=0}^{\infty} b_n x^n$ converge absolutely for $|x| < R$, and let $h(x)$ be continuous.

1. $f(x) \pm g(x) = \sum_{n=0}^{\infty} (a_n \pm b_n) x^n$ for $|x| < R$.

2. $f(x)g(x) = \left(\sum_{n=0}^{\infty} a_n x^n\right)\left(\sum_{n=0}^{\infty} b_n x^n\right) = \sum_{n=0}^{\infty} (a_0 b_n + a_1 b_{n-1} + \ldots a_n b_0) x^n$ for $|x| < R$.

3. $f(h(x)) = \sum_{n=0}^{\infty} a_n (h(x))^n$ for $|h(x)| < R$.

Notes:

8.8 Taylor Series

Example 271 **Combining Taylor series**
Write out the first 3 terms of the Taylor Series for $f(x) = e^x \cos x$ using Key Idea 32 and Theorem 78.

SOLUTION Key Idea 32 informs us that

$$e^x = 1 + x + \frac{x^2}{2!} + \frac{x^3}{3!} + \cdots \quad \text{and} \quad \cos x = 1 - \frac{x^2}{2!} + \frac{x^4}{4!} + \cdots .$$

Applying Theorem 78, we find that

$$e^x \cos x = \left(1 + x + \frac{x^2}{2!} + \frac{x^3}{3!} + \cdots\right)\left(1 - \frac{x^2}{2!} + \frac{x^4}{4!} + \cdots\right).$$

Distribute the right hand expression across the left:

$$= 1\left(1 - \frac{x^2}{2!} + \frac{x^4}{4!} + \cdots\right) + x\left(1 - \frac{x^2}{2!} + \frac{x^4}{4!} + \cdots\right) + \frac{x^2}{2!}\left(1 - \frac{x^2}{2!} + \frac{x^4}{4!} + \cdots\right)$$
$$+ \frac{x^3}{3!}\left(1 - \frac{x^2}{2!} + \frac{x^4}{4!} + \cdots\right) + \frac{x^4}{4!}\left(1 - \frac{x^2}{2!} + \frac{x^4}{4!} + \cdots\right) + \cdots$$

Distribute again and collect like terms.

$$= 1 + x - \frac{x^3}{3} - \frac{x^4}{6} - \frac{x^5}{30} + \frac{x^7}{630} + \cdots$$

While this process is a bit tedious, it is much faster than evaluating all the necessary derivatives of $e^x \cos x$ and computing the Taylor series directly.

Because the series for e^x and $\cos x$ both converge on $(-\infty, \infty)$, so does the series expansion for $e^x \cos x$.

Example 272 **Creating new Taylor series**
Use Theorem 78 to create series for $y = \sin(x^2)$ and $y = \ln(\sqrt{x})$.

SOLUTION Given that

$$\sin x = \sum_{n=0}^{\infty} (-1)^n \frac{x^{2n+1}}{(2n+1)!} = x - \frac{x^3}{3!} + \frac{x^5}{5!} - \frac{x^7}{7!} + \cdots ,$$

we simply substitute x^2 for x in the series, giving

$$\sin(x^2) = \sum_{n=0}^{\infty} (-1)^n \frac{(x^2)^{2n+1}}{(2n+1)!} = x^2 - \frac{x^6}{3!} + \frac{x^{10}}{5!} - \frac{x^{14}}{7!} \cdots .$$

Notes:

Since the Taylor series for sin x has an infinite radius of convergence, so does the Taylor series for sin(x^2).

The Taylor expansion for ln x given in Key Idea 32 is centered at $x = 1$, so we will center the series for $\ln(\sqrt{x})$ at $x = 1$ as well. With

$$\ln x = \sum_{n=1}^{\infty} (-1)^{n+1} \frac{(x-1)^n}{n} = (x-1) - \frac{(x-1)^2}{2} + \frac{(x-1)^3}{3} - \cdots,$$

we substitute $\sqrt{x}$ for x to obtain

$$\ln(\sqrt{x}) = \sum_{n=1}^{\infty} (-1)^{n+1} \frac{(\sqrt{x}-1)^n}{n} = (\sqrt{x}-1) - \frac{(\sqrt{x}-1)^2}{2} + \frac{(\sqrt{x}-1)^3}{3} - \cdots.$$

While this is not strictly a power series, it is a series that allows us to study the function $\ln(\sqrt{x})$. Since the interval of convergence of ln x is $(0, 2]$, and the range of $\sqrt{x}$ on $(0, 4]$ is $(0, 2]$, the interval of convergence of this series expansion of $\ln(\sqrt{x})$ is $(0, 4]$.

Note: In Example 272, one could create a series for $\ln(\sqrt{x})$ by simply recognizing that $\ln(\sqrt{x}) = \ln(x^{1/2}) = 1/2 \ln x$, and hence multiplying the Taylor series for ln x by $1/2$. This example was chosen to demonstrate other aspects of series, such as the fact that the interval of convergence changes.

Example 273   **Using Taylor series to evaluate definite integrals**

Use the Taylor series of e^{-x^2} to evaluate $\int_0^1 e^{-x^2}\, dx$.

SOLUTION   We learned, when studying Numerical Integration, that e^{-x^2} does not have an antiderivative expressible in terms of elementary functions. This means any definite integral of this function must have its value approximated, and not computed exactly.

We can quickly write out the Taylor series for e^{-x^2} using the Taylor series of e^x:

$$e^x = \sum_{n=0}^{\infty} \frac{x^n}{n!} = 1 + x + \frac{x^2}{2!} + \frac{x^3}{3!} + \cdots$$

and so

$$e^{-x^2} = \sum_{n=0}^{\infty} \frac{(-x^2)^n}{n!}$$

$$= \sum_{n=0}^{\infty} (-1)^n \frac{x^{2n}}{n!}$$

$$= 1 - x^2 + \frac{x^4}{2!} - \frac{x^6}{3!} + \cdots.$$

We use Theorem 75 to integrate:

$$\int e^{-x^2}\,dx = C + x - \frac{x^3}{3} + \frac{x^5}{5\cdot 2!} - \frac{x^7}{7\cdot 3!} + \cdots + (-1)^n \frac{x^{2n+1}}{(2n+1)n!} + \cdots$$

This *is* the antiderivative of e^{-x^2}; while we can write it out as a series, we cannot write it out in terms of elementary functions. We can evaluate the definite integral $\int_0^1 e^{-x^2}\,dx$ using this antiderivative; substituting 1 and 0 for x and subtracting gives

$$\int_0^1 e^{-x^2}\,dx = 1 - \frac{1}{3} + \frac{1}{5\cdot 2!} - \frac{1}{7\cdot 3!} + \frac{1}{9\cdot 4!} \cdots.$$

Summing the 5 terms shown above give the approximation of 0.74749. Since this is an alternating series, we can use the Alternating Series Approximation Theorem, (Theorem 71), to determine how accurate this approximation is. The next term of the series is $1/(11\cdot 5!) \approx 0.00075758$. Thus we know our approximation is within 0.00075758 of the actual value of the integral. This is arguably much less work than using Simpson's Rule to approximate the value of the integral.

Example 274   **Using Taylor series to solve differential equations**
Solve the differential equation $y' = 2y$ in terms of a power series, and use the theory of Taylor series to recognize the solution in terms of an elementary function.

SOLUTION   We found the first 5 terms of the power series solution to this differential equation in Example 260 in Section 8.6. These are:

$$a_0 = 1, \quad a_1 = 2, \quad a_2 = \frac{4}{2} = 2, \quad a_3 = \frac{8}{2\cdot 3} = \frac{4}{3}, \quad a_4 = \frac{16}{2\cdot 3\cdot 4} = \frac{2}{3}.$$

We include the "unsimplified" expressions for the coefficients found in Example 260 as we are looking for a pattern. It can be shown that $a_n = 2^n/n!$. Thus the solution, written as a power series, is

$$y = \sum_{n=0}^{\infty} \frac{2^n}{n!} x^n = \sum_{n=0}^{\infty} \frac{(2x)^n}{n!}.$$

Using Key Idea 32 and Theorem 78, we recognize $f(x) = e^{2x}$:

$$e^x = \sum_{n=0}^{\infty} \frac{x^n}{n!} \quad \Rightarrow \quad e^{2x} = \sum_{n=0}^{\infty} \frac{(2x)^n}{n!}.$$

Notes:

Finding a pattern in the coefficients that match the series expansion of a known function, such as those shown in Key Idea 32, can be difficult. What if the coefficients in the previous example were given in their reduced form; how could we still recover the function $y = e^{2x}$?

Suppose that all we know is that

$$a_0 = 1, \quad a_1 = 2, \quad a_2 = 2, \quad a_3 = \frac{4}{3}, \quad a_4 = \frac{2}{3}.$$

Definition 39 states that each term of the Taylor expansion of a function includes an $n!$. This allows us to say that

$$a_2 = 2 = \frac{b_2}{2!}, \quad a_3 = \frac{4}{3} = \frac{b_3}{3!}, \quad \text{and} \quad a_4 = \frac{2}{3} = \frac{b_4}{4!}$$

for some values b_2, b_3 and b_4. Solving for these values, we see that $b_2 = 4$, $b_3 = 8$ and $b_4 = 16$. That is, we are recovering the pattern we had previously seen, allowing us to write

$$f(x) = \sum_{n=0}^{\infty} a_n x^n = \sum_{n=0}^{\infty} \frac{b_n}{n!} x^n$$
$$= 1 + 2x + \frac{4}{2!}x^2 + \frac{8}{3!}x^3 + \frac{16}{4!}x^4 + \cdots$$

From here it is easier to recognize that the series is describing an exponential function.

There are simpler, more direct ways of solving the differential equation $y' = 2y$. We applied power series techniques to this equation to demonstrate its utility, and went on to show how *sometimes* we are able to recover the solution in terms of elementary functions using the theory of Taylor series. Most differential equations faced in real scientific and engineering situations are much more complicated than this one, but power series can offer a valuable tool in finding, or at least approximating, the solution.

This chapter introduced sequences, which are ordered lists of numbers, followed by series, wherein we add up the terms of a sequence. We quickly saw that such sums do not always add up to "infinity," but rather converge. We studied tests for convergence, then ended the chapter with a formal way of defining functions based on series. Such "series–defined functions" are a valuable tool in solving a number of different problems throughout science and engineering.

Coming in the next chapters are new ways of defining curves in the plane apart from using functions of the form $y = f(x)$. Curves created by these new methods can be beautiful, useful, and important.

Notes:

Exercises 8.8

Terms and Concepts

1. What is the difference between a Taylor polynomial and a Taylor series?

2. What theorem must we use to show that a function is equal to its Taylor series?

Problems

Key Idea 32 gives the nth term of the Taylor series of common functions. In Exercises 3 – 6, verify the formula given in the Key Idea by finding the first few terms of the Taylor series of the given function and identifying a pattern.

3. $f(x) = e^x; \quad c = 0$

4. $f(x) = \sin x; \quad c = 0$

5. $f(x) = 1/(1-x); \quad c = 0$

6. $f(x) = \tan^{-1} x; \quad c = 0$

In Exercises 7 – 12, find a formula for the nth term of the Taylor series of $f(x)$, centered at c, by finding the coefficients of the first few powers of x and looking for a pattern. (The formulas for several of these are found in Key Idea 32; show work verifying these formula.)

7. $f(x) = \cos x; \quad c = \pi/2$

8. $f(x) = 1/x; \quad c = 1$

9. $f(x) = e^{-x}; \quad c = 0$

10. $f(x) = \ln(1+x); \quad c = 0$

11. $f(x) = x/(x+1); \quad c = 1$

12. $f(x) = \sin x; \quad c = \pi/4$

In Exercises 13 – 16, show that the Taylor series for $f(x)$, as given in Key Idea 32, is equal to $f(x)$ by applying Theorem 77; that is, show $\lim_{n \to \infty} R_n(x) = 0$.

13. $f(x) = e^x$

14. $f(x) = \sin x$

15. $f(x) = \ln x$

16. $f(x) = 1/(1-x)$ (show equality only on $(-1, 0)$)

In Exercises 17 – 20, use the Taylor series given in Key Idea 32 to verify the given identity.

17. $\cos(-x) = \cos x$

18. $\sin(-x) = -\sin x$

19. $\frac{d}{dx}\left(\sin x\right) = \cos x$

20. $\frac{d}{dx}\left(\cos x\right) = -\sin x$

In Exercises 21 – 24, write out the first 5 terms of the Binomial series with the given k-value.

21. $k = 1/2$

22. $k = -1/2$

23. $k = 1/3$

24. $k = 4$

In Exercises 25 – 30, use the Taylor series given in Key Idea 32 to create the Taylor series of the given functions.

25. $f(x) = \cos\left(x^2\right)$

26. $f(x) = e^{-x}$

27. $f(x) = \sin(2x+3)$

28. $f(x) = \tan^{-1}(x/2)$

29. $f(x) = e^x \sin x$ (only find the first 4 terms)

30. $f(x) = (1+x)^{1/2} \cos x$ (only find the first 4 terms)

In Exercises 31 – 32, approximate the value of the given definite integral by using the first 4 nonzero terms of the integrand's Taylor series.

31. $\displaystyle\int_0^{\sqrt{\pi}} \sin\left(x^2\right)\,dx$

32. $\displaystyle\int_0^{\pi^2/4} \cos\left(\sqrt{x}\right)\,dx$

A: Solutions To Selected Problems

Chapter 5

Section 5.1

1. Answers will vary.
3. Answers will vary.
5. Answers will vary.
7. velocity
9. $1/9x^9 + C$
11. $t + C$
13. $-1/(3t) + C$
15. $2\sqrt{x} + C$
17. $-\cos\theta + C$
19. $5e^\theta + C$
21. $\frac{5^t}{2\ln 5} + C$
23. $t^6/6 + t^4/4 - 3t^2 + C$
25. $e^\pi x + C$
27. (a) $x > 0$
    (b) $1/x$
    (c) $x < 0$
    (d) $1/x$
    (e) $\ln|x| + C$. Explanations will vary.
29. $5e^x + 5$
31. $\tan x + 4$
33. $5/2x^2 + 7x + 3$
35. $5e^x - 2x$
37. $\frac{2x^4 \ln^2(2) + 2^x + x\ln 2)(\ln 32 - 1) + \ln^2(2)\cos(x) - 1 - \ln^2(2)}{\ln^2(2)}$
39. No answer provided.

Section 5.2

1. Answers will vary.
3. 0
5. (a) 3
   (b) 4
   (c) 3
   (d) 0
   (e) -4
   (f) 9
7. (a) 4
   (b) 2
   (c) 4
   (d) 2
   (e) 1
   (f) 2
9. (a) π
   (b) π
   (c) 2π
   (d) 10π
11. (a) $4/\pi$
    (b) $-4/\pi$
    (c) 0
    (d) $2/\pi$
13. (a) $40/3$
    (b) $26/3$
    (c) $8/3$
    (d) $38/3$
15. (a) 3ft/s
    (b) 9.5ft
    (c) 9.5ft
17. (a) 96ft/s
    (b) 6 seconds
    (c) 6 seconds
    (d) Never; the maximum height is 208ft.
19. 5
21. Answers can vary; one solution is $a = -2, b = 7$
23. -7
25. Answers can vary; one solution is $a = -11, b = 18$
27. $-\cos x - \sin x + \tan x + C$
29. $\ln|x| + \csc x + C$

Section 5.3

1. limits
3. Rectangles.
5. $2^2 + 3^2 + 4^2 = 29$
7. $0 - 1 + 0 + 1 + 0 = 0$
9. $-1 + 2 - 3 + 4 - 5 + 6 = 3$
11. $1 + 1 + 1 + 1 + 1 + 1 = 6$
13. Answers may vary; $\sum_{i=0}^{8}(i^2 - 1)$
15. Answers may vary; $\sum_{i=0}^{4}(-1)^i e^i$
17. 1045
19. -8525
21. 5050
23. 155
25. 24
27. 19
29. $\pi/3 + \pi/(2\sqrt{3}) \approx 1.954$
31. 0.388584
33. (a) Exact expressions will vary; $\frac{(1+n)^2}{4n^2}$.
    (b) $121/400, 10201/40000, 1002001/4000000$
    (c) $1/4$
35. (a) 8.
    (b) 8, 8, 8
    (c) 8
37. (a) Exact expressions will vary; $100 - 200/n$.

(b) 80, 98, 499/5

(c) 100

39. $F(x) = 5\tan x + 4$

41. $G(t) = 4/6t^6 - 5/4t^4 + 8t + 9$

43. $G(t) = \sin t - \cos t - 78$

Section 5.4

1. Answers will vary.
3. T
5. 20
7. 0
9. 1
11. $(5 - 1/5)/\ln 5$
13. -4
15. $16/3$
17. $45/4$
19. $1/2$
21. $1/2$
23. $1/4$
25. 8
27. 0
29. Explanations will vary. A sketch will help.
31. $c = \pm 2/\sqrt{3}$
33. $c = 64/9 \approx 7.1$
35. $2/pi$
37. $16/3$
39. $1/(e-1)$
41. 400ft
43. -1ft
45. -64ft/s
47. 2ft/s
49. $27/2$
51. $9/2$
53. $F'(x) = (3x^2 + 1)\frac{1}{x^3+x}$
55. $F'(x) = 2x(x^2 + 2) - (x+2)$

Section 5.5

1. F
3. They are superseded by the Trapezoidal Rule; it takes an equal amount of work and is generally more accurate.
5. (a) 250
   (b) 250
   (c) 250
7. (a) $2 + \sqrt{2} + \sqrt{3} \approx 5.15$
   (b) $2/3(3 + \sqrt{2} + 2\sqrt{3}) \approx 5.25$
   (c) $16/3 \approx 5.33$
9. (a) 0.2207
   (b) 0.2005
   (c) $1/5$

11. (a) $9/2(1 + \sqrt{3}) \approx 12.294$
    (b) $3 + 6\sqrt{3} \approx 13.392$
    (c) $9\pi/2 \approx 14.137$

13. Trapezoidal Rule: 3.0241
    Simpson's Rule: 2.9315

15. Trapezoidal Rule: 3.0695
    Simpson's Rule: 3.14295

17. Trapezoidal Rule: 2.52971
    Simpson's Rule: 2.5447

19. Trapezoidal Rule: 3.5472
    Simpson's Rule: 3.6133

21. (a) $n = 150$ (using max $(f''(x)) = 1$)
    (b) $n = 18$ (using max $(f^{(4)}(x)) = 7$)

23. (a) $n = 5591$ (using max $(f''(x)) = 300$)
    (b) $n = 46$ (using max $(f^{(4)}(x)) = 24$)

25. (a) Area is 25.0667 cm²
    (b) Area is 250,667 yd²

Chapter 6

Section 6.1

1. Chain Rule.
3. $\frac{1}{8}(x^3 - 5)^8 + C$
5. $\frac{1}{18}(x^2 + 1)^9 + C$
7. $\frac{1}{2}\ln|2x + 7| + C$
9. $\frac{2}{3}(x+3)^{3/2} - 6(x+3)^{1/2} + C = \frac{2}{3}(x-6)\sqrt{x+3} + C$
11. $2e^{\sqrt{x}} + C$
13. $-\frac{1}{2x^2} - \frac{1}{x} + C$
15. $\frac{\sin^3(x)}{3} + C$
17. $-\tan(4-x) + C$
19. $\frac{\tan^3(x)}{3} + C$
21. $\tan(x) - x + C$
23. The key is to multiply $\csc x$ by 1 in the form $(\csc x + \cot x)/(\csc x + \cot x)$.
25. $\frac{e^{x^3}}{3} + C$
27. $x - e^{-x} + C$
29. $\frac{27^x}{\ln 27} + C$
31. $\frac{1}{2}\ln^2(x) + C$
33. $\frac{1}{6}\ln^2(x^3) + C$
35. $\frac{x^2}{2} + 3x + \ln|x| + C$
37. $\frac{x^3}{3} - \frac{x^2}{2} + x - 2\ln|x+1| + C$
39. $\frac{3}{2}x^2 - 8x + 15\ln|x+1| + C$
41. $\sqrt{7}\tan^{-1}\left(\frac{x}{\sqrt{7}}\right) + C$
43. $14\sin^{-1}\left(\frac{x}{\sqrt{5}}\right) + C$
45. $\frac{5}{4}\sec^{-1}(|x|/4) + C$
47. $\frac{\tan^{-1}\left(\frac{x-1}{\sqrt{7}}\right)}{\sqrt{7}} + C$

49. $3\sin^{-1}\left(\frac{x-4}{5}\right) + C$

51. $-\frac{1}{3(x^3+3)} + C$

53. $-\sqrt{1-x^2} + C$

55. $-\frac{2}{3}\cos^{\frac{3}{2}}(x) + C$

57. $\frac{7}{3}\ln|3x+2| + C$

59. $\ln|x^2 + 7x + 3| + C$

61. $-\frac{x^2}{2} + 2\ln|x^2 - 7x + 1| + 7x + C$

63. $\tan^{-1}(2x) + C$

65. $\frac{1}{3}\sin^{-1}\left(\frac{3x}{4}\right) + C$

67. $\frac{19}{5}\tan^{-1}\left(\frac{x+6}{5}\right) - \ln|x^2 + 12x + 61| + C$

69. $\frac{x^2}{2} - \frac{9}{2}\ln|x^2 + 9| + C$

71. $-\tan^{-1}(\cos(x)) + C$

73. $\ln|\sec x + \tan x| + C$ (integrand simplifies to sec x)

75. $\sqrt{x^2 - 6x + 8} + C$

77. $352/15$

79. $1/5$

81. $\pi/2$

83. $\pi/6$

Section 6.2

1. T

3. Determining which functions in the integrand to set equal to "u" and which to set equal to "dv".

5. $-e^{-x} - xe^{-x} + C$

7. $-x^3\cos x + 3x^2\sin x + 6x\cos x - 6\sin x + C$

9. $x^3 e^x - 3x^2 e^x + 6xe^x - 6e^x + C$

11. $1/2 e^x(\sin x - \cos x) + C$

13. $1/13 e^{2x}(2\sin(3x) - 3\cos(3x)) + C$

15. $-1/2\cos^2 x + C$

17. $x\tan^{-1}(2x) - \frac{1}{4}\ln|4x^2 + 1| + C$

19. $\sqrt{1-x^2} + x\sin^{-1}x + C$

21. $-\frac{x^2}{4} + \frac{1}{2}x^2\ln|x| + 2x - 2x\ln|x| + C$

23. $\frac{1}{2}x^2\ln(x^2) - \frac{x^2}{2} + C$

25. $2x + x(\ln|x|)^2 - 2x\ln|x| + C$

27. $x\tan(x) + \ln|\cos(x)| + C$

29. $\frac{2}{5}(x-2)^{5/2} + \frac{4}{3}(x-2)^{3/2} + C$

31. $\sec x + C$

33. $-x\csc x - \ln|\csc x + \cot x| + C$

35. $2\sin(\sqrt{x}) - 2\sqrt{x}\cos(\sqrt{x}) + C$

37. $2\sqrt{x}e^{\sqrt{x}} - 2e^{\sqrt{x}} + C$

39. π

41. 0

43. $1/2$

45. $\frac{3}{4e^2} - \frac{5}{4e^4}$

47. $1/5\left(e^\pi + e^{-\pi}\right)$

Section 6.3

1. F

3. F

5. $\frac{1}{4}\sin^4(x) + C$

7. $\frac{1}{6}\cos^6 x - \frac{1}{4}\cos^4 x + C$

9. $-\frac{1}{9}\sin^9(x) + \frac{3\sin^7(x)}{7} - \frac{3\sin^5(x)}{5} + \frac{\sin^3(x)}{3} + C$

11. $\frac{1}{2}\left(-\frac{1}{8}\cos(8x) - \frac{1}{2}\cos(2x)\right) + C$

13. $\frac{1}{2}\left(\frac{1}{4}\sin(4x) - \frac{1}{10}\sin(10x)\right) + C$

15. $\frac{1}{2}\left(\sin(x) + \frac{1}{3}\sin(3x)\right) + C$

17. $\frac{\tan^5(x)}{5} + C$

19. $\frac{\tan^6(x)}{6} + \frac{\tan^4(x)}{4} + C$

21. $\frac{\sec^5(x)}{5} - \frac{\sec^3(x)}{3} + C$

23. $\frac{1}{3}\tan^3 x - \tan x + x + C$

25. $\frac{1}{2}(\sec x\tan x - \ln|\sec x + \tan x|) + C$

27. $\frac{2}{5}$

29. $32/315$

31. $2/3$

33. $16/15$

Section 6.4

1. backwards

3. (a) $\tan^2\theta + 1 = \sec^2\theta$
   (b) $9\sec^2\theta$.

5. $\frac{1}{2}\left(x\sqrt{x^2+1} + \ln|\sqrt{x^2+1} + x|\right) + C$

7. $\frac{1}{2}\left(\sin^{-1}x + x\sqrt{1-x^2}\right) + C$

9. $\frac{1}{2}x\sqrt{x^2-1} - \frac{1}{2}\ln|x + \sqrt{x^2-1}| + C$

11. $x\sqrt{x^2+1/4} + \frac{1}{4}\ln|2\sqrt{x^2+1/4} + 2x| + C = \frac{1}{2}x\sqrt{4x^2+1} + \frac{1}{4}\ln|\sqrt{4x^2+1} + 2x| + C$

13. $4\left(\frac{1}{2}x\sqrt{x^2-1/16} - \frac{1}{32}\ln|4x + 4\sqrt{x^2-1/16}|\right) + C = \frac{1}{2}x\sqrt{16x^2-1} - \frac{1}{8}\ln|4x + \sqrt{16x^2-1}| + C$

15. $3\sin^{-1}\left(\frac{x}{\sqrt{7}}\right) + C$ (Trig. Subst. is not needed)

17. $\sqrt{x^2-11} - \sqrt{11}\sec^{-1}(x/\sqrt{11}) + C$

19. $\sqrt{x^2-3} + C$ (Trig. Subst. is not needed)

21. $-\frac{1}{\sqrt{x^2+9}} + C$ (Trig. Subst. is not needed)

23. $\frac{1}{18}\frac{x+2}{x^2+4x+13} + \frac{1}{54}\tan^{-1}\left(\frac{x+2}{2}\right) + C$

25. $\frac{1}{7}\left(-\frac{\sqrt{5-x^2}}{x} - \sin^{-1}(x/\sqrt{5})\right) + C$

27. $\pi/2$

29. $2\sqrt{2} + 2\ln(1+\sqrt{2})$

31. $9\sin^{-1}(1/3) + \sqrt{8}$ Note: the new lower bound is $\theta = \sin^{-1}(-1/3)$ and the new upper bound is $\theta = \sin^{-1}(1/3)$. The final answer comes with recognizing that $\sin^{-1}(-1/3) = -\sin^{-1}(1/3)$ and that $\cos(\sin^{-1}(1/3)) = \cos(\sin^{-1}(-1/3)) = \sqrt{8}/3$.

Section 6.5

1. rational

3. $\frac{A}{x} + \frac{B}{x-3}$

5. $\frac{A}{x-\sqrt{7}} + \frac{B}{x+\sqrt{7}}$

7. $3\ln|x-2| + 4\ln|x+5| + C$

9. $\frac{1}{3}(\ln|x+2| - \ln|x-2|) + C$

11. $-\frac{4}{x+8} - 3\ln|x+8| + C$

13. $-\ln|2x-3| + 5\ln|x-1| + 2\ln|x+3| + C$

15. $x + \ln|x-1| - \ln|x+2| + C$

17. $2x + C$

19. $-\frac{3}{2}\ln|x^2 + 4x + 10| + x + \frac{\tan^{-1}\left(\frac{x+2}{\sqrt{6}}\right)}{\sqrt{6}} + C$

21. $2\ln|x-3| + 2\ln|x^2 + 6x + 10| - 4\tan^{-1}(x+3) + C$

23. $\frac{1}{2}\left(3\ln|x^2 + 2x + 17| - 4\ln|x-7| + \tan^{-1}\left(\frac{x+1}{4}\right)\right) + C$

25. $\frac{1}{2}\ln|x^2 + 10x + 27| + 5\ln|x+2| - 6\sqrt{2}\tan^{-1}\left(\frac{x+5}{\sqrt{2}}\right) + C$

27. $5\ln(9/4) - \frac{1}{3}\ln(17/2) \approx 3.3413$

29. $1/8$

Section 6.6

1. Because $\cosh x$ is always positive.

3. $\coth^2 x - \operatorname{csch}^2 x = \left(\frac{e^x + e^{-x}}{e^x - e^{-x}}\right)^2 - \left(\frac{2}{e^x - e^{-x}}\right)^2$

$= \frac{(e^{2x} + 2 + e^{-2x}) - (4)}{e^{2x} - 2 + e^{-2x}}$

$= \frac{e^{2x} - 2 + e^{-2x}}{e^{2x} - 2 + e^{-2x}}$

$= 1$

5. $\cosh^2 x = \left(\frac{e^x + e^{-x}}{2}\right)^2$

$= \frac{e^{2x} + 2 + e^{-2x}}{4}$

$= \frac{1}{2}\frac{(e^{2x} + e^{-2x}) + 2}{2}$

$= \frac{1}{2}\left(\frac{e^{2x} + e^{-2x}}{2} + 1\right)$

$= \frac{\cosh 2x + 1}{2}.$

7. $\frac{d}{dx}[\operatorname{sech} x] = \frac{d}{dx}\left[\frac{2}{e^x + e^{-x}}\right]$

$= \frac{-2(e^x - e^{-x})}{(e^x + e^{-x})^2}$

$= -\frac{2(e^x - e^{-x})}{(e^x + e^{-x})(e^x + e^{-x})}$

$= -\frac{2}{e^x + e^{-x}} \cdot \frac{e^x - e^{-x}}{e^x + e^{-x}}$

$= -\operatorname{sech} x \tanh x$

9. $\int \tanh x\, dx = \int \frac{\sinh x}{\cosh x}\, dx$

Let $u = \cosh x$; $du = (\sinh x)dx$

$= \int \frac{1}{u}\, du$

$= \ln|u| + C$

$= \ln(\cosh x) + C.$

11. $2\sinh 2x$

13. $\coth x$

15. $x \cosh x$

17. $\frac{3}{\sqrt{9x^2 + 1}}$

19. $\frac{1}{1 - (x+5)^2}$

21. $\sec x$

23. $y = 3/4(x - \ln 2) + 5/4$

25. $y = x$

27. $1/2 \ln(\cosh(2x)) + C$

29. $1/2 \sinh^2 x + C$ or $1/2 \cosh^2 x + C$

31. $x\cosh(x) - \sinh(x) + C$

33. $\cosh^{-1}(x^2/2) + C = \ln(x^2 + \sqrt{x^4 - 4}) + C$

35. $\frac{1}{16}\tan^{-1}(x/2) + \frac{1}{32}\ln|x-2| + \frac{1}{32}\ln|x+2| + C$

37. $\tan^{-1}(e^x) + C$

39. $x\tanh^{-1}x + 1/2\ln|x^2 - 1| + C$

41. 0

43. 2

Section 6.7

1. $0/0, \infty/\infty, 0 \cdot \infty, \infty - \infty, 0^0, 1^\infty, \infty^0$

3. F

5. derivatives; limits

7. Answers will vary.

9. $-5/3$

11. $-\sqrt{2}/2$

13. 0

15. a/b

17. $1/2$

19. 0

21. ∞

23. 0

25. -2

27. 0

29. 0

31. ∞

33. ∞

35. 0

37. 1

39. 1

41. 1

43. 1

45. 1

47. 2

49. $-\infty$

51. 0

Section 6.8

1. The interval of integration is finite, and the integrand is continuous on that interval.

3. converges; could also state < 10.

5. $p > 1$

7. $e^5/2$

9. $1/3$

11. $1/\ln 2$

13. diverges

15. 1

17. diverges

19. diverges

21. diverges

23. 1

25. 0

27. $-1/4$

29. -1

31. diverges

33. $1/2$

35. converges; Limit Comparison Test with $1/x^{3/2}$.

37. converges; Direct Comparison Test with xe^{-x}.

39. converges; Direct Comparison Test with xe^{-x}.

41. diverges; Direct Comparison Test with $x/(x^2 + \cos x)$.

43. converges; Limit Comparison Test with $1/e^x$.

Chapter 7

Section 7.1

1. T

3. Answers will vary.

5. $16/3$

7. π

9. $2\sqrt{2}$

11. 4.5

13. $2 - \pi/2$

15. $1/6$

17. On regions such as $[\pi/6, 5\pi/6]$, the area is $3\sqrt{3}/2$. On regions such as $[-\pi/2, \pi/6]$, the area is $3\sqrt{3}/4$.

19. $5/3$

21. $9/4$

23. 1

25. 4

27. $219{,}000$ ft^2

Section 7.2

1. T

3. Recall that "dx" does not just "sit there;" it is multiplied by $A(x)$ and represents the thickness of a small slice of the solid. Therefore dx has units of in, giving $A(x)\,dx$ the units of in^3.

5. $175\pi/3$ units^3

7. $\pi/6$ units^3

9. $35\pi/3$ units^3

11. $2\pi/15$ units^3

13. (a) $512\pi/15$
    (b) $256\pi/5$
    (c) $832\pi/15$
    (d) $128\pi/3$

15. (a) $104\pi/15$
    (b) $64\pi/15$
    (c) $32\pi/5$

17. (a) 8π
    (b) 8π
    (c) $16\pi/3$
    (d) $8\pi/3$

19. The cross–sections of this cone are the same as the cone in Exercise 18. Thus they have the same volume of $250\pi/3$ units^3.

21. Orient the solid so that the x-axis is parallel to long side of the base. All cross–sections are trapezoids (at the far left, the trapezoid is a square; at the far right, the trapezoid has a top length of 0, making it a triangle). The area of the trapezoid at x is $A(x) = 1/2(-1/2x + 5 + 5)(5) = -5/4x + 25$. The volume is 187.5 units^3.

Section 7.3

1. T

3. F

5. $9\pi/2$ units^3

7. $\pi^2 - 2\pi$ units^3

9. $48\pi\sqrt{3}/5$ units^3

11. $\pi^2/4$ units^3

13. (a) $4\pi/5$
    (b) $8\pi/15$
    (c) $\pi/2$
    (d) $5\pi/6$

15. (a) $4\pi/3$
    (b) $\pi/3$
    (c) $4\pi/3$
    (d) $2\pi/3$

17. (a) $2\pi(\sqrt{2} - 1)$
    (b) $2\pi(1 - \sqrt{2} + \sinh^{-1}(1))$

Section 7.4

1. T

3. $\sqrt{2}$

5. $4/3$

7. $109/2$

9. $12/5$

11. $-\ln(2 - \sqrt{3}) \approx 1.31696$

13. $\int_0^1 \sqrt{1 + 4x^2}\,dx$

15. $\int_0^1 \sqrt{1 + \frac{1}{4x}}\,dx$

17. $\int_{-1}^1 \sqrt{1 + \frac{x^2}{1-x^2}}\,dx$

19. $\int_1^2 \sqrt{1 + \frac{1}{x^4}}\,dx$

21. 1.4790

23. Simpson's Rule fails, as it requires one to divide by 0. However, recognize the answer should be the same as for $y = x^2$; why?

25. Simpson's Rule fails.

27. 1.4058

29. $2\pi \int_0^1 2x\sqrt{5}\, dx = 2\pi\sqrt{5}$

31. $2\pi \int_0^1 x^3\sqrt{1+9x^4}\, dx = \pi/27(10\sqrt{10}-1)$

33. $2\pi \int_0^1 \sqrt{1-x^2}\sqrt{1+x/(1-x^2)}\, dx = 4\pi$

Section 7.5

1. In SI units, it is one joule, i.e., one Newton-meter, or kg·m/s²·m. In Imperial Units, it is ft-lb.

3. Smaller.

5. (a) 2450 j
   (b) 1568 j

7. 735 j

9. 11,100 ft-lb

11. 125 ft-lb

13. 12.5 ft-lb

15. 0.2625 = 21/80 j

17. 45 ft-lb

19. 953, 284 j

21. 192,767 ft-lb. Note that the tank is oriented horizontally. Let the origin be the center of one of the circular ends of the tank. Since the radius is 3.75 ft, the fluid is being pumped to $y = 4.75$; thus the distance the gas travels is $h(y) = 4.75 - y$. A differential element of water is a rectangle, with length 20 and width $2\sqrt{3.75^2 - y^2}$. Thus the force required to move that slab of gas is $F(y) = 40 \cdot 45.93 \cdot \sqrt{3.75^2 - y^2}\, dy$. Total work is $\int_{-3.75}^{3.75} 40 \cdot 45.93 \cdot (4.75 - y)\sqrt{3.75^2 - y^2}\, dy$. This can be evaluated without actual integration; split the integral into $\int_{-3.75}^{3.75} 40 \cdot 45.93 \cdot (4.75)\sqrt{3.75^2 - y^2}\, dy + \int_{-3.75}^{3.75} 40 \cdot 45.93 \cdot (-y)\sqrt{3.75^2 - y^2}\, dy$. The first integral can be evaluated as measuring half the area of a circle; the latter integral can be shown to be 0 without much difficulty. (Use substitution and realize the bounds are both 0.)

23. (a) approx. 577,000 j
    (b) approx. 399,000 j
    (c) approx 110,000 j (By volume, half of the water is between the base of the cone and a height of 3.9685 m. If one rounds this to 4 m, the work is approx 104,000 j.)

25. 617,400 j

Section 7.6

1. Answers will vary.

3. 499.2 lb

5. 6739.2 lb

7. 3920.7 lb

9. 2496 lb

11. 602.59 lb

13. (a) 2340 lb
    (b) 5625 lb

15. (a) 1597.44 lb
    (b) 3840 lb

17. (a) 56.42 lb
    (b) 135.62 lb

19. 5.1 ft

Chapter 8

Section 8.1

1. Answers will vary.

3. Answers will vary.

5. $2, \frac{8}{3}, \frac{8}{3}, \frac{32}{15}, \frac{64}{45}$

7. $\frac{1}{3}, 2, \frac{81}{5}, \frac{512}{3}, \frac{15625}{7}$

9. $a_n = 3n + 1$

11. $a_n = 10 \cdot 2^{n-1}$

13. 1/7

15. 0

17. diverges

19. converges to 0

21. diverges

23. converges to e

25. converges to 0

27. converges to 2

29. bounded

31. bounded

33. neither bounded above or below

35. monotonically increasing

37. never monotonic

39. Let $\{a_n\}$ be given such that $\lim_{n\to\infty} |a_n| = 0$. By the definition of the limit of a sequence, given any $\varepsilon > 0$, there is a m such that for all $n > m$, $||a_n| - 0| < \varepsilon$. Since $||a_n| - 0| = |a_n - 0|$, this directly implies that for all $n > m$, $|a_n - 0| < \varepsilon$, meaning that $\lim_{n\to\infty} a_n = 0$.

41. Left to reader

Section 8.2

1. Answers will vary.

3. One sequence is the sequence of terms $\{a_n\}$. The other is the sequence of nth partial sums, $\{S_n\} = \{\sum_{i=1}^n a_i\}$.

5. F

7. (a) $1, \frac{5}{4}, \frac{49}{36}, \frac{205}{144}, \frac{5269}{3600}$
   (b) Plot omitted

9. (a) 1, 3, 6, 10, 15
   (b) Plot omitted

11. (a) $\frac{1}{3}, \frac{4}{9}, \frac{13}{27}, \frac{40}{81}, \frac{121}{243}$
    (b) Plot omitted

13. (a) 0.1, 0.11, 0.111, 0.1111, 0.11111
    (b) Plot omitted

15. $\lim_{n\to\infty} a_n = \infty$; by Theorem 63 the series diverges.

17. $\lim_{n\to\infty} a_n = 1$; by Theorem 63 the series diverges.

19. $\lim_{n\to\infty} a_n = e$; by Theorem 63 the series diverges.

21. Converges

23. Converges

25. Converges

27. Converges

29. Diverges

31. (a) $S_n = \left(\frac{n(n+1)}{2}\right)^2$
    (b) Diverges

33. (a) $S_n = 5\frac{1-1/2^n}{1/2}$
    (b) Converges to 10.

35. (a) $S_n = \frac{1-(-1/3)^n}{4/3}$
    (b) Converges to 3/4.

37. (a) With partial fractions, $a_n = \frac{3}{2}\left(\frac{1}{n} - \frac{1}{n+2}\right)$. Thus $S_n = \frac{3}{2}\left(\frac{3}{2} - \frac{1}{n+1} - \frac{1}{n+2}\right)$.
    (b) Converges to 9/4

39. (a) $S_n = \ln\left(1/(n+1)\right)$
    (b) Diverges (to $-\infty$).

41. (a) $a_n = \frac{1}{n(n+3)}$; using partial fractions, the resulting telescoping sum reduces to $S_n = \frac{1}{3}\left(1 + \frac{1}{2} + \frac{1}{3} - \frac{1}{n+1} - \frac{1}{n+2} - \frac{1}{n+3}\right)$
    (b) Converges to 11/18.

43. (a) With partial fractions, $a_n = \frac{1}{2}\left(\frac{1}{n-1} - \frac{1}{n+1}\right)$. Thus $S_n = \frac{1}{2}\left(3/2 - \frac{1}{n} - \frac{1}{n+1}\right)$.
    (b) Converges to 3/4.

45. (a) The nth partial sum of the odd series is $1 + \frac{1}{3} + \frac{1}{5} + \cdots + \frac{1}{2n-1}$. The nth partial sum of the even series is $\frac{1}{2} + \frac{1}{4} + \frac{1}{6} + \cdots + \frac{1}{2n}$. Each term of the even series is less than the corresponding term of the odd series, giving us our result.
    (b) The nth partial sum of the odd series is $1 + \frac{1}{3} + \frac{1}{5} + \cdots + \frac{1}{2n-1}$. The nth partial sum of 1 plus the even series is $1 + \frac{1}{2} + \frac{1}{4} + \cdots + \frac{1}{2(n-1)}$. Each term of the even series is now greater than or equal to the corresponding term of the odd series, with equality only on the first term. This gives us the result.
    (c) If the odd series converges, the work done in (a) shows the even series converges also. (The sequence of the nth partial sum of the even series is bounded and monotonically increasing.) Likewise, (b) shows that if the even series converges, the odd series will, too. Thus if either series converges, the other does.
    Similarly, (a) and (b) can be used to show that if either series diverges, the other does, too.
    (d) If both the even and odd series converge, then their sum would be a convergent series. This would imply that the Harmonic Series, their sum, is convergent. It is not. Hence each series diverges.

Section 8.3

1. continuous, positive and decreasing

3. The Integral Test (we do not have a continuous definition of $n!$ yet) and the Limit Comparison Test (same as above, hence we cannot take its derivative).

5. Converges

7. Diverges

9. Converges

11. Converges

13. Converges; compare to $\sum_{n=1}^{\infty}\frac{1}{n^2}$, as $1/(n^2 + 3n - 5) \leq 1/n^2$ for all $n > 1$.

15. Diverges; compare to $\sum_{n=1}^{\infty}\frac{1}{n}$, as $1/n \leq \ln n/n$ for all $n \geq 2$.

17. Diverges; compare to $\sum_{n=1}^{\infty}\frac{1}{n}$. Since $n = \sqrt{n^2} > \sqrt{n^2 - 1}$, $1/n \leq 1/\sqrt{n^2 - 1}$ for all $n \geq 2$.

19. Diverges; compare to $\sum_{n=1}^{\infty}\frac{1}{n}$:
$$\frac{1}{n} = \frac{n^2}{n^3} < \frac{n^2 + n + 1}{n^3} < \frac{n^2 + n + 1}{n^3 - 5},$$
for all $n \geq 1$.

21. Diverges; compare to $\sum_{n=1}^{\infty}\frac{1}{n}$. Note that
$$\frac{n}{n^2 - 1} = \frac{n^2}{n^2 - 1} \cdot \frac{1}{n} > \frac{1}{n},$$
as $\frac{n^2}{n^2-1} > 1$, for all $n \geq 2$.

23. Converges; compare to $\sum_{n=1}^{\infty}\frac{1}{n^2}$.

25. Diverges; compare to $\sum_{n=1}^{\infty}\frac{\ln n}{n}$.

27. Diverges; compare to $\sum_{n=1}^{\infty}\frac{1}{n}$.

29. Diverges; compare to $\sum_{n=1}^{\infty}\frac{1}{n}$. Just as $\lim_{n\to 0}\frac{\sin n}{n} = 1$,
$$\lim_{n\to\infty}\frac{\sin(1/n)}{1/n} = 1.$$

31. Converges; compare to $\sum_{n=1}^{\infty}\frac{1}{n^{3/2}}$.

33. Converges; Integral Test

35. Diverges; the nth Term Test and Direct Comparison Test can be used.

37. Converges; the Direct Comparison Test can be used with sequence $1/3^n$.

39. Diverges; the nth Term Test can be used, along with the Integral Test.

41. (a) Converges; use Direct Comparison Test as $\frac{a_n}{n} < n$.
    (b) Converges; since original series converges, we know $\lim_{n\to\infty} a_n = 0$. Thus for large n, $a_n a_{n+1} < a_n$.
    (c) Converges; similar logic to part (b) so $(a_n)^2 < a_n$.
    (d) May converge; certainly $na_n > a_n$ but that does not mean it does not converge.
    (e) Does not converge, using logic from (b) and n^{th} Term Test.

Section 8.4

1. algebraic, or polynomial.

3. Integral Test, Limit Comparison Test, and Root Test

5. Converges

7. Converges

A.7

9. The Ratio Test is inconclusive; the p-Series Test states it diverges.

11. Converges

13. Converges; note the summation can be rewritten as $\sum_{n=1}^{\infty} \frac{2^n n!}{3^n n!}$, from which the Ratio Test can be applied.

15. Converges

17. Converges

19. Diverges

21. Diverges. The Root Test is inconclusive, but the nth-Term Test shows divergence. (The terms of the sequence approach e^2, not 0, as $n \to \infty$.)

23. Converges

25. Diverges; Limit Comparison Test

27. Converges; Ratio Test or Limit Comparison Test with $1/3^n$.

29. Diverges; n^{th}-Term Test or Limit Comparison Test with 1.

31. Diverges; Direct Comparison Test with $1/n$

33. Converges; Root Test

Section 8.5

1. The signs of the terms do not alternate; in the given series, some terms are negative and the others positive, but they do not necessarily alternate.

3. Many examples exist; one common example is $a_n = (-1)^n/n$.

5. (a) converges
   (b) converges (p-Series)
   (c) absolute

7. (a) diverges (limit of terms is not 0)
   (b) diverges
   (c) n/a; diverges

9. (a) converges
   (b) diverges (Limit Comparison Test with $1/n$)
   (c) conditional

11. (a) diverges (limit of terms is not 0)
    (b) diverges
    (c) n/a; diverges

13. (a) diverges (terms oscillate between ± 1)
    (b) diverges
    (c) n/a; diverges

15. (a) converges
    (b) converges (Geometric Series with $r = 2/3$)
    (c) absolute

17. (a) converges
    (b) converges (Ratio Test)
    (c) absolute

19. (a) converges
    (b) diverges (p-Series Test with $p = 1/2$)
    (c) conditional

21. $S_5 = -1.1906$; $S_6 = -0.6767$;
$$-1.1906 \leq \sum_{n=1}^{\infty} \frac{(-1)^n}{\ln(n+1)} \leq -0.6767$$

23. $S_6 = 0.3681$; $S_7 = 0.3679$;
$$0.3681 \leq \sum_{n=0}^{\infty} \frac{(-1)^n}{n!} \leq 0.3679$$

25. $n = 5$

27. Using the theorem, we find $n = 499$ guarantees the sum is within 0.001 of $\pi/4$. (Convergence is actually faster, as the sum is within ε of $\pi/24$ when $n \geq 249$.)

Section 8.6

1. 1

3. 5

5. $1 + 2x + 4x^2 + 8x^3 + 16x^4$

7. $1 + x + \frac{x^2}{2} + \frac{x^3}{6} + \frac{x^4}{24}$

9. (a) $R = \infty$
   (b) $(-\infty, \infty)$

11. (a) $R = 1$
    (b) $(2, 4]$

13. (a) $R = 2$
    (b) $(-2, 2)$

15. (a) $R = 1/5$
    (b) $(4/5, 6/5)$

17. (a) $R = 1$
    (b) $(-1, 1)$

19. (a) $R = \infty$
    (b) $(-\infty, \infty)$

21. (a) $R = 1$
    (b) $[-1, 1]$

23. (a) $R = 0$
    (b) $x = 0$

25. (a) $f'(x) = \sum_{n=1}^{\infty} n^2 x^{n-1}$; $(-1, 1)$
    (b) $\int f(x)\, dx = C + \sum_{n=0}^{\infty} \frac{n}{n+1} x^{n+1}$; $(-1, 1)$

27. (a) $f'(x) = \sum_{n=1}^{\infty} \frac{n}{2^n} x^{n-1}$; $(-2, 2)$
    (b) $\int f(x)\, dx = C + \sum_{n=0}^{\infty} \frac{1}{(n+1)2^n} x^{n+1}$; $[-2, 2)$

29. (a) $f'(x) = \sum_{n=1}^{\infty} \frac{(-1)^n x^{2n-1}}{(2n-1)!} = \sum_{n=0}^{\infty} \frac{(-1)^{n+1} x^{2n+1}}{(2n+1)!}$; $(-\infty, \infty)$
    (b) $\int f(x)\, dx = C + \sum_{n=0}^{\infty} \frac{(-1)^n x^{2n+1}}{(2n+1)!}$; $(-\infty, \infty)$

31. $1 + 3x + \frac{9}{2}x^2 + \frac{9}{2}x^3 + \frac{27}{8}x^4$

33. $1 + x + x^2 + x^3 + x^4$

35. $0 + x + 0x^2 - \frac{1}{6}x^3 + 0x^4$

Section 8.7

1. The Maclaurin polynomial is a special case of Taylor polynomials. Taylor polynomials are centered at a specific x-value; when that x-value is 0, it is a Maclauring polynomial.

3. $p_2(x) = 6 + 3x - 4x^2$.

5. $p_3(x) = 1 - x + \frac{1}{2}x^3 - \frac{1}{6}x^3$

7. $p_8(x) = x + x^2 + \frac{1}{2}x^3 + \frac{1}{6}x^4 + \frac{1}{24}x^5$

9. $p_4(x) = \frac{2x^4}{3} + \frac{4x^3}{3} + 2x^2 + 2x + 1$

11. $p_4(x) = x^4 - x^3 + x^2 - x + 1$

13. $p_4(x) = 1 + \frac{1}{2}(-1+x) - \frac{1}{8}(-1+x)^2 + \frac{1}{16}(-1+x)^3 - \frac{5}{128}(-1+x)^4$

15. $p_6(x) = \frac{1}{\sqrt{2}} - \frac{-\frac{\pi}{4}+x}{\sqrt{2}} - \frac{(-\frac{\pi}{4}+x)^2}{2\sqrt{2}} + \frac{(-\frac{\pi}{4}+x)^3}{6\sqrt{2}} + \frac{(-\frac{\pi}{4}+x)^4}{24\sqrt{2}} - \frac{(-\frac{\pi}{4}+x)^5}{120\sqrt{2}} - \frac{(-\frac{\pi}{4}+x)^6}{720\sqrt{2}}$

17. $p_5(x) = \frac{1}{2} - \frac{x-2}{4} + \frac{1}{8}(x-2)^2 - \frac{1}{16}(x-2)^3 + \frac{1}{32}(x-2)^4 - \frac{1}{64}(x-2)^5$

19. $p_3(x) = \frac{1}{2} + \frac{1+x}{2} + \frac{1}{4}(1+x)^2$

21. $p_3(x) = x - \frac{x^3}{6}$; $p_3(0.1) = 0.09983$. Error is bounded by $\pm \frac{1}{4!} \cdot 0.1^4 \approx \pm 0.000004167$.

23. $p_2(x) = 3 + \frac{1}{6}(-9+x) - \frac{1}{216}(-9+x)^2$; $p_2(10) = 3.16204$. The third derivative of $f(x) = \sqrt{x}$ is bounded on $(8, 11)$ by 0.003. Error is bounded by $\pm \frac{0.003}{3!} \cdot 1^3 = \pm 0.0005$.

25. The nth derivative of $f(x) = e^x$ is bounded by 3 on intervals containing 0 and 1. Thus $|R_n(1)| \leq \frac{3}{(n+1)!} 1^{(n+1)}$. When $n = 7$, this is less than 0.0001.

27. The nth derivative of $f(x) = \cos x$ is bounded by 1 on intervals containing 0 and $\pi/3$. Thus $|R_n(\pi/3)| \leq \frac{1}{(n+1)!}(\pi/3)^{(n+1)}$. When $n = 7$, this is less than 0.0001. Since the Maclaurin polynomial of $\cos x$ only uses even powers, we can actually just use $n = 6$.

29. The nth term is $\frac{1}{n!}x^n$.

31. The nth term is x^n.

33. The nth term is $(-1)^n \frac{(x-1)^n}{n}$.

35. $3 + 15x + \frac{75}{2}x^2 + \frac{375}{6}x^3 + \frac{1875}{24}x^4$

Section 8.8

1. A Taylor polynomial is a **polynomial**, containing a finite number of terms. A Taylor series is a **series**, the summation of an infinite number of terms.

3. All derivatives of e^x are e^x which evaluate to 1 at $x = 0$.
The Taylor series starts $1 + x + \frac{1}{2}x^2 + \frac{1}{3!}x^3 + \frac{1}{4!}x^4 + \cdots$;
the Taylor series is $\sum_{n=0}^{\infty} \frac{x^n}{n!}$

5. The nth derivative of $1/(1-x)$ is $f^{(n)}(x) = (n)!/(1-x)^{n+1}$, which evaluates to $n!$ at $x = 0$.
The Taylor series starts $1 + x + x^2 + x^3 + \cdots$;
the Taylor series is $\sum_{n=0}^{\infty} x^n$

7. The Taylor series starts
$0 - (x - \pi/2) + 0x^2 + \frac{1}{6}(x - \pi/2)^3 + 0x^4 - \frac{1}{120}(x - \pi/2)^5$;
the Taylor series is $\sum_{n=0}^{\infty} (-1)^{n+1} \frac{(x-\pi/2)^{2n+1}}{(2n+1)!}$

9. $f^{(n)}(x) = (-1)^n e^{-x}$; at $x = 0, f^{(n)}(0) = -1$ when n is odd and $f^{(n)}(0) = 1$ when n is even.
The Taylor series starts $1 - x + \frac{1}{2}x^2 - \frac{1}{3!}x^3 + \cdots$;
the Taylor series is $\sum_{n=0}^{\infty} (-1)^n \frac{x^n}{n!}$.

11. $f^{(n)}(x) = (-1)^{n+1} \frac{n!}{(x+1)^{n+1}}$; at $x = 1, f^{(n)}(1) = (-1)^{n+1} \frac{n!}{2^{n+1}}$
The Taylor series starts
$\frac{1}{2} + \frac{1}{4}(x-1) - \frac{1}{8}(x-1)^2 + \frac{1}{16}(x-1)^3 \cdots$;
the Taylor series is $\sum_{n=0}^{\infty} (-1)^{n+1} \frac{(x-1)^n}{2^{n+1}}$.

13. Given a value x, the magnitude of the error term $R_n(x)$ is bounded by
$$|R_n(x)| \leq \frac{\max |f^{(n+1)}(z)|}{(n+1)!} |x^{(n+1)}|,$$
where z is between 0 and x.
If $x > 0$, then $z < x$ and $f^{(n+1)}(z) = e^z < e^x$. If $x < 0$, then $x < z < 0$ and $f^{(n+1)}(z) = e^z < 1$. So given a fixed x value, let $M = \max\{e^x, 1\}$; $f^{(n)}(z) < M$. This allows us to state
$$|R_n(x)| \leq \frac{M}{(n+1)!} |x^{(n+1)}|.$$
For any x, $\lim_{n \to \infty} \frac{M}{(n+1)!} |x^{(n+1)}| = 0$. Thus by the Squeeze Theorem, we conclude that $\lim_{n \to \infty} R_n(x) = 0$ for all x, and hence
$$e^x = \sum_{n=0}^{\infty} \frac{x^n}{n!} \quad \text{for all } x.$$

15. Given a value x, the magnitude of the error term $R_n(x)$ is bounded by
$$|R_n(x)| \leq \frac{\max |f^{(n+1)}(z)|}{(n+1)!} |(x-1)^{(n+1)}|,$$
where z is between 1 and x.
Note that $|f^{(n+1)}(x)| = \frac{n!}{x^{n+1}}$.
We consider the cases when $x > 1$ and when $x < 1$ separately.
If $x > 1$, then $1 < z < x$ and $f^{(n+1)}(z) = \frac{n!}{z^{n+1}} < n!$. Thus
$$|R_n(x)| \leq \frac{n!}{(n+1)!} |(x-1)^{(n+1)}| = \frac{(x-1)^{n+1}}{n+1}.$$
For a fixed x,
$$\lim_{n \to \infty} \frac{(x-1)^{n+1}}{n+1} = 0.$$
If $0 < x < 1$, then $x < z < 1$ and $f^{(n+1)}(z) = \frac{n!}{z^{n+1}} < \frac{n!}{x^{n+1}}$. Thus
$$|R_n(x)| \leq \frac{n!/x^{n+1}}{(n+1)!} |(x-1)^{(n+1)}| = \frac{x^{n+1}}{n+1}(1-x)^{n+1}.$$
Since $0 < x < 1$, $x^{n+1} < 1$ and $(1-x)^{n+1} < 1$. We can then extend the inequality from above to state
$$|R_n(x)| \leq \frac{x^{n+1}}{n+1}(1-x)^{n+1} < \frac{1}{n+1}.$$
As $n \to \infty$, $1/(n+1) \to 0$. Thus by the Squeeze Theorem, we conclude that $\lim_{n \to \infty} R_n(x) = 0$ for all x, and hence
$$\ln x = \sum_{n=1}^{\infty} (-1)^{n+1} \frac{(x-1)^n}{n} \quad \text{for all } 0 < x \leq 2.$$

17. Given $\cos x = \sum_{n=0}^{\infty} (-1)^n \frac{x^{2n}}{(2n)!}$,
$\cos(-x) = \sum_{n=0}^{\infty} (-1)^n \frac{(-x)^{2n}}{(2n)!} = \sum_{n=0}^{\infty} (-1)^n \frac{x^{2n}}{(2n)!} = \cos x$, as all powers in the series are even.

19. Given $\sin x = \sum_{n=0}^{\infty} (-1)^n \frac{x^{2n+1}}{(2n+1)!}$,
$\frac{d}{dx}(\sin x) = \frac{d}{dx}\left(\sum_{n=0}^{\infty} (-1)^n \frac{x^{2n+1}}{(2n+1)!}\right) = \sum_{n=0}^{\infty} (-1)^n \frac{(2n+1)x^{2n}}{(2n+1)!} = \sum_{n=0}^{\infty} (-1)^n \frac{x^{2n}}{(2n)!} = \cos x$. (The summation still starts at $n = 0$ as there was no constant term in the expansion of $\sin x$).

21. $1 + \dfrac{x}{2} - \dfrac{x^2}{8} + \dfrac{x^3}{16} - \dfrac{5x^4}{128}$

23. $1 + \dfrac{x}{3} - \dfrac{x^2}{9} + \dfrac{5x^3}{81} - \dfrac{10x^4}{243}$

25. $\displaystyle\sum_{n=0}^{\infty}(-1)^n \dfrac{(x^2)^{2n}}{(2n)!} = \sum_{n=0}^{\infty}(-1)^n \dfrac{x^{4n}}{(2n)!}$.

27. $\displaystyle\sum_{n=0}^{\infty}(-1)^n \dfrac{(2x+3)^{2n+1}}{(2n+1)!}$.

29. $x + x^2 + \dfrac{x^3}{3} - \dfrac{x^5}{30}$

31. $\displaystyle\int_0^{\sqrt{\pi}} \sin(x^2)\,dx \approx \int_0^{\sqrt{\pi}} \left(x^2 - \dfrac{x^6}{6} + \dfrac{x^{10}}{120} - \dfrac{x^{14}}{5040}\right) dx =$ 0.8877

Index

!, 397
Absolute Convergence Theorem, 448
absolute maximum, 123
absolute minimum, 123
Absolute Value Theorem, 401
acceleration, 73, 642
Alternating Harmonic Series, 419, 446, 459
Alternating Series Test
    for series, 442
a_N, 660, 670
analytic function, 480
angle of elevation, 647
antiderivative, 189
arc length, 370, 519, 545, 639, 664
arc length parameter, 664, 666
asymptote
    horizontal, 49
    vertical, 47
a_T, 660, 670
average rate of change, 627
average value of a function, 769
average value of function, 236

Binomial Series, 480
Bisection Method, 42
boundary point, 682
bounded sequence, 404
    convergence, 405
bounded set, 682

center of mass, 783–785, 787, 814
Chain Rule, 97
    multivariable, 713, 716
    notation, 103
circle of curvature, 669
closed, 682
closed disk, 682
concave down, 144
concave up, 144
concavity, 144, 516
    inflection point, 145
    test for, 145
conic sections, 490
    degenerate, 490
    ellipse, 493
    hyperbola, 496
    parabola, 490
Constant Multiple Rule
    of derivatives, 80
    of integration, 193
    of series, 419

constrained optimization, 745
continuous function, 37, 688
    properties, 40, 689
    vector–valued, 630
contour lines, 676
convergence
    absolute, 446, 448
    Alternating Series Test, 442
    conditional, 446
    Direct Comparison Test, 429
        for integration, 339
    Integral Test, 426
    interval of, 454
    Limit Comparison Test, 430
        for integration, 341
    n^{th}–term test, 422
    of geometric series, 414
    of improper int., 334, 339, 341
    of monotonic sequences, 407
    of p-series, 415
    of power series, 453
    of sequence, 400, 405
    of series, 411
    radius of, 454
    Ratio Comparison Test, 435
    Root Comparison Test, 438
critical number, 125
critical point, 125, 740–742
cross product
    and derivatives, 635
    applications, 597
        area of parallelogram, 598
        torque, 600
        volume of parallelepiped, 599
    definition, 593
    properties, 595, 596
curvature, 666
    and motion, 670
    equations for, 668
    of circle, 668, 669
    radius of, 669
curve
    parametrically defined, 503
    rectangular equation, 503
    smooth, 509
curve sketching, 152
cusp, 509
cycloid, 625
cylinder, 555

decreasing function, 136

finding intervals, 137
strictly, 136
definite integral, 201
    and substitution, 270
    properties, 203
derivative
    acceleration, 74
    as a function, 64
    at a point, 60
    basic rules, 78
    Chain Rule, 97, 103, 713, 716
    Constant Multiple Rule, 80
    Constant Rule, 78
    differential, 181
    directional, 720, 722, 723, 726, 727
    exponential functions, 103
    First Deriv. Test, 139
    Generalized Power Rule, 98
    higher order, 81
        interpretation, 82
    hyperbolic funct., 316
    implicit, 106, 718
    interpretation, 71
    inverse function, 117
    inverse hyper., 319
    inverse trig., 120
    Mean Value Theorem, 132
    mixed partial, 696
    motion, 74
    multivariable differentiability, 705, 710
    normal line, 61
    notation, 64, 81
    parametric equations, 513
    partial, 692, 700
    Power Rule, 78, 91, 111
    power series, 457
    Product Rule, 85
    Quotient Rule, 88
    Second Deriv. Test, 148
    Sum/Difference Rule, 80
    tangent line, 60
    trigonometric functions, 90
    vector–valued functions, 631, 632, 635
    velocity, 74
differentiable, 60, 705, 710
differential, 181
    notation, 181
Direct Comparison Test
    for integration, 339
    for series, 429
directional derivative, 720, 722, 723, 726, 727
directrix, 490, 555
Disk Method, 355
displacement, 230, 626, 639
distance
    between lines, 611
    between point and line, 611
    between point and plane, 620
    between points in space, 552
    traveled, 650

divergence
    Alternating Series Test, 442
    Direct Comparison Test, 429
        for integration, 339
    Integral Test, 426
    Limit Comparison Test, 430
        for integration, 341
    n^{th}–term test, 422
    of geometric series, 414
    of improper int., 334, 339, 341
    of p-series, 415
    of sequence, 400
    of series, 411
    Ratio Comparison Test, 435
    Root Comparison Test, 438
dot product
    and derivatives, 635
    definition, 580
    properties, 581, 582
double integral, 762, 763
    in polar, 773
    properties, 766

eccentricity, 495, 499
elementary function, 240
ellipse
    definition, 493
    eccentricity, 495
    parametric equations, 509
    reflective property, 496
    standard equation, 494
extrema
    absolute, 123, 740
    and First Deriv. Test, 139
    and Second Deriv. Test, 148
    finding, 126
    relative, 124, 740, 741
Extreme Value Theorem, 124, 745
extreme values, 123

factorial, 397
First Derivative Test, 139
floor function, 38
fluid pressure/force, 388, 390
focus, 490, 493, 496
Fubini's Theorem, 763
function
    of three variables, 679
    of two variables, 675
    vector–valued, 623
Fundamental Theorem of Calculus, 228, 229
    and Chain Rule, 232

Gabriel's Horn, 376
Generalized Power Rule, 98
geometric series, 413, 414
gradient, 722, 723, 726, 727, 737
    and level curves, 723
    and level surfaces, 737

Harmonic Series, 419

Head To Tail Rule, 570
Hooke's Law, 381
hyperbola
- definition, 496
- eccentricity, 499
- parametric equations, 509
- reflective property, 499
- standard equation, 497

hyperbolic function
- definition, 313
- derivatives, 316
- identities, 316
- integrals, 316
- inverse, 317
  - derivative, 319
  - integration, 319
  - logarithmic def., 318

implicit differentiation, 106, 718
improper integration, 334, 337
increasing function, 136
- finding intervals, 137
- strictly, 136
indefinite integral, 189
indeterminate form, 2, 48, 327, 328
inflection point, 145
initial point, 566
initial value problem, 194
Integral Test, 426
integration
- arc length, 370
- area, 201, 754, 755
- area between curves, 233, 346
- average value, 236
- by parts, 275
- by substitution, 257
- definite, 201
  - and substitution, 270
  - properties, 203
  - Riemann Sums, 224
- displacement, 230
- distance traveled, 650
- double, 762
- fluid force, 388, 390
- Fun. Thm. of Calc., 228, 229
- general application technique, 345
- hyperbolic funct., 316
- improper, 334, 337, 339, 341
- indefinite, 189
- inverse hyper., 319
- iterated, 753
- Mean Value Theorem, 235
- multiple, 753
- notation, 190, 201, 229, 753
- numerical, 240
  - Left/Right Hand Rule, 240, 247
  - Simpson's Rule, 245, 247, 248
  - Trapezoidal Rule, 243, 247, 248
- of multivariable functions, 751
- of power series, 457
- of trig. functions, 263
- of trig. powers, 286, 291
- of vector–valued functions, 637
- partial fraction decomp., 306
- Power Rule, 194
- Sum/Difference Rule, 194
- surface area, 374, 521, 546
- trig. subst., 297
- triple, 800, 811, 813
- volume
  - cross-sectional area, 353
  - Disk Method, 355
  - Shell Method, 362, 366
  - Washer Method, 357, 366
- work, 378
interior point, 682
Intermediate Value Theorem, 42
interval of convergence, 454
iterated integration, 753, 762, 763, 800, 811, 813
- changing order, 757
- properties, 766, 807

L'Hôpital's Rule, 324, 326
lamina, 779
Left Hand Rule, 210, 215, 240
Left/Right Hand Rule, 247
level curves, 676, 723
level surface, 680, 737
limit
- Absolute Value Theorem, 401
- at infinity, 49
- definition, 10
- difference quotient, 6
- does not exist, 4, 32
- indeterminate form, 2, 48, 327, 328
- L'Hôpital's Rule, 324, 326
- left handed, 30
- of infinity, 46
- of multivariable function, 683, 684, 690
- of sequence, 400
- of vector–valued functions, 629
- one sided, 30
- properties, 18, 684
- pseudo-definition, 2
- right handed, 30
- Squeeze Theorem, 22
Limit Comparison Test
- for integration, 341
- for series, 430
lines, 604
- distances between, 611
- equations for, 606
- intersecting, 607
- parallel, 607
- skew, 607
logarithmic differentiation, 113

Maclaurin Polynomial, *see* Taylor Polynomial
- definition, 466
Maclaurin Series, *see* Taylor Series

definition, 477
magnitude of vector, 566
mass, 779, 780, 814
    center of, 783
maximum
    absolute, 123, 740
    and First Deriv. Test, 139
    and Second Deriv. Test, 148
    relative/local, 124, 740, 743
Mean Value Theorem
    of differentiation, 132
    of integration, 235
Midpoint Rule, 210, 215
minimum
    absolute, 123, 740
    and First Deriv. Test, 139, 148
    relative/local, 124, 740, 743
moment, 785, 787, 814
monotonic sequence, 405
multiple integration, *see* iterated integration
multivariable function, 675, 679
    continuity, 688–690, 706, 711
    differentiability, 705, 706, 710, 711
    domain, 675, 679
    level curves, 676
    level surface, 680
    limit, 683, 684, 690
    range, 675, 679

Newton's Method, 160
norm, 566
normal line, 61, 513, 733
normal vector, 615
n^{th}–term test, 422
numerical integration, 240
    Left/Right Hand Rule, 240, 247
    Simpson's Rule, 245, 247
        error bounds, 248
    Trapezoidal Rule, 243, 247
        error bounds, 248

open, 682
open ball, 690
open disk, 682
optimization, 173
    constrained, 745
orthogonal, 584, 733
    decomposition, 588
orthogonal decomposition of vectors, 588
orthogonal projection, 586
osculating circle, 669

p-series, 415
parabola
    definition, 490
    general equation, 491
    reflective property, 493
parallel vectors, 574
Parallelogram Law, 570
parametric equations
    arc length, 519

concavity, 516
definition, 503
finding $\frac{d^2y}{dx^2}$, 517
finding $\frac{dy}{dx}$, 513
normal line, 513
surface area, 521
tangent line, 513
partial derivative, 692, 700
    high order, 700
    meaning, 694
    mixed, 696
    second derivative, 696
    total differential, 704, 710
perpendicular, *see* orthogonal
planes
    coordinate plane, 554
    distance between point and plane, 620
    equations of, 616
    introduction, 554
    normal vector, 615
    tangent, 736
point of inflection, 145
polar
    coordinates, 525
    function
        arc length, 545
        gallery of graphs, 532
        surface area, 546
    functions, 528
        area, 541
        area between curves, 543
        finding $\frac{dy}{dx}$, 538
        graphing, 528
polar coordinates, 525
    plotting points, 525
Power Rule
    differentiation, 78, 85, 91, 111
    integration, 194
power series, 452
    algebra of, 482
    convergence, 453
    derivatives and integrals, 457
projectile motion, 647, 648, 661

quadric surface
    definition, 558
    ellipsoid, 560
    elliptic cone, 559
    elliptic paraboloid, 559
    gallery, 559–561
    hyperbolic paraboloid, 561
    hyperboloid of one sheet, 560
    hyperboloid of two sheets, 561
    sphere, 560
    trace, 558
Quotient Rule, 88

$\mathbb{R}$, 566
radius of convergence, 454
radius of curvature, 669

Ratio Comparison Test
    for series, 435
rearrangements of series, 447, 448
related rates, 166
Riemann Sum, 210, 214, 217
    and definite integral, 224
Right Hand Rule, 210, 215, 240
right hand rule
    of Cartesian coordinates, 552
Rolle's Theorem, 132
Root Comparison Test
    for series, 438

saddle point, 742, 743
Second Derivative Test, 148, 743
sensitivity analysis, 709
sequence
    Absolute Value Theorem, 401
    positive, 429
sequences
    boundedness, 404
    convergent, 400, 405, 407
    definition, 397
    divergent, 400
    limit, 400
    limit properties, 403
    monotonic, 405
series
    absolute convergence, 446
    Absolute Convergence Theorem, 448
    alternating, 441
        Approximation Theorem, 444
    Alternating Series Test, 442
    Binomial, 480
    conditional convergence, 446
    convergent, 411
    definition, 411
    Direct Comparison Test, 429
    divergent, 411
    geometric, 413, 414
    Integral Test, 426
    interval of convergence, 454
    Limit Comparison Test, 430
    Maclaurin, 477
    n^{th}–term test, 422
    p-series, 415
    partial sums, 411
    power, 452, 453
        derivatives and integrals, 457
    properties, 419
    radius of convergence, 454
    Ratio Comparison Test, 435
    rearrangements, 447, 448
    Root Comparison Test, 438
    Taylor, 477
    telescoping, 416, 417
Shell Method, 362, 366
signed area, 201
signed volume, 762, 763
Simpson's Rule, 245, 247

    error bounds, 248
smooth, 634
smooth curve, 509
speed, 642
sphere, 553
Squeeze Theorem, 22
Sum/Difference Rule
    of derivatives, 80
    of integration, 194
    of series, 419
summation
    notation, 211
    properties, 213
surface area, 792
    solid of revolution, 374, 521, 546
surface of revolution, 556, 557

tangent line, 60, 513, 538, 633
    directional, 730
tangent plane, 736
Taylor Polynomial
    definition, 466
    Taylor's Theorem, 469
Taylor Series
    common series, 482
    definition, 477
    equality with generating function, 479
Taylor's Theorem, 469
telescoping series, 416, 417
terminal point, 566
total differential, 704, 710
    sensitivity analysis, 709
total signed area, 201
trace, 558
Trapezoidal Rule, 243, 247
    error bounds, 248
triple integral, 800, 811, 813
    properties, 807

unbounded sequence, 404
unbounded set, 682
unit normal vector
    a_N, 660
    and acceleration, 659, 660
    and curvature, 670
    definition, 657
    in $\mathbb{R}^2$, 659
unit tangent vector
    and acceleration, 659, 660
    and curvature, 666, 670
    a_T, 660
    definition, 655
    in $\mathbb{R}^2$, 659
unit vector, 572
    properties, 574
    standard unit vector, 576
    unit normal vector, 657
    unit tangent vector, 655

vector–valued function
    algebra of, 624

    arc length, 639
    average rate of change, 627
    continuity, 630
    definition, 623
    derivatives, 631, 632, 635
    describing motion, 642
    displacement, 626
    distance traveled, 650
    graphing, 623
    integration, 637
    limits, 629
    of constant length, 637, 646, 647, 656
    projectile motion, 647, 648
    smooth, 634
    tangent line, 633
vectors, 566
    algebra of, 569
    algebraic properties, 572
    component form, 567
    cross product, 593, 595, 596
    definition, 566
    dot product, 580–582
    Head To Tail Rule, 570
    magnitude, 566
    norm, 566
    normal vector, 615
    orthogonal, 584
    orthogonal decomposition, 588
    orthogonal projection, 586
    parallel, 574
    Parallelogram Law, 570
    resultant, 570
    standard unit vector, 576
    unit vector, 572, 574
    zero vector, 570
velocity, 73, 642
volume, 762, 763, 798

Washer Method, 357, 366
work, 378, 590

Differentiation Rules

1. $\dfrac{d}{dx}(cx) = c$
2. $\dfrac{d}{dx}(u \pm v) = u' \pm v'$
3. $\dfrac{d}{dx}(u \cdot v) = uv' + u'v$
4. $\dfrac{d}{dx}\left(\dfrac{u}{v}\right) = \dfrac{vu' - uv'}{v^2}$
5. $\dfrac{d}{dx}(u(v)) = u'(v)v'$
6. $\dfrac{d}{dx}(c) = 0$
7. $\dfrac{d}{dx}(x) = 1$
8. $\dfrac{d}{dx}(x^n) = nx^{n-1}$
9. $\dfrac{d}{dx}(e^x) = e^x$
10. $\dfrac{d}{dx}(a^x) = \ln a \cdot a^x$
11. $\dfrac{d}{dx}(\ln x) = \dfrac{1}{x}$
12. $\dfrac{d}{dx}(\log_a x) = \dfrac{1}{\ln a} \cdot \dfrac{1}{x}$
13. $\dfrac{d}{dx}(\sin x) = \cos x$
14. $\dfrac{d}{dx}(\cos x) = -\sin x$
15. $\dfrac{d}{dx}(\csc x) = -\csc x \cot x$
16. $\dfrac{d}{dx}(\sec x) = \sec x \tan x$
17. $\dfrac{d}{dx}(\tan x) = \sec^2 x$
18. $\dfrac{d}{dx}(\cot x) = -\csc^2 x$
19. $\dfrac{d}{dx}(\sin^{-1} x) = \dfrac{1}{\sqrt{1-x^2}}$
20. $\dfrac{d}{dx}(\cos^{-1} x) = \dfrac{-1}{\sqrt{1-x^2}}$
21. $\dfrac{d}{dx}(\csc^{-1} x) = \dfrac{-1}{|x|\sqrt{x^2-1}}$
22. $\dfrac{d}{dx}(\sec^{-1} x) = \dfrac{1}{|x|\sqrt{x^2-1}}$
23. $\dfrac{d}{dx}(\tan^{-1} x) = \dfrac{1}{1+x^2}$
24. $\dfrac{d}{dx}(\cot^{-1} x) = \dfrac{-1}{1+x^2}$
25. $\dfrac{d}{dx}(\cosh x) = \sinh x$
26. $\dfrac{d}{dx}(\sinh x) = \cosh x$
27. $\dfrac{d}{dx}(\tanh x) = \text{sech}^2 x$
28. $\dfrac{d}{dx}(\text{sech}\, x) = -\text{sech}\, x \tanh x$
29. $\dfrac{d}{dx}(\text{csch}\, x) = -\text{csch}\, x \coth x$
30. $\dfrac{d}{dx}(\coth x) = -\text{csch}^2 x$
31. $\dfrac{d}{dx}(\cosh^{-1} x) = \dfrac{1}{\sqrt{x^2-1}}$
32. $\dfrac{d}{dx}(\sinh^{-1} x) = \dfrac{1}{\sqrt{x^2+1}}$
33. $\dfrac{d}{dx}(\text{sech}^{-1} x) = \dfrac{-1}{x\sqrt{1-x^2}}$
34. $\dfrac{d}{dx}(\text{csch}^{-1} x) = \dfrac{-1}{|x|\sqrt{1+x^2}}$
35. $\dfrac{d}{dx}(\tanh^{-1} x) = \dfrac{1}{1-x^2}$
36. $\dfrac{d}{dx}(\coth^{-1} x) = \dfrac{1}{1-x^2}$

Integration Rules

1. $\displaystyle\int c \cdot f(x)\, dx = c \int f(x)\, dx$
2. $\displaystyle\int f(x) \pm g(x)\, dx = \int f(x)\, dx \pm \int g(x)\, dx$
3. $\displaystyle\int 0\, dx = C$
4. $\displaystyle\int 1\, dx = x + C$
5. $\displaystyle\int x^n\, dx = \dfrac{1}{n+1}x^{n+1} + C,\ n \neq -1$
6. $\displaystyle\int e^x\, dx = e^x + C$
7. $\displaystyle\int a^x\, dx = \dfrac{1}{\ln a} \cdot a^x + C$
8. $\displaystyle\int \dfrac{1}{x}\, dx = \ln|x| + C$
9. $\displaystyle\int \cos x\, dx = \sin x + C$
10. $\displaystyle\int \sin x\, dx = -\cos x + C$
11. $\displaystyle\int \tan x\, dx = -\ln|\cos x| + C$
12. $\displaystyle\int \sec x\, dx = \ln|\sec x + \tan x| + C$
13. $\displaystyle\int \csc x\, dx = -\ln|\csc x + \cot x| + C$
14. $\displaystyle\int \cot x\, dx = \ln|\sin x| + C$
15. $\displaystyle\int \sec^2 x\, dx = \tan x + C$
16. $\displaystyle\int \csc^2 x\, dx = -\cot x + C$
17. $\displaystyle\int \sec x \tan x\, dx = \sec x + C$
18. $\displaystyle\int \csc x \cot x\, dx = -\csc x + C$
19. $\displaystyle\int \cos^2 x\, dx = \dfrac{1}{2}x + \dfrac{1}{4}\sin(2x) + C$
20. $\displaystyle\int \sin^2 x\, dx = \dfrac{1}{2}x - \dfrac{1}{4}\sin(2x) + C$
21. $\displaystyle\int \dfrac{1}{x^2 + a^2}\, dx = \dfrac{1}{a}\tan^{-1}\left(\dfrac{x}{a}\right) + C$
22. $\displaystyle\int \dfrac{1}{\sqrt{a^2 - x^2}}\, dx = \sin^{-1}\left(\dfrac{x}{a}\right) + C$
23. $\displaystyle\int \dfrac{1}{x\sqrt{x^2 - a^2}}\, dx = \dfrac{1}{a}\sec^{-1}\left(\dfrac{|x|}{a}\right) + C$
24. $\displaystyle\int \cosh x\, dx = \sinh x + C$
25. $\displaystyle\int \sinh x\, dx = \cosh x + C$
26. $\displaystyle\int \tanh x\, dx = \ln(\cosh x) + C$
27. $\displaystyle\int \coth x\, dx = \ln|\sinh x| + C$
28. $\displaystyle\int \dfrac{1}{\sqrt{x^2 - a^2}}\, dx = \ln\left|x + \sqrt{x^2 - a^2}\right| + C$
29. $\displaystyle\int \dfrac{1}{\sqrt{x^2 + a^2}}\, dx = \ln\left|x + \sqrt{x^2 + a^2}\right| + C$
30. $\displaystyle\int \dfrac{1}{a^2 - x^2}\, dx = \dfrac{1}{2}\ln\left|\dfrac{a + x}{a - x}\right| + C$
31. $\displaystyle\int \dfrac{1}{x\sqrt{a^2 - x^2}}\, dx = \dfrac{1}{a}\ln\left(\dfrac{x}{a + \sqrt{a^2 - x^2}}\right) + C$
32. $\displaystyle\int \dfrac{1}{x\sqrt{x^2 + a^2}}\, dx = \dfrac{1}{a}\ln\left|\dfrac{x}{a + \sqrt{x^2 + a^2}}\right| + C$

The Unit Circle

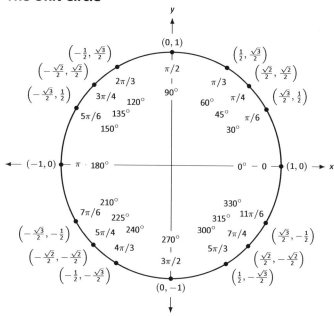

Definitions of the Trigonometric Functions

Unit Circle Definition

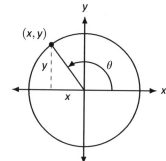

$$\sin\theta = y \qquad \cos\theta = x$$

$$\csc\theta = \frac{1}{y} \qquad \sec\theta = \frac{1}{x}$$

$$\tan\theta = \frac{y}{x} \qquad \cot\theta = \frac{x}{y}$$

Right Triangle Definition

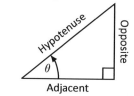

$$\sin\theta = \frac{O}{H} \qquad \csc\theta = \frac{H}{O}$$

$$\cos\theta = \frac{A}{H} \qquad \sec\theta = \frac{H}{A}$$

$$\tan\theta = \frac{O}{A} \qquad \cot\theta = \frac{A}{O}$$

Common Trigonometric Identities

Pythagorean Identities

$\sin^2 x + \cos^2 x = 1$

$\tan^2 x + 1 = \sec^2 x$

$1 + \cot^2 x = \csc^2 x$

Cofunction Identities

$\sin\left(\frac{\pi}{2} - x\right) = \cos x \qquad \csc\left(\frac{\pi}{2} - x\right) = \sec x$

$\cos\left(\frac{\pi}{2} - x\right) = \sin x \qquad \sec\left(\frac{\pi}{2} - x\right) = \csc x$

$\tan\left(\frac{\pi}{2} - x\right) = \cot x \qquad \cot\left(\frac{\pi}{2} - x\right) = \tan x$

Double Angle Formulas

$\sin 2x = 2\sin x \cos x$

$\cos 2x = \cos^2 x - \sin^2 x$

$\qquad = 2\cos^2 x - 1$

$\qquad = 1 - 2\sin^2 x$

$\tan 2x = \dfrac{2\tan x}{1 - \tan^2 x}$

Sum to Product Formulas

$\sin x + \sin y = 2\sin\left(\dfrac{x+y}{2}\right)\cos\left(\dfrac{x-y}{2}\right)$

$\sin x - \sin y = 2\sin\left(\dfrac{x-y}{2}\right)\cos\left(\dfrac{x+y}{2}\right)$

$\cos x + \cos y = 2\cos\left(\dfrac{x+y}{2}\right)\cos\left(\dfrac{x-y}{2}\right)$

$\cos x - \cos y = -2\sin\left(\dfrac{x+y}{2}\right)\sin\left(\dfrac{x-y}{2}\right)$

Power-Reducing Formulas

$\sin^2 x = \dfrac{1 - \cos 2x}{2}$

$\cos^2 x = \dfrac{1 + \cos 2x}{2}$

$\tan^2 x = \dfrac{1 - \cos 2x}{1 + \cos 2x}$

Even/Odd Identities

$\sin(-x) = -\sin x$

$\cos(-x) = \cos x$

$\tan(-x) = -\tan x$

$\csc(-x) = -\csc x$

$\sec(-x) = \sec x$

$\cot(-x) = -\cot x$

Product to Sum Formulas

$\sin x \sin y = \dfrac{1}{2}\big(\cos(x-y) - \cos(x+y)\big)$

$\cos x \cos y = \dfrac{1}{2}\big(\cos(x-y) + \cos(x+y)\big)$

$\sin x \cos y = \dfrac{1}{2}\big(\sin(x+y) + \sin(x-y)\big)$

Angle Sum/Difference Formulas

$\sin(x \pm y) = \sin x \cos y \pm \cos x \sin y$

$\cos(x \pm y) = \cos x \cos y \mp \sin x \sin y$

$\tan(x \pm y) = \dfrac{\tan x \pm \tan y}{1 \mp \tan x \tan y}$

Areas and Volumes

Triangles

$h = a \sin \theta$

Area = $\frac{1}{2}bh$

Law of Cosines:
$c^2 = a^2 + b^2 - 2ab \cos \theta$

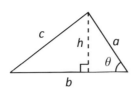

Right Circular Cone

Volume = $\frac{1}{3}\pi r^2 h$

Surface Area =
$\pi r \sqrt{r^2 + h^2} + \pi r^2$

Parallelograms

Area = bh

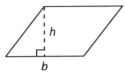

Right Circular Cylinder

Volume = $\pi r^2 h$

Surface Area =
$2\pi rh + 2\pi r^2$

Trapezoids

Area = $\frac{1}{2}(a+b)h$

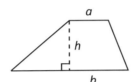

Sphere

Volume = $\frac{4}{3}\pi r^3$

Surface Area = $4\pi r^2$

Circles

Area = πr^2

Circumference = $2\pi r$

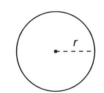

General Cone

Area of Base = A

Volume = $\frac{1}{3}Ah$

Sectors of Circles

θ in radians

Area = $\frac{1}{2}\theta r^2$

$s = r\theta$

General Right Cylinder

Area of Base = A

Volume = Ah

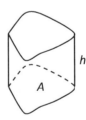

Algebra

Factors and Zeros of Polynomials
Let $p(x) = a_n x^n + a_{n-1} x^{n-1} + \cdots + a_1 x + a_0$ be a polynomial. If $p(a) = 0$, then a is a *zero* of the polynomial and a solution of the equation $p(x) = 0$. Furthermore, $(x - a)$ is a *factor* of the polynomial.

Fundamental Theorem of Algebra
An *n*th degree polynomial has *n* (not necessarily distinct) zeros. Although all of these zeros may be imaginary, a real polynomial of odd degree must have at least one real zero.

Quadratic Formula
If $p(x) = ax^2 + bx + c$, and $0 \leq b^2 - 4ac$, then the real zeros of p are $x = (-b \pm \sqrt{b^2 - 4ac})/2a$

Special Factors
$x^2 - a^2 = (x - a)(x + a)$ $\qquad\qquad\qquad\qquad$ $x^3 - a^3 = (x - a)(x^2 + ax + a^2)$
$x^3 + a^3 = (x + a)(x^2 - ax + a^2)$ $\qquad\qquad$ $x^4 - a^4 = (x^2 - a^2)(x^2 + a^2)$
$(x + y)^n = x^n + nx^{n-1}y + \frac{n(n-1)}{2!}x^{n-2}y^2 + \cdots + nxy^{n-1} + y^n$
$(x - y)^n = x^n - nx^{n-1}y + \frac{n(n-1)}{2!}x^{n-2}y^2 - \cdots \pm nxy^{n-1} \mp y^n$

Binomial Theorem
$(x + y)^2 = x^2 + 2xy + y^2$ $\qquad\qquad\qquad$ $(x - y)^2 = x^2 - 2xy + y^2$
$(x + y)^3 = x^3 + 3x^2y + 3xy^2 + y^3$ $\qquad$ $(x - y)^3 = x^3 - 3x^2y + 3xy^2 - y^3$
$(x + y)^4 = x^4 + 4x^3y + 6x^2y^2 + 4xy^3 + y^4$ $\qquad$ $(x - y)^4 = x^4 - 4x^3y + 6x^2y^2 - 4xy^3 + y^4$

Rational Zero Theorem
If $p(x) = a_n x^n + a_{n-1} x^{n-1} + \cdots + a_1 x + a_0$ has integer coefficients, then every *rational zero* of p is of the form $x = r/s$, where r is a factor of a_0 and s is a factor of a_n.

Factoring by Grouping
$acx^3 + adx^2 + bcx + bd = ax^2(cs + d) + b(cx + d) = (ax^2 + b)(cx + d)$

Arithmetic Operations
$ab + ac = a(b + c) \qquad\qquad \frac{a}{b} + \frac{c}{d} = \frac{ad + bc}{bd} \qquad\qquad \frac{a+b}{c} = \frac{a}{c} + \frac{b}{c}$

$\dfrac{\left(\frac{a}{b}\right)}{\left(\frac{c}{d}\right)} = \left(\frac{a}{b}\right)\left(\frac{d}{c}\right) = \frac{ad}{bc} \qquad \dfrac{\left(\frac{a}{b}\right)}{c} = \frac{a}{bc} \qquad\qquad \dfrac{a}{\left(\frac{b}{c}\right)} = \frac{ac}{b}$

$a\left(\frac{b}{c}\right) = \frac{ab}{c} \qquad\qquad \frac{a-b}{c-d} = \frac{b-a}{d-c} \qquad\qquad \frac{ab + ac}{a} = b + c$

Exponents and Radicals
$a^0 = 1, \ a \neq 0 \qquad (ab)^x = a^x b^x \qquad a^x a^y = a^{x+y} \qquad \sqrt{a} = a^{1/2} \qquad \frac{a^x}{a^y} = a^{x-y} \qquad \sqrt[n]{a} = a^{1/n}$

$\left(\frac{a}{b}\right)^x = \frac{a^x}{b^x} \qquad \sqrt[n]{a^m} = a^{m/n} \qquad a^{-x} = \frac{1}{a^x} \qquad \sqrt[n]{ab} = \sqrt[n]{a}\sqrt[n]{b} \qquad (a^x)^y = a^{xy} \qquad \sqrt[n]{\frac{a}{b}} = \frac{\sqrt[n]{a}}{\sqrt[n]{b}}$

Additional Formulas

Summation Formulas:

$$\sum_{i=1}^{n} c = cn \qquad \sum_{i=1}^{n} i = \frac{n(n+1)}{2}$$

$$\sum_{i=1}^{n} i^2 = \frac{n(n+1)(2n+1)}{6} \qquad \sum_{i=1}^{n} i^3 = \left(\frac{n(n+1)}{2}\right)^2$$

Trapezoidal Rule:

$$\int_a^b f(x)\,dx \approx \frac{\Delta x}{2}\left[f(x_1) + 2f(x_2) + 2f(x_3) + \ldots + 2f(x_n) + f(x_{n+1})\right]$$

with Error $\leq \dfrac{(b-a)^3}{12n^2}\left[\max |f''(x)|\right]$

Simpson's Rule:

$$\int_a^b f(x)\,dx \approx \frac{\Delta x}{3}\left[f(x_1) + 4f(x_2) + 2f(x_3) + 4f(x_4) + \ldots + 2f(x_{n-1}) + 4f(x_n) + f(x_{n+1})\right]$$

with Error $\leq \dfrac{(b-a)^5}{180n^4}\left[\max |f^{(4)}(x)|\right]$

Arc Length:

$$L = \int_a^b \sqrt{1 + f'(x)^2}\,dx$$

Surface of Revolution:

$$S = 2\pi \int_a^b f(x)\sqrt{1 + f'(x)^2}\,dx$$

(where $f(x) \geq 0$)

$$S = 2\pi \int_a^b x\sqrt{1 + f'(x)^2}\,dx$$

(where $a, b \geq 0$)

Work Done by a Variable Force:

$$W = \int_a^b F(x)\,dx$$

Force Exerted by a Fluid:

$$F = \int_a^b w\,d(y)\,\ell(y)\,dy$$

Taylor Series Expansion for $f(x)$:

$$p_n(x) = f(c) + f'(c)(x-c) + \frac{f''(c)}{2!}(x-c)^2 + \frac{f'''(c)}{3!}(x-c)^3 + \ldots + \frac{f^{(n)}(c)}{n!}(x-c)^n$$

Maclaurin Series Expansion for $f(x)$, where $c = 0$:

$$p_n(x) = f(0) + f'(0)x + \frac{f''(0)}{2!}x^2 + \frac{f'''(0)}{3!}x^3 + \ldots + \frac{f^{(n)}(0)}{n!}x^n$$

Summary of Tests for Series:

Test	Series	Condition(s) of Convergence	Condition(s) of Divergence	Comment
nth-Term	$\sum_{n=1}^{\infty} a_n$		$\lim_{n\to\infty} a_n \neq 0$	This test cannot be used to show convergence.
Geometric Series	$\sum_{n=0}^{\infty} r^n$	$\lvert r \rvert < 1$	$\lvert r \rvert \geq 1$	Sum $= \dfrac{1}{1-r}$
Telescoping Series	$\sum_{n=1}^{\infty} (b_n - b_{n+a})$	$\lim_{n\to\infty} b_n = L$		Sum $= \left(\sum_{n=1}^{a} b_n\right) - L$
p-Series	$\sum_{n=1}^{\infty} \dfrac{1}{(an+b)^p}$	$p > 1$	$p \leq 1$	
Integral Test	$\sum_{n=0}^{\infty} a_n$	$\int_{1}^{\infty} a(n)\, dn$ is convergent	$\int_{1}^{\infty} a(n)\, dn$ is divergent	$a_n = a(n)$ must be continuous
Direct Comparison	$\sum_{n=0}^{\infty} a_n$	$\sum_{n=0}^{\infty} b_n$ converges and $0 \leq a_n \leq b_n$	$\sum_{n=0}^{\infty} b_n$ diverges and $0 \leq b_n \leq a_n$	
Limit Comparison	$\sum_{n=0}^{\infty} a_n$	$\sum_{n=0}^{\infty} b_n$ converges and $\lim_{n\to\infty} a_n/b_n \geq 0$	$\sum_{n=0}^{\infty} b_n$ diverges and $\lim_{n\to\infty} a_n/b_n > 0$	Also diverges if $\lim_{n\to\infty} a_n/b_n = \infty$
Ratio Test	$\sum_{n=0}^{\infty} a_n$	$\lim_{n\to\infty} \dfrac{a_{n+1}}{a_n} < 1$	$\lim_{n\to\infty} \dfrac{a_{n+1}}{a_n} > 1$	$\{a_n\}$ must be positive. Also diverges if $\lim_{n\to\infty} a_{n+1}/a_n = \infty$
Root Test	$\sum_{n=0}^{\infty} a_n$	$\lim_{n\to\infty} (a_n)^{1/n} < 1$	$\lim_{n\to\infty} (a_n)^{1/n} > 1$	$\{a_n\}$ must be positive. Also diverges if $\lim_{n\to\infty} (a_n)^{1/n} = \infty$

Made in the USA
Middletown, DE
07 January 2019